T0177481

THE NATURE OF COMPLEX NETWORKS

The Nature of Complex Networks

Sergey N. Dorogovtsev and José F. F. Mendes

University of Aveiro, Portugal

OXFORD
UNIVERSITY PRESS

OXFORD
UNIVERSITY PRESS

Great Clarendon Street, Oxford, OX2 6DP,
United Kingdom

Oxford University Press is a department of the University of Oxford.
It furthers the University's objective of excellence in research, scholarship,
and education by publishing worldwide. Oxford is a registered trade mark of
Oxford University Press in the UK and in certain other countries.

Published in the United States of America by Oxford University Press
198 Madison Avenue, New York, NY 10016, United States of America.

British Library Cataloguing in Publication Data

Data available

Library of Congress Control Number: 2022930598

ISBN 978–0–19–969511–9

DOI: 10.1093/oso/9780199695119.001.0001

Printed and bound by
CPI Group (UK) Ltd, Croydon, CR0 4YY

For those whom we love

Gentlemen! As is custom among our kind, we do not
plunge headlong into folly on the orders of a single madman,
but act according to our own collective madness!
Assassin's Creed IV: Black Flag

Preface

The authors of this book are scientific workers in the field of complex networks. This is an explosively developing multidisciplinary area, which involves practically everything, natural and artificial, in the world, and we, theoretical physicists, explore it, armed with the standard tools of statistical physics. What we are doing is uncovering the actual simplicity and beauty of complex systems including the random complex networks. The book is about the key ideas and concepts of this field which is already called the science of networks. The highly heterogeneous networked systems, which we explore, typically have no governing authority, no command centre taking decisions and issuing orders. Nonetheless, even without orchestration, these remarkably widespread systems function very effectively and reliably. Moreover, the absence of centralization in favour of distributed self-organization is vital for their existence and robustness.

Thanks to numerous tips one can appreciate the pace and scale of efforts to surveil, control, protect, and manipulate society, in essence, large social and information networks. The sweeping reset of our 'highly interconnected and interdependent world', induced and concerted by the Covid-19 pandemic (Schwab and Malleret, 2020), is forcefully amplifying these exertions. The big problem now is how to control large complex systems that is impossible to govern. A related fundamental problem is how to structure and efficiently process the collected big data. Similarly impressive and costly efforts are under way to engineer and manage large cellular, technological, deep-learning, and many other networks. These extensive works demonstrate how important it is to understand in detail the principles of organization and function of highly heterogeneous and highly structured random networked systems, that is the nature of complex networks.

In 2003 we published the first reference book in this research area, the 'Evolution of Networks: From Biological Nets to the Internet and the WWW,' which essentially focused on evolving networks. Our new, more dense and advanced, book touches upon a much wider range of networks and networked systems paying particular attention to the directions that emerged after 2002–2003. Writing a book is about asking good questions. So, keeping the book compact, we consider only the issues and key results that, we believe, will not lose their value in the future. To avoid drowning in serious mathematics we have to present a number of basic theorems from graph theory without even sketching the ideas of their proofs. Furthermore, for the sake of brevity we do not include in the book chapters devoted to numerous applications and empirical material. The reader will find detailed discussions of these matters in a number of comprehensive books and reviews, which we cite.

The researchers studying complex networks will acquire from this concise modern book a number of new issues and ideas, not yet touched upon in other reference volumes. We propose a statistical mechanics view of random networks based on the concept of statistical ensembles, but approaches and methods of modern graph theory, concerning random graphs, overlap strongly with statistical physics. Hence mathematicians have a good chance to discover interesting things in our book, even though it does not contain mathematical proofs and trades off rigour for comprehension, brevity, and relevance. We expect that this book will be useful for undergraduate, master, and PhD students and young researchers from multidisciplinary studies, physics, computer science, and applied mathematics wishing to gain a serious insight into the principles of complex networks. The book can be used as a text in university courses on complex networks. We propose to determined students not only a brief trip to the land of complex networks but an option to stay there forever. For a more elementary and concise introduction to the science of networks, we refer most impatient readers to the text of one of us, S. N. Dorogovtsev, 'Lectures on Complex Networks.'

We are deeply indebted to our friends and colleagues all over the world for their encouragement and advice, first and foremost to Gareth Baxter, who has read and commented on the manuscript, Alexander Goltsev, Ginestra Bianconi, Pavel Krapivsky, Rui Americo da Costa, Gábor Timár, João Gama Oliveira, Nahid Azimi-Tafreshi, KyoungEun Lee, António Luís Ferreira, Sooyeon Yoon, Bruno Coutinho, Marta Santos, Nuno Araujo, Ricardo Jesus, Marian Boguñá, Maria Ángeles Serrano, Dmitri Krioukov, Edgar Wright, Romualdo Pastor-Satorras, Tiago Peixoto, Jorge Pacheco, Francisco Santos, Peter Grassberger, Mark Newman, Byungnam Kahng, Zdzislaw Burda, Michel Bauer, Raissa D'Souza, Doochul Kim, Mikko Alava, Hyunggyu Park, Petter Holme, Kimmo Kaski, Hawoong Jeong, Satu Elisa Schaeffer, José Ramasco, János Kertész, Jae Dong Noh, László Barabási, Shlomo Havlin, Matteo Marsili, Bartlomiej Waclaw, Geoff Rodgers, Malte Henkel, Davide Cellai, James Gleeson, Alexander Povolotsky, Alexander

Zyuzin, Santo Fortunato, Vittoria Colizza, Alessandro Vespignani, Alex Arenas, Guido Caldarelli, Filippo Radicchi, Adilson Motter, Sid Redner, Yamir Moreno, Maxi San Miguel, Alain Barrat, Marc Barthelemy, Agata and Piotr Fronczak, Konstantin Klemm, Lenka Zdeborová, Zoltán Toroczkai, Alexei Vázquez, Florent Krzakala, Valentina Guleva, Jesus Gomez-Gardeñes, Bosiljka Tadic, Piet Van Mieghem, Dima Shepelyansky, Sergey Maslov, Ernesto Estrada, Bruno Gonçalves, Serguei Nechaev, Alexander Gorsky, Ronaldo Menezes, Hans Herrmann, Nelly Litvak, Andrei Raigorodskii, and Dafne Bandeira. We are enormously bound to Alexander Nikolaevich Samukhin (1953–2018), our friend and collaborator, who showed us the essence and logic of statistical mechanics and its mathematical apparatus. Our book would be impossible without his friendship and tuition. Finally, our warmest thanks to the editorial and production staff at Oxford University Press for their guidance, encouragement, and patience.

Aveiro
August 2021

S.N.D.
J.F.F.M.

Contents

CONTENTS

CONTENTS

1

First Insight

1.1 Statistical mechanics view of random systems

Look at the two graphs in Figure 1.1. Could you say which graph is random, left or right? The typical student's response is: 'Of course, the right one!' This immediate answer is dramatically incorrect. It is actually impossible to say whether a finite graph is random or non-random (deterministic) since one can generate any finite graph by some deterministic algorithm. The example in Figure 1.2 explains what is really a random graph. This graph has a number of different realizations (individual graphs), and each of them occurs with some associated probability. Thus a random graph (random net-

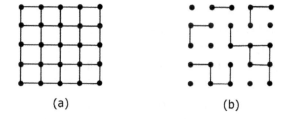

(a)   (b)

Fig. 1.1 Which graph is random, left or right?

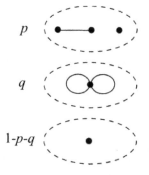

Fig. 1.2 Example of a random graph. The first, second, and third graphs are realised with probability p, q, and $1 - p - q$, respectively.

work) is a statistical ensemble of individual graphs, in which each member has its probability of realization. (Note that we do not distinguish the terms 'graph', or its generalization—'hypergraph', and 'network'.) In particular, all these probabilities may be equal, which provides a uniform ensemble. In short, a random graph is a statistical ensemble of graphs. In this picture, the result of the measurement of some characteristic of a random graph, an observable, is the average of this characteristic over the statistical ensemble accounting for the probabilities of realization of its members.[1]

This construction is identical to the foundational notion of statistical mechanics, which is just the statistical ensemble. The statistical ensemble is the complete set of all possible states (configurations) of a system. Each member of a statistical ensemble is accompanied by its statistical weight, a positive number, which is equal to the probability of realization of this member with some unimportant factor, equal for all members. In the 1870s, Ludwig Boltzmann explained that the statistical weights $\mathcal{W}_\alpha$ of the members α of a statistical ensemble are expressed in terms of the energy E_α of these members (which are states or configurations),

$$\mathcal{W}_\alpha \propto e^{-\beta E_\alpha}. \tag{1.1}$$

Here, β is a control parameter, a positive number, $\beta = 1/(k_B T)$, where k_B is the Boltzmann constant and T is temperature. The first to comprehend that thermodynamics is about probability and probability distributions, that is, statistics, was James Clerk Maxwell. Maxwell first applied the term 'statistical' to thermodynamical systems in 1871, and in 1884, Josiah Willard Gibbs coined the term 'statistical mechanics'; in 1902 he essentially completed the construction of equilibrium statistical physics elaborating the idea of statistical ensembles in his book *Elementary Principles in Statistical Mechanics*.

These efforts were directed to understanding equilibrium systems where statistical weights do not change with time. The breakthrough in understanding non-equilibrium systems (having evolving statistical weights) came from Albert Einstein (1905) and Marian Smoluchowski (1906) on molecular Brownian motion, and Smoluchowski in 1916, in which he derived a

---

[1]Note that exploring empirical material or performing simulations, a researcher often works with only a few realizations of a random network, sometimes, with a single member of a given ensemble. Inspecting this single graph, the researcher measures how frequently a given structural feature occurs within this particular realization, for example, measures the fraction of vertices with q connections. These values coincide with the corresponding averages over the full ensemble only in the infinite size limit.

linear equation for coagulation processes. The Smoluchowski equation actually provides a unified description of the evolution of statistical weights in non-equilibrium statistical ensembles. Let us write this equation in application to evolving random networks. Let $\mathcal{G}$ denote a random graph (network), G denote individual members of this ensemble, and $\mathcal{W}(G, t)$ denote their statistical weights, proportional to probabilities of realization, $\mathcal{P}(G, t) = \mathcal{W}(G, t)/\sum_{G' \in \mathcal{G}} \mathcal{W}(G', t)$. The observables in this scheme are the averages over the statistical ensemble,

$$\langle \mathcal{X} \rangle = \sum_{G \in \mathcal{G}} \mathcal{P}(G)\mathcal{X}(G) = \frac{\sum_{G \in \mathcal{G}} \mathcal{W}(G)\mathcal{X}(G)}{\sum_{G \in \mathcal{G}} \mathcal{W}(G)}. \tag{1.2}$$

We consider a stochastic process in which members of this ensemble transform into other members of the same ensemble. A graph $G \in \mathcal{G}$ transforms to graph $G' \in \mathcal{G}$ with probability $r(G', G)dt$. The Smoluchowski equation for this process is

$$\frac{\partial \mathcal{W}(G, t)}{\partial t} = \sum_{G' \in \mathcal{G}} [r(G, G')\mathcal{W}(G', t) - r(G', G)\mathcal{W}(G, t)]. \tag{1.3}$$

How does one obtain an equilibrium ensemble from this equation? It is actually not enough to demand the stationary statistical weights, $\mathcal{W}(G, t) = \mathcal{W}(G)$, for equilibrium. We must also take into account that non-dying-off chains of transitions—'flows' or 'currents'—are absent in equilibrium. That is, for any cycle, we must have the equality of the kind:

$$r(G, G')r(G', G'')r(G'', G) = r(G, G'')r(G'', G')r(G', G). \tag{1.4}$$

Furthermore, for each pair of graphs $G, G' \in \mathcal{G}$ in the ensemble, we also need the existence at least one connecting chain of transitions

$$r(G', G_n) \dots r(G_3, G_2)r(G_1, G) \neq 0. \tag{1.5}$$

There must not be separated parts of the ensemble. This guarantees that the weights are determined unambiguously: ascribing an arbitrary statistical weight to some graph, one can obtain statistical weights of all other graphs up to a constant multiple. Equation (1.4) leads to the the detailed balance condition

$$r(G, G')\mathcal{W}(G') = r(G', G)\mathcal{W}(G), \tag{1.6}$$

which provides statistical weights in terms of the transition matrix $r(G, G')$. To satisfy Eq. (1.6) or Eq. (1.4), it is sufficient to assume the factorization:

$r(G', G) = s_f(G')s_i(G)$, which gives the statistical weight of graph G as a function of only this graph, $\mathcal{W}(G) \propto s_f(G)/s_i(G)$, like in Boltzmann's formula, Eq. (1.1).

In our book we apply this standard scheme from statistical physics to random networks for which we use the terms equilibrium and non-equilibrium interchangeably. Mathematicians in graph theory employ essentially the same idea of statistical ensembles, although they do not use this wording. With this in mind, what is the difference between the two approaches used by mathematicians and physicists to the same objects—random graphs and networks? For sure, these two routes differ in language and rigour, but it is not enough to say that mathematicians prove lemmas and theorems and theoretical physicists obtain results. Rather it's about how each area of research treats infinite systems. Exploring random networks, physicists traditionally focus on the infinite size limit, as they do in condensed matter or, for example, studying gases. We often call it the thermodynamic limit. This is the starting (and very often the final) point of our study, and the approach to infinity remains unexplored or, at least, less explored. Finite size effects are more difficult to treat and appear to be of secondary importance to an impatient physicist. In contrast, mathematicians, in principle, cannot complete their job without a meticulous investigation of large but still finite graphs, a proof of the existence of the infinite size limit, and the detailed analysis of the large size asymptotics, proving upper and lower bounds.

1.2 History of networks

It is Leonhard Euler's solution of the Königsberg bridge problem in 1735 that is usually regarded as the birth point of graph theory. In Euler's time, the connections in Königsberg were as shown in Figure 1.3a, where the vertices of the graph denote lands (the left vertex is Kneiphof Island, a remarkably perfect rectangle) and edges denote bridges between these pieces of land across the Pregel River. The question was: could a pedestrian walk around Königsberg, crossing each bridge only once? In graph theory, a *walk* is an alternating sequence of vertices and edges, which begins and ends with vertices. A *trail* is a walk that has all edges distinct (though it can have repeating vertices). A trail is closed if its initial and end vertices coincide, otherwise a trail is open. Therefore, using the language of graph theory, the question was: is there a trail (open or closed) that passes all the edges of the graph in Figure 1.3a? A closed trail visiting all edges of a graph is called an *Eulerian circuit*. A graph having an Eulerian circuit is called

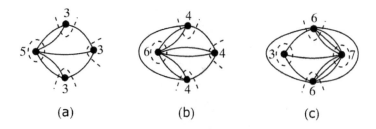

(a)   (b)   (c)

Fig. 1.3 Königsberg bridge problem. (a) Graph, representing connections between four Königsberg's land masses in Euler's time. The vertices are lands (the left vertex is Kneiphof Island), the edges are bridges between these pieces of land. The numbers are the degrees of the vertices. This graph has no trails visiting every edge. (b) Königsberg's graph at some time after Euler's epoch. Every vertex degree is even, so this graph is Eulerian. (c) Connections between land masses in modern Kaliningrad. The right vertex became an island, some of old bridges disappear, and new bridges were constructed. Two of the vertices have odd degree, and so there is an open trail passing through every edge. The trail starts at one of these two vertices and finishes at the other.

a *Eulerian graph*. An open trail visiting all edges of a graph is called an *Eulerian trail* (tour). We need one more definition: the number of the edge ends attached to a vertex is its *degree*. Euler's theorem says that a connected graph is Eulerian iff each vertex has an even degree, and so such tours were impossible in Königsberg. (In a *connected graph*, there exists a walk between each two vertices.) In fact, Euler proved that this is the necessary condition. (The proof is quite easy, the reader can immediately reproduce it, and, of course, the real breakthrough was formulating the problem in terms of a graph.) In 1873, Carl Hierholzer proved that this condition is also sufficient. Furthermore, a connected graph has an open trail visiting all edges if and only if only two of the vertices have odd degree. As is natural, this trail runs between these vertices.

The set of bridges in Königsberg (now Kaliningrad) evolved with time, Figure 1.3. In Euler's time, indeed, there was no trail passing every bridge. Afterwards, at some moment, the graph became Eulerian. Today there exists an open trail passing every bridge—an Eulerian trail.

In 1759 Euler made a big advance in solving the knight's tour problem, one of the oldest in mathematics. The problem appeared in a piece of Sanskrit poetry by Rudrata in the ninth century and it probably dates back to the sixth century when chess originated. A knight's tour is a series of moves of a knight visiting every square of a chessboard exactly once. A

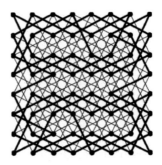

Fig. 1.4 8×8 knight's graph and one of its 13,267,364,410,532 Hamiltonian cycles, found by Euler.

knight's tour is closed if its start coincides with its end and open otherwise. Euler found one of the closed knight's tours on the 8 × 8 chessboard (Figure 1.4), which shows this tour on the knight's graph whose vertices are the chessboard's squares and the edges are the knight's legal moves.[2] In graph theory, a *path* is an open walk having no repeated vertices (and so having no repeated edges). A *cycle* is a closed walk having no repeated vertices (and edges).[3] Graphs without cycles are called *trees*. A path visiting all the vertices of a graph is called a *Hamiltonian path*. A cycle that passes through all the vertices of a graph is called a *Hamiltonian cycle*, and a graph having a Hamiltonian cycle is called a *Hamiltonian graph*. Therefore, an open knight's tour is a Hamiltonian path, a closed knight's tour is a Hamiltonian cycle, and the 8 × 8 knight's graph turns out to be Hamiltonian.

The Hamiltonian path, cycle, and graphs were named after William Rowan Hamilton. In 1856 Hamilton invented the Icosian game, a puzzle which he sold for 25 guineas to a game manufacturer in Dublin and which was widely marketed in Europe. Imagine a dodecahedron, a Platonic solid with 12 pentagonal faces, 20 vertices, and 30 edges, with each of the vertices labelled by some capital city. The puzzle is to find a closed route along the edges of the dodecahedron which passes every city only once.[4] Figure 1.5

---

[2] It is known now that the total number of closed knight's tours is huge, 13,267,364,410,532 different tours. As one could expect, the number of different open knight's tours is even greater, 19,591,828,170,979,904.

[3] Physicists systematically call vertices nodes, edges links, and cycles loops.

[4] Hamilton devised his puzzle while studying algebraic structures related to the symmetries of the regular icosahedron, a Platonic solid, dual to a dodecahedron, with 20 equilateral triangle faces, 12 vertices, and 30 edges. The term 'dual polyhedra' means that the vertices of one polyhedra correspond to the faces of the other, and each edge

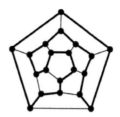

Fig. 1.5 One of the 60 different solutions of the Icosian game invented by William Hamilton. The graph is the plane projection of the dodecahedral graph.

Fig. 1.6 Another representation of the dodecahedral graph. The Hamiltonian cycle goes around the ring.

shows this route in the plane projection of the dodecahedral graph made of the edges and vertices of the dodecahedron. By definition, this closed route is a Hamiltonian cycle, and the graph is Hamiltonian. Figure 1.6 represents the dodecahedral graph in a different form, in which the Hamiltonian cycle goes around the ring. Note that any Hamiltonian graph can be represented in a similar form. The Hamiltonian paths are not just of abstract interest. They are intensively used in DNA sequencing.

The starting point of the exploration of random networks can also be precisely indicated. In 1951–1952, two applied mathematicians, Ray J. Solomonoff and Anatol Rapoport, published a series of papers in which they considered a random graph of n vertices with a fixed probability p that two vertices are connected.[5] It is often referred to as the $G(n, p)$ model.[6] In the infinite size limit, these equilibrium random graphs have virtually no ver-

---

between two vertices of the first polyhedra correspond to the joint edge of the two faces of the second.

[5]These papers were published in *The Bulletin of Mathematical Biophysics* and did not attract much attention at the time. Ray Solomonoff is also well known by putting forward the concept of algorithmic complexity (1960), which was independently developed by Andrey Kolmogorov, and it was named Kolmogorov's complexity.

[6]In the following sections, we shall use the notation N for the number of vertices in a graph and E for the number of its edges.

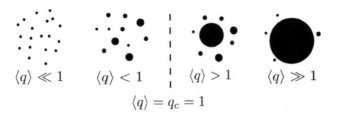

$\langle q \rangle \ll 1$ $\langle q \rangle < 1$ $\langle q \rangle > 1$ $\langle q \rangle \gg 1$

$$\langle q \rangle = q_c = 1$$

Fig. 1.7 Structure of an infinite random graph, $G(n,p)$ model, discovered by Solomonoff and Rapoport.

tices with high degrees when the average degree of a vertex is finite, that is, when a graph is sparse. Solomonoff and Rapoport discovered a fundamental phase transition in this graph. They observed two distinct phases of the random graph when it was very large (actually infinite) (Figure 1.7). When the average degree $\langle q \rangle \cong 1/p$ was smaller than 1, the graph consisted of a set of finite clusters—*finite connected components*. When the average degree exceeded 1, one of the clusters was infinite in the sense that it contained a finite fraction of all vertices—the *giant connected component*—and the remaining clusters were finite. As the average degree increased up to values much greater than 1, almost all vertices of the graph appeared in the giant connected component. A continuous phase transition occurred at the critical point where the average degree was 1. This picture turns out to be typical for large sparse random networks, and the birth of a giant connected component is the key point for a random graph as its structure dramatically differs below and above this critical point.

In the second half of the 1950s Paul Erdős and Alfréd Rényi introduced another graph model, $G(n,m)$, a uniform ensemble with fixed numbers n of vertices and m of edges, and as a matter of fact established random graph theory. Similarly to $G(n,p)$, the ensemble $G(n,m)$ is equilibrium. The ensembles $G(n,p)$ and $G(n,m)$ are equivalent when n is large and the graphs are sparse, so these classical random graphs are often called the Erdős–Rényi random graph. These random graphs are the reference objects in graph theory and the science of networks, extensively explored over many decades. One can treat them as a basic paradigm for random networks.

A crucial step was made by prominent social psychologist Stanley Milgram in 1967. Milgram asked what the typical separation of vertices is in social networks measured along the chains of social contacts, and how are these networks far from finite-dimensional lattices? He also queried how small the world of social relationships is, and his measurements, estimating the typical separation from above, led him to the conclusion that, while residing in the big world, with large geographic distances, we are confined in a surprisingly small social world with at most about 'six degrees of separation' between each other.[7] Most of random networks, which we discuss in this book, are small worlds.

In the first half of the 1960s Paul Baran in the U.S. and Donald Watts Davies in the UK drew the first sketches of complex designs for computer networks and elaborated the basic concepts of reliable communications systems, including the idea of packet switching. Visually, these networks were quite far from the classical random graphs studied by Solomonoff and Rapoport and Erdős and Rényi. Mathematicians started searching for more general random network constructions in the 1970s. In 1978, Edward Anton Bender and Earl Rodney Canfield published a paper entitled 'The asymptotic number of labelled graphs with given degree sequences', where they introduced equilibrium random networks with essentially richer architectures than the classical random graphs. Informally, these were networks, maximally uniform for arbitrary fixed degrees of their vertices. Béla Bollobás strictly formulated this generalization of the $G(n,m)$ model and called it the *configuration model* (1980).

Approximately at the same time, in 1976, Derek de Solla Price proposed a model that generated growing networks having a marked number of vertices with many edges. The Price model described the evolution of networks of scientific citations by introducing preferential citing of already highly cited papers. In this way Price explained empirical power-law distributions of the number of citations collected by scientific papers. In fact, similar ideas emerged much earlier, in another context. In 1925, George Udny Yule used a similar mechanism as a possible explanation for the observed power-law distribution of the number of species in genera. It is known that new species emerge due to mutations. Yule supposed that the evolution of a set of genera is a growth process determined by (i) the emergence of new genera with a single species and (ii) the emergence of new species in already existing genera. Yule took into account two contributing factors. First, there is a small chance that mutations create a new genus, and second, mutations in genera with more species are more frequent and so the probability of the emergence of a new species is greater. If the frequency of the emergence of a new species is a linear function of the number of species in a genus, then this

---

[7]Milgram's 1967 paper in *Psychology Today* with the results of his experiment was entitled 'The small world problem'.

growth leads to a power-law distribution. The evolution processes incorporating this mechanism were named *Yule processes*. In 1955, the prominent economist Herbert Alexander Simon developed the quantitative theory of these processes. Theoretical physicists Paul John Flory and Walter Hugo Stockmayer explored such growth in the physics of polymers in the 1940s. In application to networks, after Albert-László Barabási and Réka Albert, this mechanism is called *preferential attachment*.

Very probably, Albert and Barabási coined the term 'complex networks' in their renowned 2002 review 'Statistical mechanics of complex networks', associating it with the networks having more sophisticated architectures than the classical random graphs.[8] It became clear afterwards that classical random graphs are actually not that simple and that they demonstrate in some form many of the effects occurring in complex networks. It is impossible to focus on complex networks without considering classical random graphs, so our book is about all these networks.

1.3 Small and large worlds

The key characteristic of a large network (in particular, a large lattice) is its space dimensionality, that is, the minimal number of coordinates allowing one to specify any vertex in the network. We start with a simplistic consideration, leaving details for further chapters. Consider an infinite regular, for example, hypercubic, finite-dimensional lattice with edges of unit length. Select some its vertex and all the vertices at distance ℓ or smaller from it. The distance is measured as the length of the shortest path between two vertices counting the number of edges in the path. Then the number of vertices $M(\ell)$ in this ball increases as a power of ℓ,

$$M(\ell) \sim \ell^D \tag{1.7}$$

for large ℓ, and

$$D = \frac{d \ln M(\ell)}{d \ln \ell}. \tag{1.8}$$

[8]Note that the notion of complex networks understood in this way is distinct from complex systems. According to Giorgio Parisi (2002), 'A system is complex if its behaviour crucially depends on the details of the system. ... In other words, the behaviour of the system may be extremely sensitive to the form of the equations of motion and a small variation in the equations of motion leads to a large variation in the behaviour of the system.' Mark Newman (2011) proposed another definition: 'A complex system is a system composed of many interacting parts, often called agents, which displays collective behaviour that does not follow trivially from the behaviours of the individual parts.'

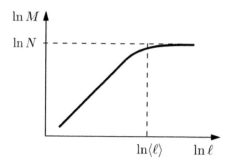

Fig. 1.8 Log-log plot of the dependence $M(\ell)$ for finite-dimensional networks. Here $\langle \ell \rangle$ is the average shortest path length and N is the number of vertices in a network.

The space dimension D, introduced in this way, coincides with the *Hausdorff dimension* in many interesting for us networks, and so we loosely use the latter term as physicists often do (Durhuus, 2009). For a finite network, we can introduce the distribution $\mathcal{P}(\ell)$ of the lengths of the shortest paths connecting the pairs of vertices, the first moment of this distribution, that is, the average shortest path distance $\langle \ell \rangle$, and the diameter of the network, $d = \max(\ell)$, that is, the maximum separation of two vertices in the network. For a large, though finite, lattice of N vertices, the power-law dependence $M(\ell)$, Eq. (1.7) approaches the upper limit N when $\ell \sim d \sim \langle \ell \rangle$ (Figure 1.8), so

$$N \sim d^D \sim \langle \ell \rangle^D. \tag{1.9}$$

Therefore, the mean separation of vertices in finite-dimensional lattices is large in the sense that it increases with N in a power-law way,

$$\langle \ell \rangle(N) \sim N^{1/D}. \tag{1.10}$$

Figure 1.8 demonstrates that distinguishing a high-dimensional network from an infinite-dimensional one is a difficult task—possible only for very large networks.

Next, let us consider a basic, non-random network, a tree, whose vertices are markedly less separated from each other than in the finite-dimensional lattices. Our example is a *Cayley tree*. This is a symmetric tree that has all vertices except *leaves* (vertices with degree 1) of the same degree $k \geq 2$ (Figure 1.9).[9] All leaves in the Cayley tree is at distance equal to one

---

[9] This is how the Cayley trees are defined by physicists. The Cayley trees in graph theory are defined without condition of symmetry.

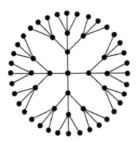

Fig. 1.9 A Cayley tree with $k = 4$.

half of it diameter, $\ell_{\max}$, from the central vertex, the *root*, so they form a boundary. We assume that our Cayley tree is sufficiently large. Let us choose some vertex at a distance greater than ℓ from the boundary. An elementary estimate gives the number $M(\ell)$ of vertices at a distance ℓ or closer from the chosen vertex,

$$M(\ell) \sim (k - 1)^{\ell}, \tag{1.11}$$

for large ℓ. Here we ignore a factor independent on ℓ on the left-hand side of this formula. Therefore, we get the following expression for the average separation of vertices in the Cayley tree:

$$\langle \ell \rangle (N) \sim \frac{\ln N}{\ln(k - 1)}, \tag{1.12}$$

that is a logarithmic function of N in contrast to a power law in Eq. (1.10). The logarithm in Eq. (1.12) can be interpreted as the limit $D \to \infty$ of the power law in Eq. (1.10). Thus, a Cayley tree has a dramatically smaller average separation of its vertices than a finite-dimensional lattice of the same size, that is, with the same (large) number N of vertices. So one can reasonably call finite-dimensional graphs *large worlds* and infinite-dimensional ones, similar to the Cayley tree, *small worlds*.

Average separations not only sharply differ from each other in these two classes of networks, but also the entire shortest path length distributions $\mathcal{P}(\ell)$ are very different. One can immediately obtain these distributions for one-dimensional lattices, namely, the chain and the ring of N vertices. Figure 1.10 shows the resulting distributions for large N. We see that in both examples, the width of the distribution is of the order of the average shortest path length, $\langle \ell \rangle$. One can check that the large width of shortest path length distributions is generic for all finite-dimensional networks and lattices.

The distribution $\mathcal{P}(\ell)$ for an infinite-dimensional network looks quite differently (Figure 1.11 for the Cayley tree with $k = 4$). The figure shows that

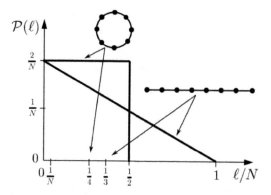

Fig. 1.10 Shortest path length distributions $P(\ell)$ for a chain and for a ring of N vertices. The average separation of vertices $\langle \ell \rangle$ is $N/3$ and $N/4$ in the large chain and the large ring, respectively.

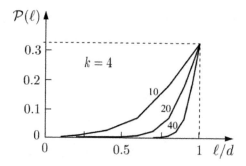

Fig. 1.11 Shortest path length distribution $P(\ell)$ for the Cayley tree with $k = 4$, see Figure 1.9. The curves are for three values of the diameter, $d = 10, 20$, and 40, which correspond to network sizes $N = 485$, $11{,}897$, and $6{,}973{,}568{,}801$, respectively.

for large Cayley trees, the shortest path length distribution clubs near the diameter of a network, and the width of this distribution is of the order of 1. That is, the majority of vertices in this network are almost equally separated from each other when it is large. Such a quasi-equidistant separation appears to be generic for small words.

1.4   First structural characteristics of networks

One can introduce local and global structural characteristics of a graph. Let us start from the local ones, which are for individual vertices and use only local information. We assume that networks are *undirected*, that is, their edges are not oriented. The first natural characteristic is the *degree* of a given vertex i in a graph. This is the number q_i of the edge ends attached to this vertex. For example, if a vertex has only a self-loop (1-cycle), then its degree equals 2, although this vertex is incident to a single edge. The next quantity, *local clustering*, characterizes the set of connections between the nearest neighbours of a vertex i. This is the ratio of the number t_i of edges between the nearest neighbours of vertex i and the maximum possible number of such connections, namely,

$$C_i = \frac{t_i}{q_i(q_i - 1)/2},\qquad(1.13)$$

so $C_i \leq 1$. Here, t_i is also the number of triangles attached to vertex i, and the denominator $q_i(q_i - 1)/2$ is the maximum possible number of such triangles (cycles of length 3).[10]

Introducing the shortest path distance ℓ_{ij} between two vertices, i and j, which depends on how connections are organized in a part of this graph between i and j, enables us to define the *diameter* $d = \max_{i,j} \ell_{ij}$ characterizing the entire graph. The diameter is an example of a global structural characteristic. Note that this characteristic is useful only if the biggest connected component in a graph is large compared to other its connected components.

The statistics of degrees in a random network is provided by the degree distribution,

$$P(q) = \left\langle \frac{N(q)}{N} \right\rangle = \left\langle \frac{1}{N} \sum_i \delta(q_i, q) \right\rangle,\qquad(1.14)$$

where N is the number of vertices in a graph—a member of the statistical ensemble (this number can be different in different members of the ensembles), $N(q)$ is the number of vertices of degree q in this graph, and $\langle \ldots \rangle$ denotes averaging over the ensemble, explained in Section 1.1.[11] $\delta(q_i, q) \equiv \delta_{q_i, q}$ is the Kronecker symbol. The moments of this distribution are

$$\langle q^n \rangle = \left\langle \frac{1}{N} \sum_i q_i^n \right\rangle = \sum_q P(q) q^n.\qquad(1.15)$$

Other observables can be obtained in a similar way. Notably, there are two options for clustering. Based on Eq. (1.13), we can introduce (i) the *average clustering*

[10]The definition, Eq. (1.13), is not applicable to vertices of degrees 0 and 1 (isolated vertices and leaves). For these vertices, it is natural to set $C_i = 0$.

[11]For example, the graph in Figure 1.2 has the following degree distribution: $P(0) = 1 - 2p/3 - q$, $P(1) = 2p/3$, $P(4) = q$.

$$\langle C \rangle = \left\langle \frac{1}{N} \sum_i \frac{t_i}{q_i(q_i - 1)/2} \right\rangle \qquad (1.16)$$

and (ii) the *clustering coefficient* (known as *transitivity* in sociology)

$$C = \left\langle \frac{\sum_i t_i}{\sum_i q_i(q_i - 1)/2} \right\rangle$$

$$= \left\langle 3 \frac{\text{the number of triangles in the graph}}{\text{the number of connected triples in the graph}} \right\rangle. \qquad (1.17)$$

A triple here is a vertex and two its nearest neighbours. Clearly, $\sum_i t_i$ equals three times the number of triangles (3-cycles) in the graph. One can see that $0 \leq C, \langle C \rangle \leq 1$. Clearly, the clustering of trees is zero.

In a similar way we can obtain the average shortest path length between two vertices (average separation between vertices) in a random network,

$$\langle \ell \rangle = \left\langle \frac{\sum_{(ij)} \ell_{ij}}{\mathcal{N}_{\text{connected pairs}}} \right\rangle = \left\langle \frac{\sum_{(ij)} \ell_{ij}}{\sum_a N_a(N_a - 1)/2} \right\rangle, \qquad (1.18)$$

where $\mathcal{N}_{\text{connected pairs}}$ is the number of pairs of vertices in a graph (a given member of the statistical ensemble) having at least one connecting path. The sum $\sum_{(ij)} = \frac{1}{2} \sum_{i,j}$ is over these pairs of vertices. When this graph consists of clusters (connected components) of size N_a, the number $\mathcal{N}_{\text{connected pairs}}$ equals $N_a(N_a - 1)/2$. The statistics of shortest path length in a random network is provided by the shortest path length distribution,

$$\mathcal{P}(\ell) = \left\langle \frac{\mathcal{N}(\ell)}{\mathcal{N}_{\text{connected pairs}}} \right\rangle$$

$$= \left\langle \frac{\sum_{(ij)} \delta(\ell_{ij}, \ell)}{\sum_a N_a(N_a - 1)/2} \right\rangle \xrightarrow{N \to \infty} \frac{\left\langle \sum_{(ij)} \delta(\ell_{ij}, \ell) \right\rangle}{\langle \sum_a N_a(N_a - 1)/2 \rangle}. \qquad (1.19)$$

In the following we discuss many other characteristics of random networks, and, in principle, each of them is introduced in a similar way to this section. Namely, first we define a quantity for an individual graph, and secondly, for a random network, we average this quantity over the corresponding statistical ensemble.

1.5   Equilibrium versus growing trees

The difference between equilibrium and non-equilibrium random networks may be huge even when they are based on, at first sight, similar graphs.

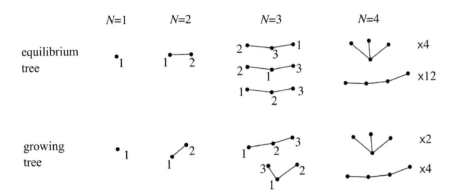

Fig. 1.12 The random equilibrium connected tree versus the random recursive tree. The upper and lower rows show the members of these two statistical ensembles for $N = 1, 2, 3$, and 4. Each of the graphs in the figure has a unit statistical weight. For the sake of compactness, at $N = 4$ we indicate the number of configurations which differ from each other only by labels.

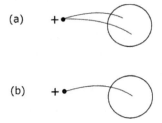

Fig. 1.13 How recursive graphs grow. (a) The growth of a general recursive graph. Each new node is attached to a number of existing vertices. Note that new edges are only between an added vertex and already existing vertices. (b) The growth of a recursive tree (we assume that the growth starts from a single vertex).

For demonstration purposes we compare two different random connected trees—equilibrium and growing—of equal size. The term *connected graph* means that there is a path between each two its vertices, that is, this graph consists of a single *connected component*. As in classical physics, the vertices of graphs in this book are distinguishable, and they are typically labelled, although, of course, one can also consider unlabelled graphs. Let us first inspect the ensemble consisting of all possible labelled connected trees with N vertices, each tree taken with equal statistical weight, say 1. This is an equilibrium ensemble. In other terms, this connected random tree is uniformly random, or one can say that this random tree is a maximally random connected tree with a given number of vertices. The number of such labelled connected trees of size N equals

$$N_{\text{labelled connected trees}} = N^{N-2}, \tag{1.20}$$

which is well-known Cayley's formula (Cayley, 1889). The number of all unlabelled connected trees with N vertices is, of course, much smaller. It asymptotically approaches the following number

$$N_{\text{unlabelled connected trees}} \cong 2.955765286^N N^{-5/2} \tag{1.21}$$

for large N (Otter, 1948; Plotkin and Rosenthal, 1994). The top row in Figure 1.12 shows these uniform random trees for $N = 1, 2, 3$, and 4.

For the sake of comparison, let us now consider the *random recursive tree*, which is the basic reference model of a growing random network. Figure 1.13a explains the process generating general recursive graphs. A recursive network grows in the following way: add a new vertex and attach it to a number of already existing vertices. Then repeat again, and again, and again. Importantly, no edges emerge between existing vertices, which makes these networks easy for analysis. Let us now suppose that in the recursive process, each new vertex has only one edge, and the growth starts from a single vertex (Figure 1.13b). Then this graph is a recursive tree (i.e., it has no cycles). One should specify how a vertex for attachment is selected from the set of existing vertices. In the random recursive tree this selection is made in the simplest way, namely, uniformly at random. That is, at each step, choose uniformly at random a vertex from existing ones and attach a new vertex to it. This construction provides us with a maximally random tree with a given number of vertices under only one restriction: this tree is obtained by sequential addition of labelled vertices—the causality restriction. We will consider this important random graph in more detail in Section 5.1.

The bottom row in Figure 1.12 shows the members of this non-equilibrium statistical ensemble for $N = 1, 2, 3$, and 4. (We label the vertices according to their age.) One can see that some of the labelled trees from the top row do not emerge in the bottom one due to the causality constraint. For example, the chain $1-3-2$ does not occur in the recursive tree. Moreover, for large N, only a tiny fraction of all possible labelled trees are realized. One can easily find the number of labelled recursive trees with N vertices and its large N asymptotics,

$$N_{\text{labelled recursive trees}} = (N-1)! \cong \sqrt{\frac{2\pi}{N}} e^{-N} N^N \tag{1.22}$$

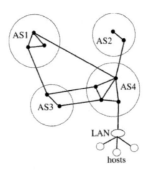

Fig. 1.14 Multilayer organization of the Internet. Open dots show host computers of users, LAN is a local area network, filled dots are routers, and ASs are autonomous systems.

(Drmota, 2009; Drmota, 2008),[12] which is exponentially smaller than the corresponding number for the equilibrium connected trees, Eq. (1.20). If we remove the labels of vertices, then both ensembles, the recursive trees and equilibrium ones, will contain the same unlabelled trees. Figure 1.12 demonstrates that the statistical weights of the members of these two ensembles, equilibrium and non-equilibrium, differ already at $N = 4$. This difference leads to a marked distinction between these two random trees and their characteristics. In particular, while the degree distribution $P(q)$ of the equilibrium random trees with large N equals $e^{-1}/(q-1)!$ (Burda, Correia, and Krzywicki, 2001; Bialas, Burda, Jurkiewicz, and Krzywicki, 2003), decaying as q^{-q} for large q, the degree distribution of the recursive random trees is exponential, $P(q) = 2^{-q}$, that is, it decays much slower.

We shall find a far more dramatic difference if we inspect the size dependence of the average shortest path distances, $\langle \ell \rangle(N)$ for these two random trees. One can easily write the equation for the evolution of $\langle \ell \rangle(N)$ in the random recursive tree. For that we compare the total lengths of the shortest paths between all pairs of vertices in this network at 'times' N and $N+1$,

$$\langle \ell \rangle(N)\frac{(N+1)N}{2} = \langle \ell \rangle(N)\frac{N(N-1)}{2} + 1 + [\langle \ell \rangle(N) + 1](N-1). \quad (1.23)$$

The first term on the right-hand side is the sum of the lengths of the shortest paths between all pairs of vertices at 'time' N. The term on the left-hand

[12]The probability of generating each labelled tree of N vertices, allowed by this process, is the same: $1 \times 1 \times \dfrac{1}{2} \times \dfrac{1}{3} \times \ldots \times \dfrac{1}{(N-1)} = \dfrac{1}{(N-1)!}$, which explains Eq. (1.22). Thus in each of these two statistical ensembles all members (labelled graphs) are equiprobable.

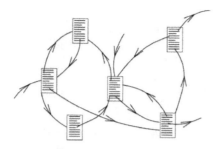

Fig. 1.15 Connections in the WWW. Notice reciprocal edges.

side is this total length at 'time' $N+1$. The term $1+[\langle\ell\rangle(N)+1](N-1)$ on the right-hand side is the difference due to the new attached vertex. Solving this equation we get $\langle\ell\rangle(N) \cong 2\ln N$ when N is large, that is, the random recursive tree is a small world. The calculation of the average shortest path length for the equilibrium random connected trees is more demanding mathematically (for details, see Durhuus, 2009). The result is $\langle\ell\rangle(N) \sim \sqrt{N}$, that is, the Hausdorff dimension of this tree equals 2.[13] Consequently, this equilibrium random tree is a large world unlike the random recursive tree. One could hardly imagine a greater difference between two random networks.

1.6 Real-world networks

Let us briefly outline how real networks look. Here are only a few demonstrative examples. The first is the Internet, the network of computers interconnected by wires and wireless links. This undirected multilayer network includes hosts (computers of users), servers (computers or programs providing a network service), and routers arranging the traffic. Three layers of the Internet (see Figure 1.14) can be considered: (i) the network of computers (hosts), (ii) the network of routers, and (iii) the network of the autonomously administered subnetworks (autonomous systems, AS). The routers have geographic locations, while the autonomous systems can span vast areas, overlapping each other. All three networks are sparse and loopy. Their average degrees $\langle q \rangle$ are about 10^1, the empirical degree distributions

---

[13]The Hausdorff dimension of equilibrium connected trees with the degree distribution $P(q) \sim q^{-\gamma}$ has the values

$$D_H = \begin{cases} 2, & \gamma \geq 3, \\ \dfrac{\gamma-1}{\gamma-2} > 2, & 2 < \gamma < 3 \end{cases}$$

(Burda, Correia, and Krzywicki, 2001), see Section 6.2.

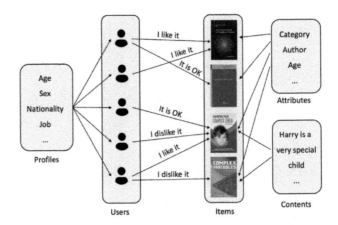

Fig. 1.16 Recommender network. Adapted from Lü, Medo, Yeung, Zhang, Zhang, and Zhou (2012).

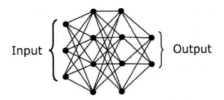

Fig. 1.17 A small deep learning network.

are power laws, $P(q) \sim q^{-\gamma}$, where exponent γ is close to 2.1 or 2.2 in a sufficiently broad range of degrees. Networks with power-law degree distributions are often called *scale-free*. In contrast to the Internet, the World Wide Web (WWW, Web) is a directed network. Its vertices are documents (web pages) and directed edges are hyperlinks between these documents (Figure 1.15). Notably, up to 60% of all connections in the WWW are reciprocal, and there are a lot of short cycles and triangles. The empirical distribution of the number of incoming connections of a page is power-law, with the exponent close to that of the Internet, while the distribution of the number of outgoing links of a page decays faster than any power law.

Recommender systems have a very different network structure. These networks contain vertices of two kinds: items (products) and users (buyers) (Figure 1.16). The edges in these nets, which connect users to items (buyers to products), can indicate purchases, likes, dislikes, reactions, inspections, recommendations, and so on.

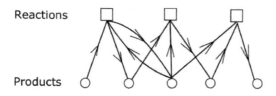

Fig. 1.18 Chemical reaction network.

The next example of complex network architectures is provided by specific algorithms—deep learning networks (a particular type of artificial neural networks) containing many layers of nodes. These nodes are non-linear transformations of signals coming from other nodes. In basic feedforward deep learning networks all possible connections between consecutive layers are present (Figure 1.17).

Networks of chemical reactions have two kinds of vertices—products and chemical reactions (Figure 1.18). Each reaction vertex is connected with the vertices participating in this reaction by directed edges. Some of these edges can be reciprocal (or bidirectional). Typically, connections within each of these two layers—products and chemical reactions—are absent. In principle, such links are also possible. For example, one chemical reaction producing heat can influence another reaction.

Finally, let us try to guess the rational structural organization of the huge arrays of information about individuals and their contacts and interactions, which are so zealously collected all over the world according to the tips of Mr Snowden and many others. Figure 1.19 schematically shows a possible network structure of such an array. Each vertex in this network is a set of

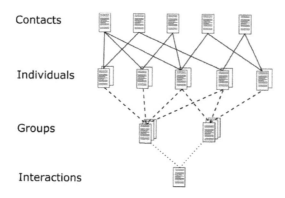

Fig. 1.19 Hypothetical network structure of big data about individuals and groups.

files of information. The vertices are (at least) of four different types: (i) for individuals, (ii) for their contacts (each contact can involve any number of participants), (iii) for groups, communities, cells, teams, etc., and (iv) for various interactions between these groups. Pieces of information stored in different files can overlap with each other. The resulting complex multilayer architecture provides a wide range options for data processing.

2

Graphs

2.1 Types of graphs

Let us put aside random graphs, statistical ensembles, and probabilities for a while in favour of individual graphs while we introduce a few necessary basic notions.[1] First we discuss undirected graphs.

A *hypergraph* is a set of vertices in which some multi-subsets are tied (Figure 2.1).[2] These ties are called edges (or, often, hyperedges). So a hypergraph is a pair of coupled sets, vertices and edges: $G = (V, E)$, $V = \{v_1, v_2, \ldots, v_{|V|}\}$, $V = \{e_1, e_2, \ldots, e_{|E|}\}$. Here $|V|$ and $|E|$ are the numbers of the vertices and edges of the hypergraph.[3] A *simple hypergraph*

[1] For the sake of compactness, we often omit the labels of vertices.
[2] A *multiset* is a set in which some elements can occur more than once.
[3] For the sake of compactness, we often denote $|V| \equiv N$ and $|E| \equiv E$.

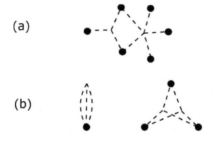

(a)

(b)

Fig. 2.1 Simple hypergraph (a) and multi-hypergraph (b).

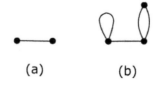

(a)   (b)

Fig. 2.2 Simple graph (a) and multigraph (b).

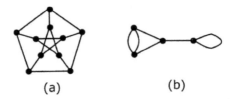

(a)   (b)

Fig. 2.3 Examples of regular graphs. (a) Petersen graph—a simple 3-regular graph. (b) A small 3-regular multi-graph.

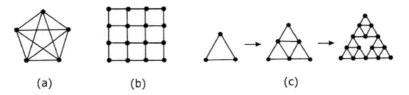

(a)   (b)   (c)

Fig. 2.4 (a) Complete graph, (b) lattice, (c) fractal (Sierpinski gasket). At each step, the diameter d of the Sierpinski gasket doubles while the numbers of its edges and vertices (asymptotically) triple, $N \cong (3/2)d^{\ln 3/\ln 2}$. So its Hausdorff dimension is $D_H = \ln 3/\ln 2$.

(Figure 2.1a) does not contain self-loops (1-cycles) and repeated edges, in contrast to a *multi-hypergraph*, (Figure 2.1b). The *m-uniform hypergraphs* (or simply *m-hypergraph*) contain only edges with m ends (edges of *cardinality m* or, often, *m-order edges*). The maximum number of edges in the simple m-hypergraph with N vertices equals $\binom{N}{m}$. The maximum number of edges in the simple nonuniform hypergraph with N vertices equals $\sum_{m=1}^{N} \binom{N}{m} = 2^N - 1$ (here we include even 1-edges).

Graphs are 2-uniform hypergraphs, that is, each edge of a graph has exactly two ends. We mostly speak about *simple graphs*, that is, the graphs that have no 1- and 2-cycles (cycles of length 1 and 2, i.e., edges having the same end vertices) (Figure 2.2a). The number of labelled simple graphs of N vertices equals

$$N_{\text{labelled simple graphs}} = 2^{N(N-1)/2} \overset{N\to\infty}{\cong} N_{\text{connected labelled simple graphs}}, \quad (2.1)$$

and only a small fraction, about $2/N^3$, of labelled simple graphs of size N are not connected. The number of large unlabelled simple graphs asymptotically equals

$$N_{\text{unlabelled simple graphs}} \cong N_{\text{connected unlabelled simple graphs}} \cong \frac{2^{N(N-1)/2}}{N!} \quad (2.2)$$

24

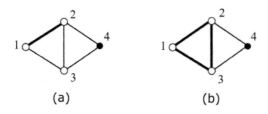

Fig. 2.5 (a) One of subgraphs with vertices 1, 2, and 3. (b) The subgraph induced by the same vertices.

with the same small fraction, about $2/N^3$, of unconnected graphs. Compare these formulas with those for labelled and unlabelled connected trees, Eqs. (1.20) and (1.21). The *multigraphs* have 1- and/or 2-cycles (Figure 2.2b).[4] A *regular graph* has all vertices with equal degrees. A *q-regular graph* has all vertices of degree q (Figure 2.3). Figure 2.4 provides a few examples of simple graphs, including a *complete graph* (fully connected graph), a lattice, and a fractal. The sum of the degrees of vertices of a graph equals twice the number of its edges,

$$\sum_i q_i = 2E. \tag{2.3}$$

For hypergraphs, the direct generalization of this relation also looks simple:

$$\sum_i q_i = \sum_m m E_m, \tag{2.4}$$

where E_m is the number of edges of cardinality m in a hypergraph.

A *subgraph* is a piece of a graph. The subgraph $G' = (V', E')$ of a graph $G' = (V, E)$ is formed by subsets of vertices $V' \subseteq V$ and of edges $E' \subseteq E$ such that each edge in E' connects two of the vertices of V' (Figure 2.5a). In the subgraph $G' = (V', E')$ induced by vertices V' (*vertex-induced subgraph*), the subset E' consists of all the edges of the graph G that connect vertices V' (Figure 2.5b). A complete subgraph is called a *clique*. The *n-clique* is the complete subgraph of n vertices.

The *clique (or primal) graph* of a hypergraph is a graph having the same vertices as the hypergraph and edges interconnecting all pairs of end vertices for each hyperedge (Figure 2.6). Clearly, this representation is not unique: different hypergraphs may have the same clique graph.

---

[4]Sometimes they are called *pseudographs*.

Fig. 2.6 Hypergraph with two hyperedges of cardinality 3 and 4 and its clique (or primal) graph.

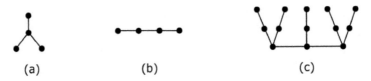

(a)　　　　　　　(b)　　　　　　　(c)

Fig. 2.7 (a) Star, (b) chain, (c) brush.

Graphs can consist of a number of *connected components*. These are disjoint subgraphs, each having at least one path between any two of its vertices. A *connected graph* consists of a single connected component.

Trees are simple connected graphs having no cycles.[5] *Stars* (trees with the maximum number of leaves), *chains* (trees with the minimum number of leaves, two), *brushes*, etc., provide simple examples of trees (Figure 2.7). When one vertex of a tree is special (*root*), the tree is called *rooted*. The numbers of vertices, N, and edges, E, in a tree are related in a simple way,

$$N = E + 1. \tag{2.5}$$

That is, the total degree of the vertices of a tree is $\sum_{i=1}^{N} q_i = 2(N-1)$. In a *forest*, all connected components are trees. If the number of these components is M, then $N = E + M$.

A tree subgraph T of an undirected graph G, having all its vertices, that is, $|T| = |G|$, is called a *spanning tree* of the graph. The number of spanning trees[6] of a (labelled) complete graph with N vertices equals N^{N-2}, which readily follows from Cayley's formula, Eq. (1.20).

It is easy to turn a graph with cycles into a tree, for example, by deleting edges. One can show that the minimal number of edges that must be removed from an undirected graph with N vertices, E edges, and M connected components to eliminate all its cycles equals

$$r = E - N + M. \tag{2.6}$$

---

[5] Arthur Cayley coined this term in 1857.

[6] This number is also called also called the *complexity of a graph*.

This number is called the *cycle (or circuit) rank*, or *cyclomatic number* (the term and notion introduced by Gustav Kirchhoff). This is the number of *independent cycles* in an undirected graph, where the set of independent cycles, by definition, consists of all cycles that cannot be made by combining other cycles of this set.[7] Any other cycle in a graph can be represented as a combination of two or more of these independent cycles.

A wide class of simple graphs allow embedding in topological surfaces—sphere, torus, etc.—without intersecting their edges with each other. In particular, *planar graphs (planars)* can be embedded in the plane in this way. Let such finite graph contain a single component. Then the embedding surface is divided into regions surrounded by edges of the graph, *faces*. The numbers of vertices, N, edges, E, and faces, F, in this graph satisfy Euler's renowned formula:

$$N - E + F = \chi = 2 - 2g. \tag{2.7}$$

Here χ is the Euler characteristic of the topological surface in which the graph is embedded. For a sphere with g genera (holes, or handles), $\chi = 2 - 2g$.[8] If in addition, p discs are cut from this sphere, then $\chi = 2 - 2g - p$. For a sphere, $\chi = 2$, for a torus, $\chi = 0$, for a disk $\chi = 1$, for a cylinder and for a Möbius strip, $\chi = 0$, and so on.[9] For a planar graph, $N - E + F = 2$, if we count the surrounding infinite area as a face. Then a tree, which is certainly a planar graph, has a single 'face', and we return to Eq. (2.5). The reader can easily check all these relations for smallest possible graphs.

Two (hyper)graphs are *isomorphic* iff their vertices are connected in the same way, that is, there exists a permutation of the vertices of the first graph such that the adjacent vertices of this graph, after the permutation, become the adjacent vertices of the second. So the difference between two

---

[7]Note that the set of independent cycles of a graph is not unique. For example, for the graph in Figure 2.5, this set can consist of the cycles $(1, 2), (2, 3), (3, 1)$ and $(2, 4), (4, 3), (3, 2)$ or of the cycles $(1, 2), (2, 3), (3, 1)$ and $(1, 2), (2, 4), (4, 3), (3, 1)$, or, finally, of the cycles $(2, 4), (4, 3), (3, 2)$ and $(1, 2), (2, 4), (4, 3), (3, 1)$.

[8]The number g is also the second Betti number, b_2. In simple terms, the d-th *Betti number b_d* is the number of cavities (holes, voids) in d-dimensional topologically non-equivallent subspaces of a topological space. For example, for a torus, $b_0 = 1$ (single component), $b_1 = 2$ (due to two nonequivalent circles on the surface of the torus) $b_2 = 1$ (the hole in the torus). For an undirected graph, b_1 is the cycle rank, Eq. (2.6).

[9]For the sake of completeness, let us mention cross-cups (self-intersection of a one-sided surface), the topological features which can be imagined in the following way. Join together two points of a sphere; clasp the resulting tube along a line from this point up to the hilt; reconnect crosswise the two surfaces along the zip. For a sphere with g genera, p discs removed, and c cross-cups, $\chi = 2 - 2g - p - c$.

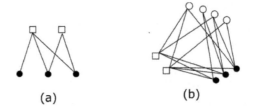

(a)   (b)

Fig. 2.8 Bipartite (a) and tri-partite (b) graphs.

Fig. 2.9 A small directed graph.

isomorphic (hyper)graphs is only in labelling their vertices. For graphs embedded in topological surfaces and graphs with finite maximum degree, one can determine in a polynomial time whether two graphs are isomorphic or not.

Figure 2.8 explains *bipartite* and, more generally, *multipartite graphs*. In p-partite (simple) graphs, there are $p \geq 2$ different types of vertices, and edges connect only vertices of different types. Notice that the multipartite graphs have no triangles. The hypergraphs can be represented as the bipartite graphs if we treat the hyperedges as the vertices of the second kind.

In *directed graphs* (also called *digraphs*), at least some of their edges are directed (directed edge has a tail and a head) (Figure 2.9). The basic characteristic of a given vertex i in a directed graph is the triple (q_i, q_o, q) (*in-degree, out-degree, degree*), that is the numbers of incoming, outgoing, and undirected edges adjacent to the vertex. In *oriented graphs*, all edges are directed. It is also possible to introduce *directed hypergraphs*. For that, one has to define a *oriented hyperedge* in the following way. The ends of such a hyperedge are divided into two groups (tail and head), which determines the direction of the hyperedge (Banerjee and Char, 2017). More generally, one can introduce a list of source and target end vertices for each hyperedge, that is, orientation, which leads to a *multi-directed hypergraph*. In particular, if edges have cardinality 2, this reduces to a better-known *bidirected graphs*.[10]

Usually *weighed graphs* are defined as the graphs whose edges have ascribed weights, typically real numbers (Figure 2.10). More generally, weighted

---

[10]In fact, in graph theory, bidirected graphs are defined with edges of cardinality 1 and 2, so edges with a single end vertex, source, or target, are also possible. Defined in this way bidirected graphs are actually hypergraphs.

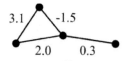

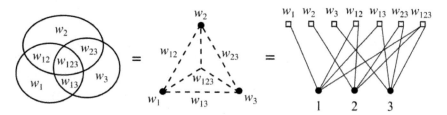

Fig. 2.10 Edge-weighted graph.

Fig. 2.11 The full set of overlaps of three masses (communities) and two its equivalent representations. Here w_i is the weight of the part of mass i that does not overlap with other masses, w_{ik} is the weight of the part of the overlap of i and k that does not overlap with the third mass; w_{123} is the weight of the overlap of all three masses.

graphs can be edge-weighted and vertex-weighted. Similarly, one can introduce directed weighted graphs and *weighed hypergraphs*. Figure 2.11 shows an example of a weighted hypergraph describing the overlaps of three masses (or communities) and its equivalent representation by a bipartite graph.

2.2   Graphicality

Suppose that we have a given set of some (typically local) structural features. The *graphicality* is about whether a graph with this set of features can be realized or not. If we focus on the simplest set of local features, namely the sequence of vertex degrees $q_1, q_2, \ldots, q_N$, then immediately we see that a graph with a single vertex of degree 1 is impossible. A *graphic sequence* is the sequence of non-negative integers such that there exists a graph with this degree sequence. The Erdős–Gallai theorem provides a criterion that a sequence of numbers is graphic. Namely, a sequence $q_1 \geq q_2 \geq \ldots \geq q_N$, $N \geq 1$, is graphic iff two conditions hold:

(1) $\sum_{i=1}^{N} q_i$ is even;
(2) the following N inequalities are satisfied:

$$\sum_{i=1}^{k} q_i \leq k(k-1) + \sum_{i=k+1}^{N} \min(q_i, k), \quad \text{for any } 1 \leq k \leq N \qquad (2.8)$$

29

(Erdős and Gallai, 1960).

Based on the following observation, there may be alternative, and somewhat more practical, algorithms for checking if a sequence is graphic. For example, if we remove some element j from a graphic sequence $q_1, q_2, \ldots, q_N$, then a sequence of non-negative integers $q'_1, q'_2, \ldots, q'_{N-1}$ with $\sum_{i=1}^{N-1} q'_i = \sum_{i \neq j}^{N} q_i - q_j$ is also graphic, and vice versa. Progressively applying this idea to shorter and shorter sequences, one can always come to sequences so short that the problem becomes trivial. One of possible ways to perform such a reduction is realized in the Havel–Hakimi algorithm that uses the equivalence: the sequence $q_1 \geq q_2 \geq \ldots \geq q_N$ ($N \geq 3$, $q_1 \geq 1$) is graphic iff the sequence $q_2 - 1, q_3 - 1, \ldots, q_{q_1+1} - 1, q_{q_1}, \ldots, q_N$ is graphic (Havel, 1955; Hakimi, 1962). Similar graphicality criteria and algorithms were derived for directed, bipartite, and some other graphs.

2.3 Representations of graphs

By definition, a graph is represented by a set of pairs of labels; each pair contains two labels of the end vertices of the corresponding edge. For a directed edge, this pair is ordered. There are two other convenient ways to mathematically represent a graph: (i) an incidence matrix indicating the pairs: an edge—a vertex incident to it, and (ii) an adjacency matrix indicating the pairs of adjacent vertices.

2.3.1 Incidence matrix

The *incidence matrix* B of a graph with N vertices and E edges has N rows and E columns. For an oriented graph,[11] this $N \times E$ matrix has the elements:[12]

- $B_{ij} = -1$ if vertex i is the source vertex of edge j;
- $B_{ij} = +1$ if vertex i is the target vertex of edge j;
- $B_{ij} = 2$ if edge j is a self-loop at vertex i;
- the remaining entries are zero.

So, for a simple oriented graph, the sum of elements in each column of this matrix is zero, $\sum_i B_{ij} = 0$ (an edge has one source and one target).

---

[11]Recall that, by definition, in an oriented graph, all edges are directed.

[12]Recall that B_{ij} is the entrance at the ith row and jth column of the matrix. Note that often the incidence matrix is defined with exchanged rows and columns compared to our definition, that is, as B^T. Sometimes, with such alternative definition, it is called the edge incidence matrix.

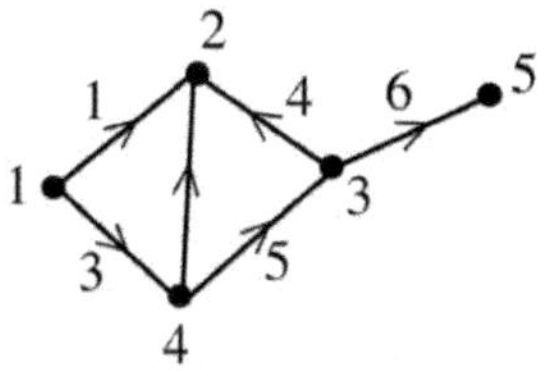

$$= \begin{array}{c} \\ 1 \\ 2 \\ 3 \\ 4 \\ 5 \end{array} \begin{array}{cccccc} 1 & 2 & 3 & 4 & 5 & 6 \\ \left(\begin{array}{cccccc} -1 & 0 & -1 & 0 & 0 & 0 \\ +1 & +1 & 0 & +1 & 0 & 0 \\ 0 & 0 & 0 & -1 & +1 & -1 \\ 0 & -1 & +1 & 0 & -1 & 0 \\ 0 & 0 & 0 & 0 & 0 & +1 \end{array} \right) \end{array}$$

Fig. 2.12 One of possible orientations of a simple graph of 5 vertices and 6 edges and the corresponding 5×6 incidence matrix.

This allows us conveniently describe, for example, electrical potential $\mathbf{x}$ (N-component vector for the graph) at the vertices of the graph without current (x_i is the component of the vector $\mathbf{x}$ at vertex i):

$$B^T \mathbf{x} = 0, \tag{2.9}$$

where T denotes the transposition operation. This just gives E equalities for all edges, namely,

$$x_i - x_{i'} = 0 \tag{2.10}$$

for an edge connecting vertices i and i'. Furthermore, if we multiply the matrix B and a E-component vector $\mathbf{y}$ (components y_α, $\alpha = 1, 2, \ldots, E$),

$$B\mathbf{y} = 0, \tag{2.11}$$

we get the Kirchhoff law for currents flowing along the edges of the graph. Equation (2.11) provides N equalities,

$$-\sum_{j \in \overleftarrow{\partial i}} y_{ji} + \sum_{j \in \overrightarrow{\partial i}} y_{ji} = 0, \quad i = 1, \ldots, N. \tag{2.12}$$

Here the sums are actually over the edges $\alpha = (ji)$ to which vertex i is incident; the first sum is over outgoing edges and the second is over incoming. Notice the notations that we shall frequently use: ∂i is the full set of the nearest neighbours of vertex i; $\overleftarrow{\partial i}$ is the set of the second ends of the outgoing edges of vertex i; $\overrightarrow{\partial i}$ is the set of the second ends of the incoming edges of vertex i. So we have $\partial i = \overrightarrow{\partial i} + \overleftarrow{\partial i}$.

For undirected graphs, one can introduce two distinct incidence matrices. The *unoriented incidence matrix* (*incidence matrix*) $\tilde{B}$ of an undirected graph is the $N \times E$ matrix with the elements:

- $\tilde{B}_{ij} = 1$ if vertex i is a vertex incident to edge j;

31

- $\tilde{B}_{ij} = 2$ if edge j is a self-loop at vertex i;
- the remaining entries are zero.

The second, *oriented incidence matrix*, is used much more widely. For a simple undirected graph, this $N \times E$ matrix is defined in the following way. Set some direction for each of the edges of the graph and use the incidence matrix for this now-oriented graph, which gives the following entries:

- $B_{ij} = -1$ if vertex i is the source vertex of edge j;
- $B_{ij} = +1$ if vertex i is the target vertex of edge j;
- the remaining entries are zero.

Figure 2.12 shows the oriented incidence matrix of a small sample graph. With this matrix, Eqs. (2.9) and (2.11) (Kirchhoff law for currents) hold, which makes it particularly useful. The oriented incidence matrix defined in this way is, of course, not unique, but this does not lead to any inconvenience. The *nullspace (kernel)* of an oriented incidence matrix of an undirected graph is the set of E-component vectors representing all Eulerian subgraphs of this graph. The *nullity* (dimension of the nullspace) is just the cycle (or circuit) rank or cyclomatic number r, the number of independent cycles in a graph, which was introduced in Section 2.1. This number equals

$$r = E - \text{rank}(B), \tag{2.13}$$

that is, the number of columns E of the oriented incidence matrix minus its rank.[13]

One can easily see that if a graph consists of a number of connected components, $G = C_1 \bigcup C_2 \bigcup \ldots$, its incidence matrix can be transformed to the block form:

$$B(G) = \begin{pmatrix} B(C_1) & 0 & 0 \\ 0 & B(C_2) & 0 \\ 0 & 0 & \ddots \end{pmatrix}, \tag{2.14}$$

where $\mathcal{I}(C_i)$ is the incidence matrix of the component C_i, and zeroes 0 here denote the blocks of zeroes.

[13] For the sake of clarity, let us inspect the oriented incidence matrix in Figure 2.12. Its nullspace is based, for example, on two vectors, $(1, -1, -1, 0, 0, 0)$ and $(0, 1, 0, -1, -1, 0)$ representing two cycles passing through edges $1, 2, 3$ and $2, 4, 5$, respectively. The cyclomatic number $r = 2$, while $E = 6$ and $\text{rank}(B) = 4$.

2.3.2 Adjacency matrix

The *adjacency matrix of an undirected graph* of N vertices, $A(G)$, can be constructed in the following way. Start with the $N \times N$ matrix having all its entries zeroes. For each edge between vertices i and j in the graph, add 1 to the entries A_{ij} and A_{ji} of the matrix, and for each self-loop at vertex i add 2 to the entry A_{ii} of the matrix. The resulting adjacency matrix of an undirected graph of N vertices is a $N \times N$ symmetric matrix with the entries:

- $A_{ij} = n$ if there are n edges interconnecting vertices i and j, $i \neq j$;
- $A_{ii} = 2m$ if vertex i has m self-loops;
- the remaining entries are zero.

In particular, in the adjacency matrix of a simple undirected graph, $A_{ij} = 1$ if there is an edge between vertices i and j, $A_{ij} = 0$ if such edge is absent, and the diagonal elements are zero, $A_{ii} = 0$.

The *adjacency matrix of an oriented graph* of N vertices can be introduced in a similar way. Start with the $N \times N$ matrix having all its entries zeroes. For each edge going from vertex i to vertex j in the graph, add 1 to the entry A_{ij} of the matrix, and for each self-loop at vertex i add 1 to the entry A_{ii} of the matrix.[14] The resulting adjacency matrix of an oriented graph of N vertices is a $N \times N$ matrix with the entries:

- $A_{ij} = n$ if there are n edges going from vertex i to vertex j, $i \neq j$;
- $A_{ii} = m$ if vertex i has m directed self-loops;
- the remaining entries are zero.

In terms of these matrices—the incidence matrix and the adjacency matrix—using the joint notation M for both matrices, two graphs, G and G' are isomorphic when there exists a permutation matrix P such that[15]

---

[14]Substituting a pair of reciprocal edges for an undirected edge we arrive at the same matrix for both graphs. So different graphs, undirected and directed, may correspond to the same adjacency matrix.

[15]A permutation matrix for a graph of N vertices is an $N \times N$ matrix each of whose rows and columns contains a single 1 and the rest entries equal zero. For example, one of possible permutation matrices for a graph of 4 vertices looks as follows:

$$P = \begin{pmatrix} 0 & 1 & 0 & 0 \\ 0 & 0 & 0 & 1 \\ 0 & 0 & 1 & 0 \\ 1 & 0 & 0 & 0 \end{pmatrix}.$$

Notice that $P^T = P^{-1}$.

$$M' = PMP^T. \tag{2.15}$$

If a graph G consists of a number of connected components, $C_1, C_2, \ldots$, then its adjacency matrix can be transformed to the block form:

$$A(G) = \begin{pmatrix} A(C_1) & \mathbf{0} & \mathbf{0} \\ \mathbf{0} & A(C_2) & \mathbf{0} \\ \mathbf{0} & \mathbf{0} & \ddots \end{pmatrix}, \tag{2.16}$$

where $A(C_i)$ is the adjacency matrix of the component C_i, and zeroes $\mathbf{0}$ here denote blocks of zeroes.

We may treat a bipartite graph with N_1 vertices of one kind, $i = 1, 2, \ldots$, N_1, and N_2 vertices of the second kind, $i = N_1+1, N_1+2, \ldots, N_1+N_2$, as a graph of N_1+N_2 vertices. Then we arrive at the following structure of the adjacency matrix:

$$A = \begin{pmatrix} \mathbf{0} & B \\ B^T & \mathbf{0} \end{pmatrix}, \tag{2.17}$$

where the $N_1 \times N_2$ matrix B is the so-called *biadjacency or incidence matrix of a bipartite graph* and B^T is its transpose. The matrix B is also the *incidence matrix of the hypergraph*—the counterpart of the bipartite graph—with the same vertices $i = 1, 2, \ldots, N_1$ and N_2 hyperedges substituted for N_2 vertices of the second type in the bipartite graph. The matrix B^T coincides with the incidence matrix of the hypergraph that is the second counterpart of the bipartite graph. This hypergraph has the same vertices $i = N_1+1, N_1+2, \ldots, N_1+N_2$ as the bipartite graph and N_1 hyperedges substituted for N_1 vertices of the first type in the bipartite graph. One can introduce the adjacency matrix of a hypergraph (Appendix A), but for practical purposes it is usually convenient to use an incidence matrix or, equivalently, the adjacency matrix of the corresponding bipartite graph, Eq. (2.17).

The degree q_i of vertex i in an undirected graph of size N is expressed in terms of its adjacency matrix in the following way:

$$q_i = \sum_{j=1}^{N} A_{ij}, \tag{2.18}$$

and so the number of edges equals

$$E = \frac{1}{2} \sum_{i,j=1}^{N} A_{ij}. \tag{2.19}$$

For the total degree of vertices in a simple undirected graph, we have

$$Q \equiv \sum_i q_i = 2E = \sum_{ij} A_{ij} = \sum_i (A^2)_{ii} \equiv \mathrm{Tr}\, A^2. \qquad (2.20)$$

Similarly, the in- and out- degrees, $q_i^{(\mathrm{in})}$ and $q_i^{(\mathrm{out})}$, of vertex i in an oriented graph are expressed in terms of the elements of the adjacency matrix as

$$q_i^{(\mathrm{in})} = \sum_{j=1}^N A_{ji}, \quad q_i^{(\mathrm{out})} = \sum_{j=1}^N A_{ij}, \qquad (2.21)$$

and the total in- and out-degree equals

$$Q^{(\mathrm{in})} = \sum_i q_i^{(\mathrm{in})} = Q^{(\mathrm{out})} = \sum_i q_i^{(\mathrm{out})} = E = \sum_{ij} a_{ij} = \mathrm{Tr}\, A^2. \qquad (2.22)$$

While the adjacency matrix of a simple graph has only entries 0 and 1, and for a multigraph, the entries of an adjacency matrix are non-negative integers, one can also introduce a *signed graph* having an adjacency matrix with entries—integers of both signs. An adjacency matrix of a weighted graph has entries—real numbers.

2.3.3 Laplacian matrix

For relating incidence and adjacency matrices of an undirected graph, let us introduce the *degree matrix D* with elements $D_{ij} = q_i \delta_{ij}$, that is, a matrix with diagonal elements equal vertex degrees of a graph and the rest elements equal zero. Here $\delta_{ij} \equiv \delta(i,j)$ is the Kronecker symbol. It corresponds to the *identity matrix I* having diagonal elements equal 1 and the rest elements zero, i.e. $I_{ij} = \delta_{ij}$. Then the *Laplacian matrix*[16] $L \equiv D - A$ of a graph is expressed in terms of its oriented incidence matrix B:

$$L \equiv D - A = BB^T. \qquad (2.23)$$

One can see that while the matrix B depends on the orientation of the graph, the matrices L, D, and A do not. Applying L to an N-component vector $\mathbf{x}$ produces a vector with the components

$$(L\mathbf{x})_i = -\sum_{j \in \partial i} (x_j - x_i). \qquad (2.24)$$

Hence we have the discrete version of the Laplacian operator describing, in particular, diffusion processes. We discuss this important matrix in detail

---

[16]This matrix is also called the *admittance matrix*.

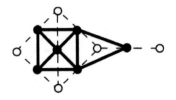

Fig. 2.13 The line graph (solid edges and filled vertices) of the undirected version of the graph from Figure 2.12 (dashed edges and empty vertices).

later. Similarly, the *signless Laplacian matrix* $Q \equiv D + A$ of a graph is expressed in terms of its incidence matrix $\tilde{B}$:

$$Q \equiv D + A = \tilde{B}\tilde{B}^T. \tag{2.25}$$

Notice the relation:

$$2D = BB^T + \tilde{B}\tilde{B}^T. \tag{2.26}$$

One can relate the incidence matrix of an undirected graph with the adjacency matrix of another graph. For that, let us introduce the *line (or edge) graph* of an undirected graph: (i) the set of vertices of the line graph corresponds to the set of edges of the undirected graphs, (ii) two vertices of the line graph are connected by an edge if the corresponding two edges of the original graph have a joint vertex. Figure 2.13 explains the line graph. With this definition, one can easily check that the $E \times E$ adjacency matrix A_{line} of the line graph of an undirected graph is expressed in terms of the $N \times E$ incidence matrix $\tilde{B}$ of the undirected graph:

$$A_{\text{line}} = \tilde{B}^T\tilde{B} - 2I, \tag{2.27}$$

where I is the $E \times E$ identity matrix.

2.4 Walks, paths, and cycles

Here we discuss how basic quantities for a graph are expressed in terms of its adjacency matrix A. For the sake of brevity, we focus on undirected graphs. First, one can easily see that the total number of walks of length ℓ connecting vertices i and j equals

$$n_{\text{walks}}(i, j; \ell) = (A^\ell)_{ij}. \tag{2.28}$$

In particular, the degree of vertex i equals[17]

[17] This also follows from the equality $(A^2)_{ii} = \sum_j A_{ij} A_{ji} = \sum_j A_{ij} = q_i$.

$$q_i = (A^2)_{ii}. \tag{2.29}$$

In fact, $n_{\text{walks}}(i, j; 2)$ is the number of joint nearest neighbours of vertices i and j. This enables one to introduce a measure for a structural dissimilarity of two vertices in a graph, namely the *Hamming distance* h_{ij} that is the number of the nearest neighbours of i or j that are not joint nearest neighbours of both vertices:

$$h_{ij} = q_i + q_j - 2n_{\text{walks}}(i, j; 2) = \sum_k (A_{ik} + A_{jk} - 2A_{ik}A_{kj}) = \sum_k (A_{ik} - A_{jk})^2. \tag{2.30}$$

Further, clearly, the shortest path length ℓ_{ij} between vertices i and j equals the smallest integer power ℓ of the adjacency matrix with a non-zero (ij) entry $(A^\ell)_{ij}$:[18]

$$(A^{\ell-1})_{ij} = 0, \quad (A^\ell)_{ij} \neq 0. \tag{2.31}$$

The total number of walks shorter or equal to ℓ between a pair of vertices is given by the corresponding entry of the matrix equal to the sum:

$$I + A + A^2 + \ldots + A^\ell. \tag{2.32}$$

If we introduce the probability z to take each step, then expression (2.32) enables us to write the following generating function

$$G(z) = I + \sum_{k=1}^{\infty} z^k A^k = (I - zA)^{-1}, \tag{2.33}$$

which we use for description of walks in networks.

The trace of an integer power of an adjacency matrix $\operatorname{Tr} A^\ell$ plays particularly important role in graph theory. Let us consider a simple undirected graph. We have already indicated that the total number of edges in a graph, $E = \frac{1}{2}\operatorname{Tr} A^2$, Eq. (2.20). One can easily see that the total number of cycles of length 3 (triangles) in this graph equals

$$N_3 = \frac{1}{6}\operatorname{Tr} A^3, \tag{2.34}$$

---

[18]In oriented graphs, a walk can visit vertices and edges more than once only if it (or its part) forms an (oriented) cycle (the term 'cycle' for an oriented graph typically suggests an oriented cycle with all its edges having the same direction). It can occur for these graphs that all entries of a power of the adjacency matrix become zero, $A^m = 0$, starting from some integer d, $m > d$. This means that such graph has no (oriented) cycles, and the maximum length of oriented path in this graph equals d.

since Tr A^3 gives the number of closed walks of length 3, where each triangle contributes 6 times—three starting vertices and two directions. The numbers of 4- and 5-cycles are

$$N_4 = \frac{1}{8}\left[\text{Tr } A^4 - 2\sum_i (A^2)_{ii}(A^2)_{ii} + \text{Tr } A^2\right] = \frac{1}{8}\left(\text{Tr } A^4 - 2\sum_i q_i^2 + \text{Tr } A^2\right),$$

$$N_5 = \frac{1}{10}\left[\text{Tr } A^5 - 5\sum_i (A^2)_{ii}(A^3)_{ii} + 5\text{Tr } A^3\right], \tag{2.35}$$

and so on.

Despite the elegance of these formulas, a number of dedicated algorithms provide structural metrics for large graphs in a much faster way. For example, the breadth-first search based algorithm can be used for finding shortest paths between a given vertex and each of the rest vertices in an unweighted graph. The algorithm explores all first nearest neighbours of a given vertex, one by one, then passes to the second-nearest neighbours, explores and lists them, and so on until the graph will be completely scanned. One can show that this can be done in $O(N + E)$ time.[19]

2.5   Triangles

Triangles of edges are of special interest as they are the shortest cycles in simple graphs. Clearly, a graph has a triangle (3-cycle) when its adjacency matrix A and its square A^2 have non-zero elements in the same position ij: $A_{ij}, (A^2)_{ij} > 0$. Equation (2.34) expresses the total number of triangles in terms of the adjacency matrix.

Let us focus on large simple graphs and introduce two densities: edges, e, and triangles, t. We define the edge density e as

$$e \equiv \frac{E}{\binom{N}{2}} = \frac{E}{N(N-1)/2}, \tag{2.36}$$

where the denominator is the number of edges in the complete graph of N vertices, and the triangle density t as

$$t \equiv \frac{N_3}{\binom{N}{3}} = \frac{N_3}{N(N-1)(N-2)/6}, \tag{2.37}$$

---

[19]The more refined *Dijkstras algorithm*, which is also called the uniform cost search, is applicable both to directed and undirected weighted graphs (weights must be positive). This algorithm provides the weighted shortest paths (where the weighted length is the sum of weights along this path) between a given vertex and each other vertex in a graph.

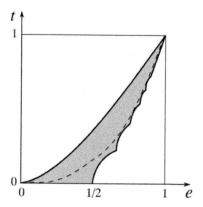

Fig. 2.14 Area of the attainable edge, e, and triangle, t, densities of large simple graphs. The upper boundary curve is $t = e^{3/2}$. The dashed curve $t = e^3$ is for the Erdős–Rényi graph. The corners of the lower boundary are on the parabola $t = 2e^2 - e$. The lower boundary consists of an infinite number of concave pieces. Adapted from Kenyon, Radin, Ren, and Sadun (2016).

where the denominator is the number of triangles in the complete graph of N vertices.[20] Notably, for any large simple graph, these two densities, of edges and of triangles, occur only within the restricted area of the (e, t) plane (Lovász, 2012) (Figure 2.14), where the boundaries of this area are highlighted. The right boundary consists of an infinite number of concave pieces which were strictly described rather recently (Razborov, 2008).

A better characteristic than t of a relative number of triangles in a graph was introduced in sociology. This is the *transitivity* or the *clustering coefficient*, C. A simple graph contains $\mathcal{T}$ connected triples of vertices,[21]

$$\mathcal{T} = \frac{1}{2} \sum_{i \neq j} (A^2)_{ij} = \frac{1}{2} \sum_i q_i(q_i - 1). \tag{2.38}$$

The transitivity (clustering coefficient) is the fraction of the connected triples of vertices that belong to triangles (in other words, the fraction of the pairs of adjacent edges, belonging to triangles):

$$C \equiv 3\frac{N_3}{\mathcal{T}} = \frac{N_3}{\sum_i q_i(q_i - 1)/6}. \tag{2.39}$$

[20]Note that the triangle density t does not coincide with the transitivity C (the clustering coefficient).

[21]For two adjacent edges (ij) and (jk), the connected triple of vertices is $(i, j, k) = (k, j, i)$.

The factor 3 appears because each triangle contains three different triples of connected vertices. Unlike the density of triangles t, the clustering coefficient accounts for the degrees of the vertices of a graph. The *local clustering C_i* (or vertex clustering) is defined for an individual vertex i of a graph. This is the fraction of the pairs of the nearest neighbours of the vertex that are the nearest neighbours of each other, or, in other words, the ratio of the number T_i of triangles attached to vertex i and the maximum possible number of such triangles occurring when the nearest neighbours of vertex i form a fully connected graph:

$$C_i = \begin{cases} 0 & \text{if } q_i < 2, \\ \dfrac{T_i}{(A^3)_{ii}/2} = \dfrac{T_i}{q_i(q_i - 1)/2} & \text{if } q_i \geq 2. \end{cases} \tag{2.40}$$

Then the *mean local clustering $\overline{C}$* is

$$\overline{C} = \frac{1}{N} \sum_i C_i. \tag{2.41}$$

As is natural, in random networks, these two characteristics of clustering, C and $\overline{C}$, Eqs. (2.39) and (2.41) respectively, should be in addition averaged over a given statistical ensemble.

Similarly to T_i for each vertex i, one introduces the *edge multiplicity m_{ij}* for each edge (ij), that is, the number of triangles to which this edge belongs (Serrano and Boguñá, 2006a). Clearly,

$$T_i = \frac{1}{2} \sum_j m_{ij}, \tag{2.42}$$

compare to Eq. (2.18) expressing the vertex degree in terms of the entries of the adjacency matrix.[22]

2.6 Cliques

Edges and triangles are particular examples of complete subgraphs of an undirected graph, cliques.[23] Finding cliques of a graph constitute a set of

---

[22] One can conveniently normalize the edge multiplicity m_{ij} dividing it by the maximal possible number of triangles in which the edge can participate, $\min(q_i, q_j) - 1$, when $\min(q_i, q_j) > 1$.

[23] As a complement to a clique, one can define an *independent set* (anti-clique, co-clique) that is a set of non-adjacent vertices in a graph. If $F \equiv (V, E_F)$, $|E_F| = |V|(|V| - 1)/2$,

computational problems, some of which are NP-complete. In general, it takes a long time $O(2^{N/4})$ or so (Robson, 2001) to find the *maximum clique* of a graph, that is, the clique with the largest possible number of vertices in the graph (*clique number*).

Moving forward we use the notion of the *maximal clique*, which is a *clique* not included in any larger clique. The maximum number of maximal cliques in a graph is $3^{N/3}$ (Moon and Moser, 1965), and all maximal cliques can be found in time $O(3^{N/3})$. A simple graph can be naturally represented as a complex of overlapping maximal cliques, where the overlaps of cliques are smaller cliques.

2.7 Betweenness centrality

A set of so-called centrality measures for an individual vertex account for the structure of the entire network. Ranking vertices with a measure of this sort allows one to indicate the 'centre of a network' understood in respect of this measure. A simple example of such a measure is the mean separation of vertex i from the rest vertices in the graph (we assume that the graph contains a single connected component):

$$\bar{\ell}_i = \frac{1}{N-1} \sum_{j \neq i} \ell_{ij}. \tag{2.43}$$

The inverse $1/\bar{\ell}_i$ is called the *closeness centrality*.

Another widely used measure, the *betweenness centrality* (physicists also call it *load*), helps to characterize the distribution of shortest paths over a network. The *vertex betweenness centrality* (or simply vertex betweenness) shows how often shortest paths in a network pass through a given vertex. The betweenness centrality $\mathcal{B}_v$ of vertex v (that is not a leave) is

$$\mathcal{B}_v \equiv \sum_{i,j \neq v} \frac{s(i,v,j)}{s(i,j)}, \tag{2.44}$$

where $s(i,j) > 0$ is the number of shortest paths between vertices i and j, where $i,j \neq v$, and $s(i,v,j)$ is the number of these paths passing through vertex v. When vertex v is a leave, its betweenness centrality is set to zero,

is the fully connected graph with the same set of vertices as a graph $G = (V, E)$, then the graph $C(G) = (V, E_F \backslash E)$ is called the complement of the graph G (or its inverse). Clearly, $C(C(G)) = G$. Then a clique in G is an independent set in the complement graph $C(G)$, and vice versa.

$\mathcal{B}_v \equiv 0$. The *edge betweenness centrality* (*edge betweenness*) is introduced in a similar way. The betweenness centrality $\mathcal{B}_{kl}$ of the edge (kl) (that is not attached to a leave) is

$$\mathcal{B}_{kl} \equiv \sum_{i,j\neq k,l} \frac{s(i, kl, j)}{s(i, j)}, \qquad (2.45)$$

where $s(i, j) > 0$ is the number of shortest paths between vertices i and j, where $i, j \neq k, l$, and $s(i, kl, j)$ is the number of these paths passing through the edge (kl). When at least one of the end vertices of the edge (kl) is a leave, its betweenness centrality is set to zero, $\mathcal{B}_{kl} \equiv 0$.[24] Imagine that every vertex in a graph sends packages to each other vertex with rate $r/(N-1)$ by shortest paths. Then betweenness centrality allows one to find the resulting traffic through given vertices and edges of a graph.[25] That is, it measures the influence of a vertex (or edge) over the information flow in the network in this situation. If each vertex only permits the passage of a flow not exceeding a unit threshold, then the vertex with the highest betweenness will be the first congested, and the congestion emerges at the rate

$$r_c = \frac{N - 1}{\max\limits_v \mathcal{B}_v}. \qquad (2.46)$$

One can easily find these numbers for all vertices and edges of an arbitrary tree since there is only one path between any two vertices in a tree. Let vertex v have degree q_v, and the number of vertices in each of these q_v branches equals $M_i^{(v)}$, $\sum_{k=1}^{q_v} M_i^{(v)} = N-1$. Then the betweenness centrality of this vertex equals

$$\mathcal{B}_v = \frac{(N - 1)^2 - \sum_{k=1}^{q_v}(M_i^{(v)})^2}{(N - 1)(N - 2)}. \qquad (2.47)$$

Now let the edge (kl) divide the tree into two pieces (branches) of M and $N - M$ vertices. Then the betweenness centrality of this edge equals

$$\mathcal{B}_{kl} = \frac{(M - 1)(N - M - 1)}{(N - 2)(N - 3)/2}. \qquad (2.48)$$

Usually, one can get the betweenness centrality of a vertex or edge in a graph with cycles only numerically.

[24]See Newman (2001*b*) and Brandes (2001) for a fast algorithm calculating betweenness centrality for all edges (or vertices) in a graph in time $O(NE)$.

[25]For more realistic models of traffic in networks based on random walks, see Holme (2003) and Tadić, Thurner, and Rodgers (2004).

2.8 Connectivity

When physicists started exploration of complex networks in the late 1990s they frequently used the term 'connectivity' for the number of edges of a vertex instead of degree, which caused a lot of confusion. In graph theory the connectivity is actually about how parts of a graph are connected, so it is actually about paths in a graph.

Let us focus on connected graphs. Clearly, a graph of N vertices described by the adjacency matrix A is connected when the matrix $A+A^2+\ldots+A^{N-1}$ has no zero entries off the diagonal. We are interested in the set of paths running between different parts of a graph. To characterize this set, it is convenient to introduce the notion of independent paths. The paths between vertices i and j are called *vertex independent paths* if they do not share vertices other than i and j. Similarly, the paths between vertices i and j are called *edge independent paths* if they do not share edges. The total number of vertex independent paths between vertices i and j and the total number of the edge independent paths between i and j show how well these two vertices are interconnected. These numbers are determined by the 'width' of the 'narrowest piece' of the graph between vertices i and j, which can be characterized in the following way. The *minimum vertex cut* for vertices i and j is defined as the minimum set of vertices, distinct from i and j, whose deletion disconnects these two vertices. The *minimum edge cut* for vertices i and j is defined as the minimum set of edges whose deletion disconnects these vertices.[26]

Menger's theorem for a finite simple undirected graph states:

(i) for each pair of nonadjacent vertices, the size of the minimum vertex cut coincides with the number of vertex independent paths between these vertices;

---

[26]One can go further and define a constant characterizing the 'narrowest' part, the 'bottleneck' of a graph. This 'bottleneck' determines important properties of a graph, including some features of graph spectra, relaxation of dynamical models and processes on a graph, synchronizability, etc. The *Cheeger constant h* of a graph of N vertices is

$$h \equiv \min_{S:\, 1\leq|S|\leq N/2} \frac{|\partial S|}{|S|}.$$

Here the minimum is over all nonempty sets S of vertices containing at most half of all vertices in the graph. The edge boundary ∂S of S is the set of edges in the graph, having one end in S. A small h indicates a pronounced 'bottleneckedness' of the graph. In particular, in large lattices, h approaches 0, while in sufficiently uniform small worlds, h is finite.

(ii) for each pair of vertices, the size of the minimum edge cut coincides with the number of edge independent paths between these two vertices. These statements are clear without proving, and they can be generalized to directed graphs (for directed paths).[27]

By using the introduced notions one can describe the connectedness of a graph. A simple undirected graph is *k-vertex-connected* when every pair of its vertices has at least k vertex independent paths between them. This graph remains connected after deletion of fewer than k vertices. A simple undirected graph is *k-edge-connected* when every pair of its vertices has at least k edge independent paths between them. This graph remains connected after deletion of fewer than k edges.

2.9 Spectra of graphs

This section briefly touches on with spectra in simple graphs. We use the basic *Perron–Frobenius theorem* stating that a real square matrix with non-negative entries has a unique largest real non-negative eigenvalue λ_1 and that all the elements of the corresponding eigenvector (leading eigenvector) can be chosen non-negative. In this situation, this eigenvalue is the *spectral radius* that is the radius of a circle on the complex plain, within which all the eigenvalues of the matrix sit. One should also remember that all N eigenvalues, λ_α, of a symmetric $N \times N$ matrix are real and their eigenvectors are orthogonal.

2.9.1 Spectra of adjacency matrices

First we discuss the spectra of adjacency matrices of undirected graphs,

$$A\mathbf{v} = \lambda\mathbf{v}. \tag{2.49}$$

Clearly, if there are no self-loops,

---

[27] Menger's theorem is a particular case of the famous *max-flow min-cut theorem* which can be formulated for undirected and directed graphs. L. R. Ford Jr. and D. R. Fulkerson proved this theorem in 1954–1955. For the flow from a source vertex s to a sink vertex t in a weighted directed graph, the theorem states that the maximum flow is equal to the minimum capacity over all edge cuts for vertices s and t. It is assumed here that a flow through an edge cannot exceed the edge weight, and, by definition, the *capacity* of an edge cut is the sum of the weights in the cut. The Ford–Fulkerson algorithm, based on this theorem, allows to find the maximum flow between two vertices.

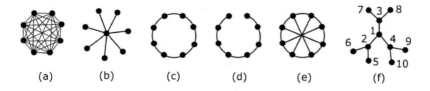

Fig. 2.15 Sample graphs whose adjacency matrix spectra are given by Eqs. (2.53)–(2.58).

$$\sum_\alpha \lambda_\alpha = 0,$$

$$\sum_\alpha \lambda_\alpha^2 = 2E,$$

$$\sum_\alpha \lambda_\alpha^3 = 6N_3, \tag{2.50}$$

where N_3 is the number of triangles in the graph. The eigenvalues are the roots of the *characteristic polynomial of a graph*

$$p_A(\lambda) \equiv \det(\lambda I - A) = \lambda^N + a_1\lambda^{N-1} + a_2\lambda^{N-2} + \ldots + a_{N-1}\lambda + a_N, \tag{2.51}$$

whose first three coefficients are

$$a_1 = 0,$$
$$a_2 = -E,$$
$$a_3 = -2N_3. \tag{2.52}$$

The following distinct small graphs (Figure 2.15) will help the reader to grasp how the spectra are organized. For the fully connected undirected simple graph of N vertices (Figure 2.15a), we have the following eigenvalues and the corresponding eigenvectors:[28,29]

$$\lambda_1 = N - 1, \quad \lambda_{2 \leq \alpha \leq N} = -1,$$
$$v_1 = (1, 1, \ldots, 1)^T,$$
$$v_\alpha = (1, 0, 0, \ldots, -1, 0, \ldots)^T, \tag{2.53}$$

[28]Since the adjacency matrices of undirected graphs are real and symmetric, all their eigenvectors can be made real.

[29]For the symmetric matrices, a right eigenvector (column vector) coincides with the transpose of the left eigenvector (row vector) corresponding to the same eigenvalue.

where the first element in the eigenvectors $\mathbf{v}_{2 \leq \alpha \leq N}$ equals 1, the α-th element equals -1, and the remaining $N-2$ elements equal 0.[30]

For the q-star with the central vertex 1 and q leaves (Figure 2.15b), the eigenvalues and the corresponding eigenvectors are

$$\lambda = \sqrt{q}, 0, 0, \ldots, 0, -\sqrt{q},$$
$$\mathbf{v}_1 = (\sqrt{q}, 1, 1, \ldots, 1)^T,$$
$$\mathbf{v}_{2 \leq \alpha \leq q} = (0, 1, 0, 0, \ldots, -1, 0, 0, \ldots)^T,$$
$$\mathbf{v}_{q+1} = (-\sqrt{q}, 1, 1, \ldots, 1)^T. \tag{2.54}$$

Here in $\mathbf{v}_{2 \leq \alpha \leq q}$, the second entry equals 1 and the $\alpha+1$-th one equals -1.[31]

For the simple undirected ring of 8 vertices (Figure 2.15c), the eigenvalues and the corresponding eigenvectors are

$$\lambda = 2, \sqrt{2}, \sqrt{2}, 0, 0, -\sqrt{2}, -\sqrt{2}, -2,$$
$$\mathbf{v}_1 = (1, 1, 1, 1, 1, 1, 1, 1)^T,$$
$$\mathbf{v}_2 = (1, \sqrt{2}, 1, 0, -1, -\sqrt{2}, -1, 0)^T, \; \ldots. \tag{2.55}$$

For the simple undirected chain of 8 vertices (Figure 2.15d), the eigenvalues and the corresponding eigenvectors are

$$\lambda \approx 1.879, 1.532, 1, 0.347, -0.347, -1, -1.532, -1.879,$$
$$\mathbf{v}_1 \approx (1, 1.879, 2.532, 2.879, 2.879, 2.532, 1.879, 1)^T,$$
$$\mathbf{v}_2 \approx (1, 1.532, 1.347, 0.532, -0.532, -1.347, -1.532, -1)^T, \; \ldots. \tag{2.56}$$

For the undirected ring of 8 vertices with opposite vertices linked (Figure 2.15e), so we have a regular graph with vertices of degree 3,[32] the eigenvalues and the corresponding eigenvectors are

$$\lambda = 3, 1, 1, \sqrt{2} - 1, \sqrt{2} - 1, -1, -\sqrt{2} - 1, -\sqrt{2} - 1,$$
$$\mathbf{v}_1 = (1, 1, 1, 1, 1, 1, 1, 1)^T,$$
$$\mathbf{v}_2 = (1, 0, -1, 0, 1, 0, -1, 0)^T, \; \ldots. \tag{2.57}$$

Finally, for the Cayley tree of 10 vertices of degree 3 (Figure 2.15f), the eigenvalues and the corresponding eigenvectors are[33]

---

[30] For the adjacency matrix spectrum, we numerate the eigenvalues, starting from the largest one till the smallest, $\lambda_1, \lambda_2, \ldots, \lambda_N$.

[31] We do not normalize eigenvectors in this section, unless it is specially indicated.

[32] In the infinite size limit, $N \to \infty$, this graph is a deterministic parody of the infinite random regular graph with vertices of degree 3.

[33] In the infinite k-Cayley tree, $\lambda_1 = k$ and $\lambda_N = -k$.

$$\lambda = \sqrt{5}, \sqrt{2}, \sqrt{2}, 0, 0, 0, 0, -\sqrt{2}, -\sqrt{2}, -\sqrt{5},$$
$$\mathbf{v}_1 = (3, \sqrt{5}, \sqrt{5}, \sqrt{5}, 1, 1, 1, 1, 1, 1)^T,$$
$$\mathbf{v}_2 = (0, \sqrt{2}, 0, -\sqrt{2}, 1, 1, 0, 0, -1, -1)^T, \ \ldots . \tag{2.58}$$

These spectra are shown in Figure 2.16.

Thus the entries of the eigenvectors corresponding to the largest and the second-largest eigenvalues of the adjacency matrix of a graph give some idea about the global structure of this graph and the position of a given vertex in it. In a chain, a star, and a Cayley tree, the entries of the $\mathbf{v}_1$ eigenvector for the central vertices are the largest, see Eqs. (2.54), (2.56), and (2.58), respectively. So the entries of the $\mathbf{v}_1$ eigenvector can be used as a measure of centrality of a vertex in an undirected graph. This measure is called the *eigenvector centrality*.

Figure 2.17 demonstrates that in directed graphs some of the entries of $\mathbf{v}_1$ equal zero:[34]

$$\lambda = 1, -\frac{1}{2} + i\frac{\sqrt{3}}{2}, -\frac{1}{2} - i\frac{\sqrt{3}}{2}, 0, 0, 0,$$
$$\mathbf{v}_1 = (2, 1, 1, 1, 1, 0)^T, \ \ldots , \tag{2.59}$$

and so in this case, the eigenvector centrality is unfeasible. The renowned *Katz centrality* is applicable even in this situation. The Katz centrality $\mathbf{K}$ (N-vector) is the solution of the equation

$$\mathbf{K} = aA\mathbf{K} + b\mathbf{1} \tag{2.60}$$

(compare to Eq. (2.49)), where $b > 0$, $a < 1/\lambda_1$, and $\mathbf{1}$ is the column vector with all N entries equal 1, $(\mathbf{1})_i = 1$, $i = 1, \ldots, N$. The value of the non-negative number b is actually not important and it can be set to 1. The solution of Eq. (2.60), $\mathbf{K}$, can be obtained by iterations, whose convergence is guaranteed by the condition $a < 1/\lambda_1$. Clearly, when $\alpha \to 1/\lambda_1 - 0$, the normalized vector $\mathbf{K}$ approaches the normalized eigenvector $\mathbf{v}_1$, which shows the relation between the Katz centrality and the principal eigenvector. The condition $a < 1/\lambda_1$ enables us to write the solution Eq. (2.60) in the following form:

$$\mathbf{K} = b(I - aA)^{-1}\mathbf{1} = b\sum_{\alpha=1}^{N} a^\alpha A^\alpha \mathbf{1}, \tag{2.61}$$

[34] These are the entries of $\mathbf{v}_1$ corresponding to the vertices of the out-component of a graph; see Section 6.5.

GRAPHS

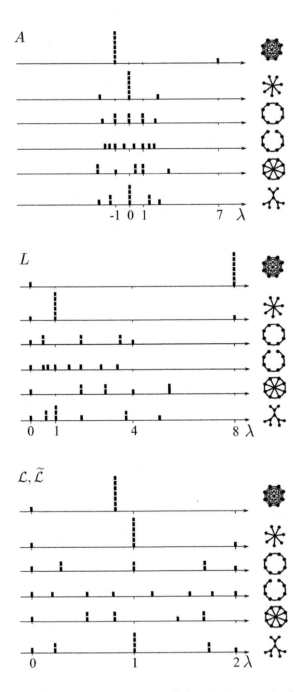

Fig. 2.16 Spectra of the adjacency matrix A, Laplacian matrix L, and symmetric normalized Laplacian matrix $\mathcal{L}$ and normalized Laplacian matrix $\widetilde{\mathcal{L}}$ (spectra of $\mathcal{L}$ and $\widetilde{\mathcal{L}}$ coincide) for the sample graphs in Figure 2.15.

48

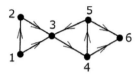

Fig. 2.17 Sample directed graph whose spectrum is given by Eq. (2.59). Vertices 3, 4, 5 and edges interconnecting them form the strongly connected component, vertices 1 and 2 provide the in-component, vertex 6 provide the out-component (see Section 6.5 for the notions of in-, out-, and strongly connected components).

which explains the meaning of the Katz centrality. Namely, the i-th component of the vector $\mathbf{K}$, $(\mathbf{K})_i = b\sum_{k=0}\sum_j a^k(A^k)_{ij}$, is the sum of the weighted contributions of all possible walks from vertex i to any vertex j, where each walk of length α gives the contribution ba^α.

2.9.2 Spectra of Laplacian matrices

Section 2.3 introduced the Laplacian matrix of a graph, $L \equiv D - A$, where D is the degree matrix of this graph. Clearly, at least one of its eigenvalues equals 0, $\lambda_1 = 0$, and the corresponding eigenvector has all entries equal, that is $\mathbf{v}_1 = \mathbf{1}$ if we do not normalize it; all eigenvalues are non-negative, $\lambda_\alpha \geq 0$,

$$\text{Tr}L = \sum_\alpha \lambda_\alpha = \sum_i q_i = 2E,$$

$$\text{Tr}L^2 = \sum_\alpha \lambda_\alpha^2 = \sum_i q_i^2 + \sum_i q_i. \tag{2.62}$$

When a graph is regular, the spectrum of L is readily obtained from the spectrum of A.[35] Returning to our examples, for the fully connected graph of N vertices (Figure 2.15a), we have

$$\lambda_1 = 0, \quad \lambda_{2\leq\alpha\leq N} = N,$$
$$\mathbf{v}_1 = (1, 1, \ldots, 1)^T,$$
$$\mathbf{v}_\alpha = (1, 0, 0, \ldots, -1, 0, \ldots)^T. \tag{2.63}$$

For the ring of 8 vertices (Figure 2.15c), the spectrum of the Laplacian matrix is

---

[35]For the Laplacian spectra, for the sake of convenience, we numerate the eigenvalues, starting from the smallest to the largest, $\lambda_1, \lambda_2, \ldots, \lambda_N$. So for q-regular graphs, $\lambda_{\alpha-1}^{(L)} = q - \lambda_\alpha^{(A)}$, $\alpha = 1, \ldots, N$, where the superscripts A and L indicate the adjacency matrix and Laplacian spectra, respectively.

$$\lambda = 0, 2 - \sqrt{2}, 2 - \sqrt{2}, 2, 2, 2 + \sqrt{2}, 2 + \sqrt{2}, 4,$$
$$\mathbf{v}_1 = (1, 1, 1, 1, 1, 1, 1, 1)^T,$$
$$\mathbf{v}_2 = (1, \sqrt{2}, 1, 0, -1, -\sqrt{2}, -1, 0)^T, \ \dots . \tag{2.64}$$

For the undirected ring of 8 vertices with opposite vertices linked (Figure 2.15e), the eigenvalues and the corresponding eigenvectors of the Laplacian matrix spectrum are

$$\lambda = 0, 2, 2, 4 - \sqrt{2}, 4 - \sqrt{2}, 4, 4 + \sqrt{2}, 4 + \sqrt{2},$$
$$\mathbf{v}_1 = (1, 1, 1, 1, 1, 1, 1, 1)^T,$$
$$\mathbf{v}_2 = (1, 0, -1, 0, 1, 0, -1, 0)^T, \ \dots . \tag{2.65}$$

For non-regular graphs, the Laplacian matrix spectra significantly differ from the spectra of the adjacency matrices of corresponding graphs. For the q-star with the central vertex 1 and q leaves (Figure 2.15b), the Laplacian eigenvalues and the corresponding eigenvectors are

$$\lambda = 0, \overbrace{1, 1, \dots, 1}^{q-1}, q + 1,$$
$$\mathbf{v}_1 = (1, 1, \dots, 1)^T,$$
$$\mathbf{v}_{2 \leq \alpha \leq q} = (0, 1, 0, 0, \dots, -1, 0, 0, \dots)^T,$$
$$\mathbf{v}_{q+1} = (q, -1, -1, \dots, -1)^T. \tag{2.66}$$

Here in $\mathbf{v}_{2 \leq \alpha \leq q}$, the second entry equals 1 and the $\alpha + 1$-th one equals -1. For the chain of 8 vertices (Figure 2.15d), the Laplacian spectrum is

$$\lambda \approx 0, 0.152, 0.586, 1.235, 2, 2.766, 3.414, 3.848,$$
$$\mathbf{v}_1 = (1, 1, 1, 1, 1, 1, 1, 1)^T,$$
$$\mathbf{v}_2 \approx (1, 0.848, 0.566, 0.199, -0.199, -0.566, -0.848, -1)^T, \ \dots . \tag{2.67}$$

Finally, for the Cayley tree of 10 vertices of degree 3 (Figure 2.15f), the eigenvalues and the corresponding eigenvectors are

$$\lambda = 0, 2 - \sqrt{3}, 2 - \sqrt{3}, 1, 1, 1, 2, 2 + \sqrt{3}, 2 + \sqrt{3}, 5,$$
$$\mathbf{v}_1 = (1, 1, 1, 1, 1, 1, 1, 1, 1, 1)^T,$$
$$\mathbf{v}_2 = (0, 1 - \sqrt{3}, -1 + \sqrt{3}, 0, -1, -1, 1, 1, 0, 0)^T, \ \dots . \tag{2.68}$$

The Laplacian matrix spectra of larger Cayley trees can be found in Erzan and Tuncer (2020). Figure 2.16 shows all these Laplacian spectra.

When a graph contains n_c connected components, the eigenvalue $\lambda = 0$ of the spectrum of L is degenerate, the degeneracy equals n_c, and each of the n_c corresponding eigenvectors has non-zero entries only for vertices of its own connected component.

The spectrum of the Laplacian matrix of a graph provides diverse information about a variety of global features of a graph. As the first example of what can be found, we indicate *Kirchhoff's matrix tree theorem*. Consider a labelled connected undirected multigraph of N vertices. The matrix tree theorem states that the number of spanning trees of this graph equals the absolute value of the determinant of any minor[36] of the Laplacian matrix of the graph, which is equal to the product of non-zero eigenvalues in the spectrum divided by N,

$$n_{\mathrm{sp}} = (-1)^{i+j} \det M_{ij} = \frac{1}{N} \prod_{k=2}^{N} \lambda_\alpha. \qquad (2.69)$$

On the other hand, for a simple graph,

$$N \sum_{k=2}^{N} \lambda_\alpha^{-1} = \sum_{i<j\in V} \Omega_{ij}, \qquad (2.70)$$

where Ω_{ij} is the *resistance distance* between vertices i and j of the graph, defined as the resistance measured between i and j after exchanging all edges to 1 ohm resistances, $\sum_{i,j\in E} \Omega_{ij} = N - 1$.

2.9.3 Spectra of normalized Laplacian matrices

If the Laplacian matrix L is about flows and diffusion, the following two matrices are about random walks. The *symmetric normalized Laplacian matrix* (also called Chung's Laplacian) is[37]

$$\mathcal{L} \equiv D^{-1/2} L D^{-1/2} = I - D^{-1/2} A D^{-1/2}, \qquad (2.71)$$

so its entries are

$$\mathcal{L}_{ij} = \begin{cases} 1 & \text{if } i = j \text{ and } q_i \neq 0 \\ -\dfrac{A_{ij}}{\sqrt{q_i q_j}} & \text{if } i \neq j \\ 0 & \text{otherwise.} \end{cases} \qquad (2.72)$$

---

[36]The minor M_{ij} of the matrix L is the latter matrix with row i and column j removed.
[37]For Laplacian matrices in weighted graphs, see Butler and Chung (2006).

The *random walk Laplacian matrix* (*normalised Laplacian matrix*) is

$$\tilde{\mathcal{L}} \equiv LD^{-1} = I - AD^{-1}, \tag{2.73}$$

where $AD^{-1} = P$ is the *transition matrix* for a random walk, giving the rate at which a random walker hops from vertex j to i:

$$\mathcal{P}_{ij} = \frac{A_{ij}}{q_j}. \tag{2.74}$$

The entries of the matrix $\tilde{\mathcal{L}}$ are

$$\tilde{\mathcal{L}}_{ij} = \begin{cases} 1 & \text{if } i = j \text{ and } q_i \neq 0 \\ -\dfrac{A_{ij}}{q_j} & \text{if } i \neq j \\ 0 & \text{otherwise.} \end{cases} \tag{2.75}$$

Since

$$\tilde{\mathcal{L}} = D^{1/2} \mathcal{L} D^{-1/2}, \tag{2.76}$$

the matrices $\mathcal{L}$ and $\tilde{\mathcal{L}}$ have the same eigenvalues, $\lambda_1 = 0$, $\lambda_\alpha \geq 0$, $\alpha = 2, \ldots, N$, lying within the window $0 \leq \lambda_\alpha \leq 2$. The sum of the eigenvalues is $\sum_\alpha \lambda_\alpha = N_{q>0}$, where $N_{q>0}$ is the number of vertices with non-zero degrees in the graph. The spectrum of $\mathcal{L}$ and $\tilde{\mathcal{L}}$ resembles the one of L, also having only non-negative eigenvalues (see our examples in Appendix B and in Figure 2.16).[38] Similarly to the matrix L, the degeneracy of the zero eigenvalue in the spectrum of $\mathcal{L}$ and $\tilde{\mathcal{L}}$ equals the number of connected components in a graph. If a graph consists of a single component, then

$$\mathbf{v}_1^{(\tilde{\mathcal{L}},\text{right})} = D\mathbf{1}, \quad \mathbf{v}_1^{(\mathcal{L})} = D^{1/2}\mathbf{1}. \tag{2.77}$$

More generally,[39]

---

[38] In particular, for the q-regular graphs, each eigenvalue in the spectrum of $\mathcal{L}$ and $\tilde{\mathcal{L}}$ is by q times smaller than the corresponding eigenvalue of L.

[39] Note that since a normalized Laplacian matrix $\tilde{\mathcal{L}}$ is non-symmetric, the right eigenvector $\mathbf{v}_\alpha^{(\tilde{\mathcal{L}},\text{right})}$ differs from the left one $\mathbf{v}_\alpha^{(\tilde{\mathcal{L}},\text{left})}$ corresponding to the same eigenvalue λ_α, but they are related,

$$(\mathbf{v}_\alpha^{(\tilde{\mathcal{L}},\text{right})})^T = D\mathbf{v}_\alpha^{(\tilde{\mathcal{L}},\text{left})},$$

where the left eigenvector is a column vector and the right one is a row vector. Assuming the normalization

$$\mathbf{v}_\alpha^{(\tilde{\mathcal{L}},\text{right})}\mathbf{v}_\alpha^{(\tilde{\mathcal{L}},\text{left})} = I,$$

$$\mathbf{v}_\alpha^{(\mathcal{L})} = D^{-1/2}\mathbf{v}_\alpha^{(\tilde{\mathcal{L}},\text{right})}, \quad \alpha = 1,\ldots,N. \tag{2.78}$$

For a connected graph of N vertices, the smallest non-zero eigenvalue λ_1 of $\mathcal{L}$ and $\tilde{\mathcal{L}}$, the *spectral gap*,[40] is related to the diameter d and the largest degree q_{max} of this graph by the following inequalities:

$$\frac{1}{d\sum_i q_i}, \frac{1}{(q_{max}+1)q_{max}^{\lceil d/2-1\rceil}} \leq \lambda_2 \leq 1 - 2\frac{\sqrt{q_{max}-1}}{q_{max}}\left(1-\frac{2}{d}\right)+\frac{2}{d} \tag{2.79}$$

(Chung, 1997), so, loosely speaking, the larger diameter, the narrower the spectral gap; see also Eq. (9.25).[41] In infinite finite-dimensional lattices, the spectral gap tends to zero and the densities of eigenvalues for both spectra (of L and of $\mathcal{L}$ and $\tilde{\mathcal{L}}$) are power-law at small λ, namely $\rho(\lambda) \sim \lambda^{D/2-1}$, where D is the space dimension of the lattice. This enables one to introduce the *spectral dimension* D_s for general infinite graphs,

$$\rho(\lambda) \sim \lambda^{D_s/2-1}, \tag{2.80}$$

if the density of eigenvalues of the L (or, equivalently, of $\mathcal{L}$ and $\tilde{\mathcal{L}}$) spectra is power-law.[42] The spectral dimension has two bounds determined by the Hausdorff dimension (Durhuus, 2009):

$$2\frac{D_H}{D_H+1} \leq D_s \leq D_H, \tag{2.81}$$

one can see that for the eigenvalue $\lambda_1 = 0$, the right and left eigenvectors equal, respectively,

$$\mathbf{v}_1^{(\tilde{\mathcal{L}},\text{right})} = \frac{1}{\sqrt{\sum_i q_i}}(q_1, q_2, \ldots, q_N)^T,$$

$$\mathbf{v}_1^{(\tilde{\mathcal{L}},\text{left})} = \frac{1}{\sqrt{\sum_i q_i}}(1, 1, \ldots, 1).$$

In turn, these eigenvectors coincide with right and left eigenvectors associated with the largest eigenvalue $\lambda_1 = 1$ of the transition matrix $\mathcal{P}$, whose eigenvalues are labelled in descending order, namely, $\mathbf{v}_1^{(\mathcal{P},\text{right})} = \mathbf{v}_1^{(\tilde{\mathcal{L}},\text{right})}$ and $\mathbf{v}_1^{(\mathcal{P},\text{left})} = \mathbf{v}_1^{(\tilde{\mathcal{L}},\text{left})}$.

[40]This eigenvalue is also called the *Fiedler eigenvalue* and the corresponding eigenvector called the *Fiedler eigenvector*.

[41]In some networks, this is not true (Motter, Zhou, and Kurths, 2005; Eldan, Rácz, and Schramm, 2017).

[42]In Chapter 10 we discuss how D_s is related to diffusion and random walks.

so the spectral dimension of a small world ($D_H = \infty$) can appear to be finite.[43] For a large graph of N vertices having $D_s < \infty$, we can roughly estimate λ_2 in the following way:

$$N \int_0^{\lambda_2} d\lambda \rho(\lambda) \sim 1, \tag{2.82}$$

which results in the estimate of the Fiedler eigenvalue:

$$\lambda_2(N) \sim N^{-2/D_s}. \tag{2.83}$$

The spectral gap λ_2 in the spectrum of $\mathcal{L}$ and $\widetilde{\mathcal{L}}$ is also related to the presence of a 'bottleneck' in a graph through the inequalities containing the Cheeger constant h:

$$\frac{1}{2}h^2 \leq 1 - \sqrt{1 - h^2} \leq \lambda_2 \leq 2h, \tag{2.84}$$

where the graph is assumed to be not complete. That is, if $\mathcal{L}$ (or, equivalently, $\widetilde{\mathcal{L}}$) has a wide spectral gap, then the graph has no bottlenecks.[44]

---

[43]In particular, the lower limit of the inequality for D_s is realized in the uniformly random trees, where $D_H = 2$ and $D_s = 4/3$ (Alexander and Orbach, 1982; Destri and Donetti, 2002). Moreover, it was conjectured that $D_s = 2D_H/(D_H + 1)$ even for more general random trees with an arbitrary D_H and that the greatest Hausdorff and spectral dimensions for random trees are realized just in the uniformly random trees (Destri and Donetti, 2002). The spectral dimension of the Web graph is $D_s \approx 1.8$, it exceeds 2 for the Internet on the Autonomous Systems' level (Hwang, Lee, and Kahng, 2012a, 2012b), while in uncorrelated networks $D_s = \infty$.

[44]In infinite lattices, the Cheeger constant approaches zero, and a spectral gap is absent. In sufficiently homogeneous small worlds, the Cheeger constant is finite, unless these networks contain long chains, and there is a spectral gap; see Section 9.2.

3
Classical Random Graphs

Here we give an insight into two basic models of equilibrium random networks. Often they are both called the Erdős–Rényi random graph, although, strictly speaking, this name is only for the second model.

3.1 $G(N, p)$ model

Let us consider a set of N vertices, each two connected by an edge with probability p. This is the $G(N, p)$ model, which is also called the Bernoulli, binomial, Poisson random graph, the Gilbert model, or the Erdős–Rényi random graph. This is a statistical ensemble containing

$$Z_N = 2^{N(N-1)/2} \tag{3.1}$$

labelled simple graphs of N vertices, in which each of these graphs, G, has the probability of realization determined by the number of edges E in this graph,

$$\mathcal{P}(G) = p^E(1-p)^{N(N-1)/2-E}, \quad \sum_G \mathcal{P}(G) = 1. \tag{3.2}$$

Since each edge of a vertex is present with probability p, the degree distribution equals

$$P(q) = \binom{N-1}{q} p^q (1-p)^{N-1-q}, \tag{3.3}$$

and so the average degree of a vertex is

$$\langle q \rangle = p(N-1). \tag{3.4}$$

In the limit of an infinite sparse network, $N \to \infty$, $\langle q \rangle \to \text{const} < \infty$, this binomial distribution approaches the Poisson

$$P(q) = e^{-\langle q \rangle} \frac{\langle q \rangle^q}{q!}. \tag{3.5}$$

Clearly, these networks are *uncorrelated* in the sense that their vertices 'know' nothing about the properties of even their nearest neighbours, let alone other vertices.

The $G(N,p)$ ensemble is actually a grand canonical one, in which the number E of edges (playing the role of particles) fluctuates, and the chemical potential, μ, is directly related to the fixed probability p. The partition function of this ensemble,

$$Z(N,\mu(p)) \propto \sum_E e^{-\mu E} Z(N,E) = \sum_E \left(\frac{p}{1-p}\right)^E Z(N,E)$$

$$= (1-p)^{-N(N-1)/2} = (1+e^{-\mu})^{N(N-1)/2}, \tag{3.6}$$

where $Z(N,E)$ is the number of all simple graphs with N vertices and E edges, and so $p/(1-p) = e^{-\mu}$ and $\mu = \ln[(1-p)/p]$. Here we used Eq. (3.2).[1] Then the average number of edges is

$$\langle E \rangle = -\partial_\mu \ln Z(N,\mu) = p\frac{N(N-1)}{2}, \tag{3.7}$$

which confirms Eq. (3.4), while the fluctuations are given by

$$\langle E^2 \rangle - \langle E \rangle^2 = -\partial_\mu^2 \ln Z(N,\mu) = p(1-p)\frac{N(N-1)}{2}. \tag{3.8}$$

So in this ensemble the relative fluctuations of the number of edges vanish as $1/N$ for large N.

The entropy

$$S = -\langle \ln \mathcal{P}(G) \rangle = -\sum_G \mathcal{P}(G) \ln \mathcal{P}(G), \tag{3.9}$$

which is the average logarithm of the number of graphs in the random network ensemble, can be written immediately,

$$S(N,p) = \frac{N(N-1)}{2}[-p \ln p - (1-p) \ln(1-p)], \tag{3.10}$$

if we take into account that all edges are independent and for each pair of vertices the contribution is $-p \ln p - (1-p) \ln(1-p)$.

Let the random graph be sparse, that is $\langle q \rangle \overset{N\to\infty}{\cong} pN \to \text{const} < \infty$. Then

$$S(N,p) \cong \frac{\langle q \rangle}{2} N \ln N, \tag{3.11}$$

and we see that this entropy is non-extensive (that is, not proportional to N) even in sparse graphs.

[1]See Bogacz, Burda, and Wacław (2006) for more detail.

3.2 Erdős–Rényi model

The second model was introduced by Paul Erdős and Alfréd Rényi in the second half of the 1950s. We denote it $G(N, E)$.[2] This is the statistical ensemble of all possible simple labelled graphs with N vertices and E edges, in which each graph has the same probability.[3] In other words, this is the maximally random (simple) network for given numbers of vertices N and edges E. The total number of these graphs is[4]

$$Z(N, E) = \binom{N(N-1)/2}{E}. \tag{3.12}$$

Recall that this number appeared in Eq. (3.6). Then the probability of realization of each graph equals

$$\mathcal{P}(G) = 1/Z(N, E), \tag{3.13}$$

and then the entropy, Eq. (3.9), of this ensemble equals $S(N, E) = \ln Z(N, E)$, and so in the limit of a large sparse graph, we get

$$S(N, E) \cong E \ln N = \frac{\langle q \rangle}{2} N \ln N, \tag{3.14}$$

which coincides with the entropy of the corresponding $G(N, p)$ random graph, Eq. (3.11), in this regime if $p = 2E/N^2$. In this sense, the two distinct statistical ensembles, the $G(N, E)$ and the $G(N, p)$ models, are equivalent. Bogacz, Burda, and Wacław (2006) provide the detailed derivation of the

---

[2]In graph theory, it is usually denoted as $G(n, m)$ or $G_{n,m}$.

[3]This is what is called a *'uniformly random'* ensemble.

[4]Compare to the number of labelled multigraphs (excluding those with self-loops) with N vertices and E edges

$$\binom{N(N-1)/2 + E - 1}{E}$$

and to the number of labelled multigraphs (self-loops are allowed) with N vertices and E edges

$$\binom{N(N+1)/2 + E - 1}{E}.$$

In fact, in the infinite size, sparse regime, $N \to \infty$, $\langle q \rangle = 2E/N \to \text{const} < \infty$, these two ensembles, of simple graphs and multigraphs differ rather insignificantly and the ratio of the total number of members (graphs to multigraphs) tends to the finite number $e^{-2\langle q \rangle(\langle q \rangle + 1)}$. Thus, in the large sparse network regime, all these ensembles should be equivalent.

same Poisson degree distribution for this ensemble as in Eq. (3.5).[5] This kind of equivalence of different statistical ensembles providing large sparse graphs with identical degree distributions is actually rare. It is violated in random networks more complex than the Erdős–Rényi random graphs (Bianconi, 2007a, 2009; Anand and Bianconi, 2009; Squartini, de Mol, den Hollander, and Garlaschelli, 2015). We use this equivalence extensively when we explore the more easily treatable $G(N,p)$ model and present results in the limit of large sparse networks, valid for both classical random graph models.

3.3 Cycles and clustering

Taking into account the independence of edges occurring with probability p we get the clustering of these random graphs:

$$C = \langle C \rangle = p \cong \frac{\langle q \rangle}{N}, \tag{3.15}$$

where the right expression, given for the limit of large sparse networks, shows that the clustering vanishes as $1/N$. This corresponds to the average number of triangles

$$\langle N_3 \rangle = \binom{N}{3} p^3 \cong \frac{1}{6} \langle q \rangle^3 \tag{3.16}$$

in these random graphs, which is finite when they are sparse. More generally, one can easily obtain the average number of k-cliques,[6]

$$\langle N_{k\text{-clique}} \rangle = \binom{N}{k} p^{\binom{k}{2}} \cong \frac{1}{k!} \langle q \rangle^{N(N-1)/2} N^{-[N(N-1)/2-k]}, \tag{3.17}$$

---

[5]Notice some discrepancies in how random network models and specific statistical ensembles are related by different authors. Bogacz, *et al.* consider the $G(N,p)$, $G(N,E)$, and configuration models as grand canonical, canonical, and microcanonical ensembles, respectively, while usually if the constraints are strictly imposed for every realization of a random network, then the ensemble is treated as microcanonical, and if the constraints are satisfied on average, then the ensemble is called canonical (Bianconi, 2018a).

[6]Equally easily one can get the number of maximal k-cliques

$$\langle N_{\text{max }k\text{-clique}} \rangle = \binom{n}{k} p^{\binom{k}{2}} (1 - p^k)^{N-k},$$

which is smaller, as is natural (Matula, 1970).

where $p^{\binom{k}{2}}$ is the probability that given k vertices form a k-clique, and $\binom{N}{k}$ is the number of ways to choose these k vertices of N vertices of the random graph. Thus $(k \geq 4)$-cliques vanish in large sparse random classical graphs.[7]

On the other hand, the average number of ℓ-cycles in these random graphs equals

$$\langle N_\ell \rangle = \frac{1}{2} \binom{N}{\ell} (\ell - 1)! p^\ell \cong \frac{1}{2\ell} \langle q \rangle^\ell. \tag{3.18}$$

Indeed, for an ℓ-cycle we need a 'ring' of ℓ vertices (and edges), which gives the factor p^ℓ. These ℓ vertices can be chosen of the N vertices of the random graph in $\binom{N}{\ell}$ ways. There are $\ell!$ permutations of these ℓ labelled vertices, which we must divide by 2ℓ, since a starting point and direction are irrelevant. Taking into account all these factors, we get Eq. (3.18). Notice that, although the average number of ℓ-cycles (including triangles) is finite in large sparse networks if we fix ℓ, there is something special about the value $\langle q \rangle = 1$. Namely, $\langle N_\ell \rangle$ practically vanishes when $\langle q \rangle < 1$,[8] and it exponentially grows with ℓ when $\langle q \rangle > 1$. Thus, we guess that a large sparse classical graph consists almost only of trees when $\langle q \rangle < 1$, and it has few finite cycles when $\langle q \rangle > 1$. In the latter the large sparse classical random graph is *locally tree-like* in the sense that with high probability a finite neighbourhood of a vertex looks like a tree. This does not mean that cycles do not matter in these networks, but rather that only long cycles (with ℓ increasing with N) should be somehow taken into account.[9] Then, if $\langle q \rangle$

---

[7]For large dense $G(N, p)$ random graphs, i.e., with a finite p, the average size of the largest clique (average clique number) equals

$$\frac{2}{\ln(1/p)} [\ln N + O(\ln(\ln N / \ln(1/p)))]$$

(Bollobás, 1985), which can be estimated from the condition $\langle N_{k\text{-clique}} \rangle = 1$.

[8]This results in the finite total number of vertices in cycles, which can be estimated as

$$\sum_{\ell \geq 3} \ell \langle N_\ell \rangle \cong \frac{\langle q \rangle^3}{2(1 - \langle q \rangle)}$$

for $\langle q \rangle < 1$.

[9]Note that when $\langle q \rangle > 1$, the total number of vertices in cycles is about

$$\sum_{\ell \geq 3} \ell \langle N_\ell \rangle \sim \sum_{\ell \geq 3} \langle q \rangle^\ell \sim N,$$

i.e., it is a finite fraction of all vertices. Here the sum is determined by the upper limit, which shows that only the long cycles really matter, and each vertex in the graph has a finite probability to participate in a long cycle.

sufficiently exceeds 1, we can estimate the average separation of vertices in a large sparse classical graph in the spirit of Eq. (1.12), namely,

$$\langle \ell \rangle \cong \frac{\ln N}{\ln \langle q \rangle}, \tag{3.19}$$

demonstrating that these random graphs are small worlds. We later explain the $\langle q \rangle$ in the denominator instead of $\langle q \rangle - 1$, which one could expect based on Eq. (1.12).

3.4 Giant connected component

Probably, the strongest result in random graph theory belongs to Ray J. Solomonoff and Anatol Rapoport who discovered in 1951 that when the average vertex degree $\langle q \rangle$ of an infinite $G(N, p)$ random graph exceeds the critical value 1, this graph contains a *giant connected component* including a finite fraction of all vertices (Solomonoff and Rapoport, 1951; Solomonoff, 1952).[10] The situation in both classical random graph models is as follows. When a classical random graph has many edgess, $\langle q \rangle > 1$, it consists of a single giant connected component and, also, numerous *finite connected components* of size much smaller than N (see Figure 1.7). On the other hand, if $\langle q \rangle < 1$, the giant connected component is absent and there are plenty of finite connected components. In infinite uncorrelated networks, including classical random graphs, finite components almost surely do not contain cycles; rather, they are trees, and so these networks have no cycles when $\langle q \rangle < 1$, as discussed in the previous section.

This qualitative picture is generic for random networks. The general properties of a network are primarily determined by whether or not a giant connected component is present. So the first question about any network should be about the presence and relative size of this component. Solomonoff and Rapoport found that a giant connected component in classical random graphs emerges exactly when a mean degree $\langle q \rangle$ surpasses 1. This happens without a jump (Figure 3.1). In that sense the birth of a giant connected

---

[10]Physicists, working in the field of condensed matter, are familiar with the percolation problem (Stauffer and Aharony, 1991). Remove at random a fraction of sites from an infinite conducting lattice so that a fraction p of sites are retained. Then below some critical value of p, the lattice is split into a set of finite disconnected clusters, and the system is isolating. On the other hand, a current can flow—'percolate'—from one border of the lattice to another if p is above this critical concentration—a percolation threshold p_c. The current flows on an infinite percolation cluster of connected sites, which is the giant connected component.

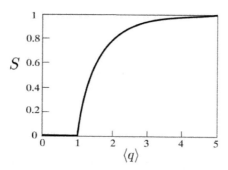

Fig. 3.1 The relative size of a giant connected component in a classical random graph versus the average degree of its vertices. Near the birth point, $S \cong 2(\langle q \rangle - 1)$.

component is a continuous phase transition, where $\langle q \rangle_c = 1$ is the critical point.[11] This is the main structural transition in a network, where network characteristics dramatically change. All these changes in uncorrelated (and many other) networks take place in the regime of a sparse network, in which the number of connections is low compared to the maximal possible number. Furthermore, Figure 3.1 demonstrates that as a giant connected component approaches the size of a classical random graph it is still in a sparse regime. In particular, the relative size of a giant connected component is already above 99% at $\langle q \rangle = 5$. Thus main qualitative changes in the architecture of networks exist in the narrow region $\langle q \rangle \ll N$. Remarkably, most studied real-world networks are sparse.

The estimate of the average shortest path distance between vertices in sparse classical random graphs, Eq. (3.18), was actually obtained for networks with a reasonably large giant connected component. This formula fails close to the birth point of a giant connected component. As $\langle q \rangle$ approaches the critical point from above, the giant connected component diminishes, its number of cycles rapidly decreases, and this component tends to a uniformly random tree whose Hausdorff dimension $D_H = 2$ and the spectral dimension $D_S = 4/3$. We discuss this special point in more detail in the following chapters.

Let us exploit the statistics of the connected components (including the giant one) in the large sparse $G(N, p)$ random graphs at different $\langle q \rangle$ using

---

[11]In physics, if there is no jump (discontinuity) of an order parameter (in our case, the relative size of a giant component) at the critical point, then it is a *continuous phase transition*. If in the critical singularity $S \propto (\langle q \rangle - \langle q \rangle_c)^\beta$, the exponent $0 < \beta \leq 1$, then this is the second-order phase transition; if $1 < \beta \leq 2$, then this is the third order phase transition; and so on.

the following approach. Here we focus only on the distribution of sizes and treat the random graph as a result of a random process. We start with $N \gg 1$ isolated vertices and progressively add edges between randomly chosen pairs of vertices.[12] If a giant component is absent, the chance to select two times vertices within the same finite component is negligible, and this process can be treated as consecutive aggregation (merging) of clusters (connected components) selected with probability proportional to their sizes.[13] It is convenient to rescale time in a such a way that the time variable t coincides with $\langle q \rangle$, that is, $t \equiv \langle q \rangle \cong 2E/N$ can be assumed continuous in large networks. Clearly, for the $G(N, p)$ model, this process can be reversed, as it is an equilibrium system.

In order to describe completely this specific aggregation process we must analyse the evolution of the size distribution of components. It is convenient to introduce the probability distribution $\mathcal{P}(s)$ for a finite connected component of s vertices to which a randomly chosen vertex belongs:

$$\mathcal{P}(s) = \frac{sn(s)}{\langle s \rangle} = \frac{sN(s)}{N},$$

(3.20)

where $n(s)$ is the size distribution of components (the probability that a uniformly randomly chosen component contains s vertices), and $\langle s \rangle$ is the average size for all components, including the giant connected component. On the other hand, $N(s)$ is the average number of components of size $s < \infty$, so $N_{\text{comp}} = \sum_s N(s)$ is the average total number of components and $\langle s \rangle \cong N/N_{\text{comp}}$. We denote the moments of the distribution $\mathcal{P}(s)$ as $\langle s^n \rangle_{\mathcal{P}} \equiv \sum_s s^n \mathcal{P}(s)$. This distribution satisfies the sum rule

$$\sum_s \mathcal{P}(s) = 1 - S.$$

(3.21)

Here, S is the relative size of the percolation cluster (giant connected component) and the sum is over $s < \infty$. Consequently, the average size of a finite component to which a randomly chosen vertex belongs is

$$\langle s \rangle_{\text{finite comp}} = \frac{\sum_s s\mathcal{P}(s)}{\sum_s \mathcal{P}(s)} = \frac{\langle s \rangle_{\mathcal{P}}}{1 - S}.$$

(3.22)

---

[12]This is often called the *Erdős–Rényi random graph process.*

[13]See Krapivsky, Redner, and Ben-Naim (2010), Chapter 5 for introduction to aggregation processes, and Ben-Naim and Krapivsky (2005) for a detailed treatment of the Erdős–Rényi process.

Note that for the sake of brevity, we often do not indicate that the distributions depend on time. For the infinite system, the evolution of the distribution $\mathcal{P}(s,t)$ is described exactly by the infinite set of evolution equations:

$$\frac{\partial \mathcal{P}(s,t)}{\partial t} = \frac{1}{2}s \sum_{u+v=s} \mathcal{P}(u,t)\mathcal{P}(v,t) - s\mathcal{P}(s,t). \tag{3.23}$$

This is actually a version of the Smoluchowski equation for our aggregation process. The reader can easily grasp how to get Eq. (3.23): first we derive the equation for $n(s,t)$ (taking into account that probability to select a component is proportional to its size s) and then pass to $\mathcal{P}(s,t)$.

We can use the generating functions technique (Appendix C), one of the fastest ways to solve bilinear equations with convolutions, à la Eq. (3.23). Let us introduce the generating function $G(z)$ of the distribution $\mathcal{P}(s)$,

$$G(z) \equiv \sum_s \mathcal{P}(s)z^s, \tag{3.24}$$

so $G(1) = 1 - S$, which we need first. Multiplying both sides of Eq. (3.23) by z^s and taking sum over s, we get the partial differential equation for $G(z,t)$:

$$\frac{\partial G(z)}{\partial t} = \frac{1}{2}z\frac{\partial G^2(z)}{\partial z} - z\frac{\partial G(z)}{\partial z} = [G(z)-1]\frac{\partial G(z)}{\partial \ln z}. \tag{3.25}$$

Its solution is

$$\ln z = (1-G)t + f(G), \tag{3.26}$$

where the function $f(G)$ is found from the initial condition $\mathcal{P}(s{=}1, t{=}0) = 1$, i.e.,

$$G(z, t=0) \equiv \sum_s \delta(s,1)z^s = z, \tag{3.27}$$

so

$$\ln z = (1-G)t + \ln G \tag{3.28}$$

and, finally, the implicit expression for the generating function $G(z,t)$ is

$$z = G(z,t)e^{[1-G(z,t)]t}. \tag{3.29}$$

Accounting for $G(1) = 1 - S$, we get the transcendental equation for S

$$S = 1 - e^{-S\langle q \rangle}, \tag{3.30}$$

where we recalled that our time t is $\langle q \rangle$. The solution of this equation, $S(\langle q \rangle)$, is shown in Figure 3.1. Near the critical point $\langle q \rangle$,

$$S \cong 2(\langle q \rangle - 1). \tag{3.31}$$

Thus, the critical exponent β for $S \sim (\langle q \rangle - \langle q \rangle_c)^\beta$ equals 1, and so this phase transition is second order.

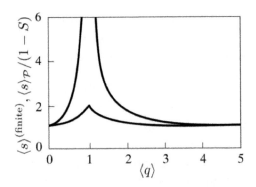

Fig. 3.2 The lower curve: the average size (number of vertices) $\langle s \rangle^{(\text{finite})}$ of a finite component in a classical random graph versus the average degree of its vertices. The upper curve: the mean size $\langle s \rangle_P / (1 - S)$ of a finite component to which a randomly chosen vertex belongs as a function of average degree $\langle q \rangle$. $\langle s \rangle_P$ diverges at the critical point as $1/|\langle q \rangle - 1|$, while $\langle s \rangle^{(\text{finite})}$ equals 2.

3.5 Finite connected components

The first question about finite components is how many there are. The total number of finite components in terms of the distribution $\mathcal{P}(s)$ follows from Eq. (3.20),

$$\mathcal{N}_{\text{finite comp}} = \sum_s N(s) = N \sum_s \frac{\mathcal{P}(s)}{s}. \tag{3.32}$$

Using Eq. (C.12),

$$\frac{N_{\text{comp}}}{N} = \sum_s \frac{\mathcal{P}(s)}{s} = \int_0^1 dz \frac{G(z)}{z}, \tag{3.33}$$

taking into account the equality directly following from Eq. (3.29),

$$\frac{d \ln G(z)}{d \ln z} = \frac{1}{1 - \langle q \rangle G(z)}. \tag{3.34}$$

Hence,

$$\frac{N_{\text{comp}}}{N} = \int_0^1 dz \frac{dG(z)}{dz} [1 - \langle q \rangle G(z)]$$

$$= \int_0^{1-S} dG[1 - \langle q \rangle G] = 1 - S - \frac{(1-S)^2}{2} \langle q \rangle. \tag{3.35}$$

Thus, the number of components decreases with $\langle q \rangle$, taking the value $N/2$ at the critical point $\langle q \rangle = 1$. At this point, the average size of a finite

64

component $\langle s \rangle^{\text{(finite)}} \cong N(1-S)/N_{\text{comp}}$ equals 2 and reaches the maximum (see Figure 3.2).

In fact, the standard term 'finite connected components' is conditional, meaning that each finite connected component contains a vanishing fraction of vertices in a network as $N \to \infty$, unlike the giant connected component. If we rank the connected components by their sizes in descending order, we get $s^{(\alpha)}$, $\alpha = 1, 2, \dots$. Mathematicians studied the asymptotics of the sizes $s^{(\alpha)}(N)$ of the αth largest components with a fixed (i.e., independent of N) α and large N in the classical random graphs and proved that:

- for $\langle q \rangle < 1 - CN^{-1/3}$, the component sizes are $s^{(\alpha \geq 1)}(N) \sim \ln N$;
- within the so-called scaling window $|\langle q \rangle - 1| < CN^{-1/3}$, $s^{(\alpha \geq 1)}(N) \sim N^{2/3}$;
- for $\langle q \rangle > 1 + CN^{-1/3}$, the giant connected component $s^{(1)}(N) \sim N$, and $s^{(\alpha > 1)}(N) \sim \ln N$,

where C is a constant (Borgs, Chayes, Kesten, and Spencer, 2001; Bollobás and Riordan, 2003).

These statements concern a vanishing fraction of all components. Physicists, when exploring continuous phase transitions, like percolation, usually ask very different questions than mathematicians. Typically we start from an infinite system. What is the nature of long range order in a given problem? What is the *order parameter*? What is the *control parameter*? What is the field? What is the susceptibility? What are the critical singularities, scaling relations, and scaling functions? What is the order of this phase transition? What are the upper and lower critical dimensions in the problem?[14] Only afterwards do we ask about finite size effects.

We have already shown that near the birth point of the giant connected component, its relative size is $S \sim (\langle q \rangle - \langle q \rangle_c)^\beta$, where $\langle q \rangle_c = 1$ and critical exponent $\beta = 1$, and so apparently, in this problem, $\langle q \rangle$ can be used as the control parameter and S is the order parameter. What about the susceptibility? Let us inspect the dependence of the first moment $\langle s \rangle_P$ of the distribution $\mathcal{P}(s)$ (probability that a uniformly randomly chosen vertex is

---

[14]In the theory of continuous phase transitions, the *upper critical dimension* D_{up} is the space dimension about which fluctuations are small, and mean-field theories are asymptotically exact in the critical region. Below and at the *lower critical dimension* D_{low}, a long range order exists only in the limit point of the control parameter, and so the phase transition disappears. For example, for the ferromagnetic Ising model on a D-dimensional lattice, $D_{\text{low}} = 1$ and $D_{\text{up}} = 4$. For the percolation problem on on a D-dimensional lattice, $D_{\text{low}} = 1$ and $D_{\text{up}} = 6$.

in a finite connected component of size s) on $\langle q \rangle$. From Eq. (3.23), we can derive the coupled set of equations for the moments $\langle s^m \rangle_{\mathcal{P}}$ and S:

$$\frac{\partial S}{\partial \langle q \rangle} = S\langle s \rangle_{\mathcal{P}},$$

$$\frac{\partial \langle s \rangle_{\mathcal{P}}}{\partial \langle q \rangle} = \langle s \rangle_{\mathcal{P}}^2 - S\langle s^2 \rangle_{\mathcal{P}}, \qquad (3.36)$$

and so on.[15] From the first two equations of the set we get the critical singularity

$$\langle s \rangle_{\mathcal{P}} \cong \frac{1}{|\langle q \rangle - 1|} \qquad (3.37)$$

(see Figure 3.2). The same directly follows from Eq. (3.34).

Let us add a vertex to a random graph, connecting it to each of the graphs vertices by an edge with a vanishingly small finite probability h and inspect how much the size of the giant component $N_{\text{giant}}(h) = NS$ increases (see Figure 3.3).[16] The response can be treated as the susceptibility:

$$\chi = \frac{\partial N_{\text{giant}}(h)}{\partial (Nh)} = \frac{N(1-S)h\langle s \rangle_{\text{finite comp}}}{Nh}$$

$$= \frac{N(1-S)h\langle s \rangle_{\mathcal{P}}/(1-S)}{Nh} = \langle s \rangle_{\mathcal{P}}. \qquad (3.38)$$

We can also explore the sum of the pair correlation functions $\langle c_{ij} \rangle$ over the vertices within all finite components, where $c_{ij} = 1$ if there is a path connecting vertices i and j (in particular, $c_{ii} = 1$) and $c_{ij} = 0$ if such a path is absent, and get the same quantity

---

[15]These equations are derived in the following way:

$$\frac{\partial \langle s^m \rangle_{\mathcal{P}}}{\partial t} = \frac{1}{2} \sum_{u,v} \sum_{s} s^{m+1} \mathcal{P}(u,t) \mathcal{P}(v,t) \delta(u+v,s) - \langle s^{m+1} \rangle_{\mathcal{P}}$$

$$= \frac{1}{2} \sum_{u,v} (u+v)^{m+1} \mathcal{P}(u,t) \mathcal{P}(v,t) - \langle s^{m+1} \rangle_{\mathcal{P}},$$

so

$$\frac{\partial \langle s^m \rangle_{\mathcal{P}}}{\partial t} = \frac{1}{2} \sum_{k=1}^{m} \binom{m+1}{k} \langle s^k \rangle_{\mathcal{P}} \langle s^{m+1-k} \rangle_{\mathcal{P}} - S\langle s^{m+1} \rangle_{\mathcal{P}}.$$

[16]Note that this is applicable both above and below the critical point. In the normal phase, such an external field induces a 'small' giant connected component of size proportional to h.

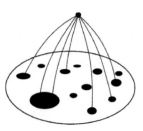

Fig. 3.3 An additional vertex attached to uniformly randomly chosen vertices in a random graph increases the giant connected component and so plays the role of an external field. The black spots are connected components in a random graph.

$$\frac{1}{N}\Big\langle \sum_{i,j\in\text{finite}} c_{ij} \Big\rangle = N\Big\langle \sum_\alpha \Big(\frac{s_\alpha}{N}\Big)^2 \Big\rangle = N\sum_s \frac{N\mathcal{P}(s)}{s}\Big(\frac{s}{N}\Big)^2$$

$$= \sum_s s\mathcal{P}(s) = \langle s\rangle_{\mathcal{P}} = \chi, \tag{3.39}$$

which provides another interpretation of the susceptibility.[17] Thus, in this problem the critical exponent γ of the susceptibility $\chi \propto |\langle q\rangle - \langle q\rangle_c|^{-\gamma}$ takes the value

$$\gamma = 1 \tag{3.40}$$

(Curie–Weiss law) as is should be in percolation above the upper critical dimension $D_{\text{up}} = 6$.

Next we ask what is the distribution $\mathcal{P}(s, \langle q\rangle)$? Knowledge of the generating function $G(z, \langle q\rangle)$, Eq. (3.29), enables us to get the explicit form of the

---

[17]We can consider the sum of the correlation functions $\langle c_{ij}\rangle$ over all the vertices in a random graph:

$$\frac{1}{N}\Big\langle \sum_{i,j} c_{ij} \Big\rangle - N\langle S\rangle^2 = \frac{1}{N}\Big\langle \sum_{i,j\in\text{finite}} c_{ij} \Big\rangle + \Big(\frac{1}{N}\Big\langle \sum_{i,j\in\text{giant}} c_{ij} \Big\rangle - N\langle S\rangle^2\Big)$$

$$= \langle s\rangle_{\mathcal{P}} + N(\langle S^2\rangle - \langle S\rangle^2).$$

The second, complementary term $N(\langle S^2\rangle - \langle S\rangle^2)$ in the final expression, coming from the summation over the vertices within the giant connected component, is also often called the 'susceptibility'. For ordinary percolation it has the same critical singularity as the first term, though in more complex percolation problems with discontinuous transitions, in particular, for the giant mutually connected component problem in interdependent networks and for k-cores, the singularities may differ from each other (Lee, Choi, Stippinger, Kertész, and Kahng, 2016a; Lee, Jo, and Kahng, 2016b). The expression of the susceptibility as the sum of pairwise correlation functions is similar to that for the susceptibility of ferromagnetic systems, given that the 1-state Potts model is equivalent to the bond percolation problem (Kasteleyn and Fortuin, 1969; Fortuin and Kasteleyn, 1972).

distribution for arbitrary $\langle q \rangle$ by implementing Eqs. (C.29) and (C.30), but we focus on the critical region as the most interesting. At the critical point $\langle q \rangle = 1$, Eq. (3.29) gives $G(z, \langle q \rangle{=}1) \cong 1 - \sqrt{2}(1 - z)^{1/2}$ near $z = 1$. According to Eq. (C.21), this suggests the power-law decay of the distribution at the critical point,

$$\mathcal{P}(s, \langle q \rangle = 1) \propto s^{-3/2}. \tag{3.41}$$

For the sake of brevity, let us derive the distribution in the critical region on the normal phase side, $\langle q \rangle = 1 - \delta$, $\delta \ll 1$. Equation (3.34) indicates that the point of singularity z^* of $G(z)$ happens when $1 - \langle q \rangle G(z^*, \langle q \rangle) = 0$, that is, $G(z^*, 1 - \delta) = 1/(1 - \delta)$. Substituting this into Eq. (3.28), we get the point of singularity, $\ln z^* = \delta^2/2$, and find a square root singularity of $G(z)$ at z^*. Equations (C.26)–(C.28) in Appendix C show that this singularity of $G(z)$ corresponds to the distribution

$$\mathcal{P}(s, \langle q \rangle) \cong \frac{1}{\sqrt{2\pi}} s^{-3/2} e^{-s\,\delta^2/2}. \tag{3.42}$$

Taking into account the singularity of $S(\langle q \rangle)$ at the critical point, Eq. (3.31), we can easily show that this form is also valid above the transition in the critical region. Thus, we have the scaling form

$$\mathcal{P}(s, \langle q \rangle) = s^{1-\tau} f(s\delta^{1/\sigma}) = \delta^{(\tau-1)/\sigma} \tilde{f}(s\delta^{1/\sigma}), \tag{3.43}$$

where $\delta = |\langle q \rangle - 1| \ll 1$ and $f(x)$ and $\tilde{f}(x)$ are scaling functions,

$$f(x) = x^{\tau-1} \tilde{f}(x) = \frac{1}{\sqrt{2\pi}} e^{-x}, \tag{3.44}$$

and τ and σ are critical exponents, taking the values $\tau = \frac{5}{2}$ and $\sigma = \frac{1}{2}$ in the classical random graphs. For a theoretical physicist, obtaining the scaling form, Eq. (3.43), and finding the scaling function and the full set of critical exponents just solves the problem of a given continuous phase transition.[18]

Substituting the scaling form of the distribution, Eq. (3.43), into the sum rule $\sum_s \mathcal{P}(s, \langle q \rangle) = 1 - S$ at $\langle q \rangle > \langle q \rangle_c$ and using the equality $\sum_s \mathcal{P}(s, \langle q \rangle_c) =$

---

[18] An elegant way to find the scaling function and the relevant critical exponent would be to substitute the scaling form of $\mathcal{P}(s, \langle q \rangle)$, Eq. (3.43), into the evolution equation for this distribution, Eq. (3.23), and solve the resulting equation for $f(x)$, which also provides the critical exponents τ and σ. The reader can find a compact technical description of this approach in da Costa, Dorogovtsev, Goltsev, and Mendes (2014, 2015). For a more comprehensive survey of the methods of theory of aggregation processes, see Leyvraz (2003).

1 at the critical point, we obtain the relative size of the giant connected component:[19]

$$S = \sum_s [\mathcal{P}(s, \langle q \rangle_c) - \mathcal{P}(s, \langle q \rangle)] \propto \delta^\beta, \tag{3.45}$$

where the critical exponent β is

$$\beta = \frac{\tau - 2}{\sigma}. \tag{3.46}$$

The critical exponent γ of the susceptibility

$$\chi = \sum_s s \mathcal{P}(s, \langle q \rangle) \propto |\delta|^{-\gamma} \tag{3.47}$$

is related to other critical exponents:

$$\gamma = \frac{3 - \tau}{\sigma}. \tag{3.48}$$

The values $\beta = 1$, $\gamma = 1$, $\tau = \frac{5}{2}$, and $\sigma = \frac{1}{2}$ found for the classical random graphs satisfy the general relations Eqs. (3.46) and (3.48). These values are common for the standard (site and bond) percolation problems above the upper critical dimension $D_{up} = 6$ (one should add to this list two critical exponents for the pairwise correlation function, $\nu = \frac{1}{2}$ and $\eta = 0$). Furthermore, three values, $\gamma = 1$, $\nu = \frac{1}{2}$, and $\eta = 0$, are valid for all cooperative models with a continuous phase transition when space dimension of a system exceeds D_{up}, that is, when a mean-field description is asymptotically exact in the neighbourhood of the critical point. The knowledge of the critical exponents enables us to obtain the key dimensions in the problem, including the Hausdorff dimension of a critical cluster in percolation (the giant component near the point of its birth). In addition, when the critical exponents are found by using mean-field theories, they also provide the value of the upper critical dimension, as shown in Appendix D.

---

[19]This is derived in the following way:

$$S \cong \int_1^\infty ds\, s^{1-\tau}[f(0) - f(s\delta^{1/\sigma})] = \delta^{(\tau-2)/\sigma} \int_{\delta^{1/\sigma}}^\infty dx\, x^{1-\tau}[f(0) - f(x)]$$

$$= \frac{\delta^{(\tau-2)/\sigma}}{\tau - 2} \left\{ -x^{2-\tau}[f(0) - f(x)]\Big|_0^\infty - \int_0^\infty dx\, x^{1-\tau} \frac{df(x)}{dx} \right\} \propto \delta^\beta.$$

The expression within the braces must be finite, which restricts the class of possible scaling functions.

4

Equilibrium Networks

4.1 The configuration model

By the late 1970s, the theory of classical random graphs was already well developed, and mathematicians started to search for more general network constructions. In 1978, Edward A. Bender and E. Rodney Canfield published a paper entitled 'The asymptotic number of labelled graphs with given degree sequences' in which they described random networks with significantly richer architectures than the Erdős–Rényi graph. Béla Bollobás strictly formulated this generalization of the Erdős–Rényi model in his 1980 paper 'A probabilistic proof of an asymptotic formula for the number of labelled random graphs' and named it the *configuration model.* This generalization turned out to be a major step toward real networks in the post-Erdős epoch.

Before introducing the configuration model, we revisit the idea of classical random graphs. Speaking in simple terms, a classical random graph is the maximally random network that is possible for a given average degree of a node, $\langle q \rangle$. The Erdős–Rényi model, $G(N, E)$, is one of the two versions of classical random graphs: a maximally random network (more rigorously, uniformly random ensemble of graphs) under two restrictions: (i) the total number of vertices N is fixed and (ii) the total number of edges, E, is fixed.

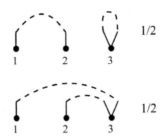

Fig. 4.1 A random network ensemble in the configuration model. In this example, the degree sequence is $q_1 = 1$, $q_2 = 1$, $q_3 = 2$; $N(1) = 2$ and $N(2) = 1$. The two members of the ensemble have equal probability of realization, $1/2$.

In fact, there is an extra (less important) restriction that the graphs in this uniform ensemble are simple, which does not produce significant difference from a similar ensemble with multigraphs in the interesting sparse network regime. In the sparse regime, these constraints result in the Poisson form of the degree distribution which markedly differs from the degree distributions of real-world networks. To make a step towards real networks, one should be able to construct a network with, at least, a real degree distribution $P(q)$, and not only a real average degree. The idea was to build the maximally random network for a given degree distribution. The configuration model provides one of the ways to achieve this goal by directly generalizing the Erdős–Rényi construction. In graph theory, the term 'sequence of degrees' usually means the vector (set) of the degrees of the vertices of a network: $\mathbf{q} \equiv (q_1, q_2, \ldots, q_N)$, which provides the number of vertices of degree q, $N(q) = \sum_i \delta(q_i, q)$, $\sum_q N(q) = N$. Let this sequence be given and, for graphicality, the total degree $\sum_i q_i$ be even. The configuration model is a statistical ensemble whose members are all possible labelled graphs, each with the same given sequence of degrees. All these members are realized with equal probability,

$$P_{\mathbf{q}}(A) = \frac{1}{Z_{\mathbf{q}}} \prod_{i=1}^{N} \delta\left(q_i, \sum_{j=1}^{N} A_{ij}\right), \tag{4.1}$$

which corresponds to the maximum possible randomness—uniform randomness (A is the adjacency matrix of a member of the ensemble). Physicists call such ensembles microcanonical. The normalization constant $Z_{\mathbf{q}}$ in Eq. (4.1) is the total number of multigraphs satisfying these constraints, that is, the number of members of this ensemble.[1]

The same can be done in the following way. Consider a set of N vertices with a degree sequence $\mathbf{q}$, where $\sum_i q_i$ is assumed to be even. Supply each of the vertices by q_i stubs of edges, $i = 1, 2, \ldots, N$; choose stubs in pairs at random and join each pair together into an edge (Figure 4.1). As a result we

---

[1]The asymptotic formula for this number was given by Bender and Canfield (1978). Here we only show a more compact formula for the number of simple graphs in this ensemble in the sparse network regime,

$$Z_{\mathbf{q}}^{(\text{simple})} \cong \frac{(\sum_i q_i - 1)!!}{\prod_i q_i!} \exp\left[-\frac{1}{2}\left(\frac{\sum_i q_i^2}{\sum_i q_i}\right)^2\right]$$

(Bianconi, 2007a, 2007b), from which one can get the entropy of the version of the configuration model, based on simple graphs. The total number of multigraphs is larger, though it is not that far from the expression shown.

have a network with a given sequence of degrees, but otherwise random.[2] If every vertex has the same number of connections, then this network is reduced to a *random regular graph*. On the other hand, if the sequence of degrees is drawn from a Poisson distribution, then, in the infinite, sparse network limit, we get approximately a classical random graph.[3] The heterogeneity of these networks is completely determined by degree distributions; correlations are absent in contrast to real-world networks. Fortunately, the majority of phenomena in complex networks can be explained qualitatively based only on the form of degree distribution, without accounting for correlations. The configuration model construction can be strightforwardly generalized to other random networks (directed, multipartite, hypergraphs, etc.) and it appeares to be remarkably handy for analytical treatment.

4.2 Local tree-likeness

Let us first consider an arbitrary undirected graph—a single realization— with the numbers $N(q)$ of vertices of degree q, $\sum_q N(q) = N$. The frequency of occurrence of vertices of degree q is given by the ratio $N(q)/N$. The total degree of the graph—double the number of edges—equals $\sum_q qN(q) = N\langle q \rangle$. Clearly, $N(q)$ vertices of degree q attract $qN(q)$ stubs ('halves of edges') of the total number $N\langle q \rangle$. Then the frequency with which end vertices of a randomly chosen edge have degree q, is $qN(q)/(N\langle q \rangle)$. In other words, select an edge at random, choose at random one of its end vertices, then the probability that it has q connections equals $qN(q)/(N\langle q \rangle)$.[4]

For random networks, this important statement is formulated as follows. In an arbitrary (uncorrelated or correlated) random network with a degree distribution $P(q)$, the degree distribution of an end vertex of a randomly chosen edge is

---

[2]Note that the term 'at random' without further specification usually means 'uniformly at random'. To build a particular realization of this network, make the following: (i) create the full set of vertices with a given sequence of stub bunches; (ii) from all these stubs, choose at random a pair of stubs and join them together; (iii) from the rest of the stubs, choose at random a pair of stubs and join them together, and so on until no stubs remain.

[3]The difference is that the configuration model is a statistical ensemble of multigraphs, while the classical random graphs are based on simple graphs. We shall see that for sparse uncorrelated networks with rapidly decaying degree distributions, this difference is not essential.

[4]We can also look at this problem in a slightly different way. Let us consider the full multiset of all $2E = N\langle q \rangle$ nearest neighbours of all vertices in an arbitrary graph. Then the frequency with which vertices in this multiset have degree q is equal to $qN(q)/(N\langle q \rangle)$.

$P(q)$

$$\frac{qP(q)}{\langle q \rangle}$$

Fig. 4.2 End vertices of a randomly chosen edge in a network have a different statistics of connections from the degree distribution of this network.

$$\widetilde{P}(q) = \frac{qP(q)}{\langle q \rangle} \qquad (4.2)$$

(Figure 4.2). Therefore, the connections of end vertices of edges are organized in a different way from those of uniformly randomly chosen vertices. In particular, we have $\langle q^2 \rangle / \langle q \rangle$ for the average degree of an end vertex of a randomly chosen edge, which is greater than the mean degree $\langle q \rangle$ of a vertex in the network in almost all situations.[5]

Relations similar to those in the two previous paragraphs, can be also written for directed graphs and for random directed networks. In particular, in an arbitrary (uncorrelated or correlated) random directed network with an in/out-degree distribution $P(q_i, q_o)$, the degree distributions of the source and target end vertices of a randomly chosen edge are, respectively,

$$\widetilde{P}_s(q_i, q_o) = \frac{q_o P(q_i, q_o)}{\langle q \rangle / 2},$$

$$\widetilde{P}_t(q_i, q_o) = \frac{q_i P(q_i, q_o)}{\langle q \rangle / 2}, \qquad (4.3)$$

where $\langle q_i \rangle = \langle q_o \rangle = \langle q \rangle / 2$, $q \equiv q_i + q_o$.

Let us now return to undirected uncorrelated networks. Let the degree distribution of the configuration model network be $P(q)$. What is the degree distribution of a nearest neighbour of a uniformly randomly chosen vertex? Clearly, for uncorrelated networks, this is exactly the distribution of an end vertex of a randomly chosen edge, Eq. (4.2), and the joint degree–degree distribution of the end vertices of an edge in these networks equals

$$P(q, q') = \frac{qP(q) \, q'P(q')}{\langle q \rangle^2}. \qquad (4.4)$$

The average degree of any neighbour of a vertex equals

---

[5]The equality $\langle q^2 \rangle / \langle q \rangle = \langle q \rangle$ takes place only if a network has no vertices other than the bare ones or the dead ends.

$$\widetilde{\langle q \rangle} = \frac{\langle q^2 \rangle}{\langle q \rangle}, \tag{4.5}$$

which is almost always greater than the average degree $\langle q \rangle$ of a vertex in the network. Consequently, the nearest neighbours of a vertex are, on average, better connected than the vertex itself, which leads to a number of important consequences. The rest of this book is mostly about these consequences. The effect is particularly strong when the degree distribution has a fat tail allowing even a huge difference, $\langle q^2 \rangle / \langle q \rangle \gg \langle q \rangle$.

While discussing classical random graphs, we noticed that the sparse networks are most interesting since the point of the birth of a giant connected component falls in this regime. Let us focus on this sparse situation in the configuration model and first find the average number of self-loops. In the configuration model, the probability that two stubs are connected by an edge equals $1/(2E - 1) \cong 1/(N\langle q \rangle)$. The number of choices for connection of two stubs of a given vertex of degree q equals $\binom{q}{2}$, so we readily have

$$\langle N_{\text{self-loops}} \rangle \cong \frac{1}{2} \frac{\langle q(q-1) \rangle}{\langle q \rangle}, \tag{4.6}$$

where we neglected double, etc., self-loops at vertices, since their contribution is much smaller, of the order of $1/N$. Similarly, one can get for the average number of multiple edges,

$$\langle N_{\text{multiple edges}} \rangle \cong \frac{1}{4} \left(\frac{\langle q(q-1) \rangle}{\langle q \rangle} \right)^2, \tag{4.7}$$

that is, the average number of cycles of length 2, and so on, for longer cycles, as we investigate later. In general, in the infinite uncorrelated networks, if the second moment of the degree distribution $\langle q \rangle$ does not diverge, then for each finite cycle length ℓ, the average number of the ℓ-cycles asymptotically goes to a constant; we have already observed it in the particular case of the classical random graphs. So uncorrelated networks are locally tree-like, which greatly simplifies their analytical treatment, since the statistics of trees are much easier than for loopy networks.[6]

The local tree-likeness still typically assumes an infinite number of infinite cycles in these networks, which can dramatically influence their long-range properties. Recall a Cayley tree, a finite fraction of whose vertices are

---

[6]In fact, much wider range of networks than uncorrelated ones are locally tree-like. In particular, numerous correlated generalizations of the configuration model belong to this class.

leaves. As it is a tree, removing any arbitrary small (yet finite) fraction of vertices or edges from it, we destroy a giant connected component effectively cutting off branches from the tree, and so the percolation phase transition is absent like in a one-dimensional chain. To get a long-range order and a phase transition, we use the following trick. Consider a sufficiently large Cayley tree of diameter $\mathcal{D}_{CT}$ and a central part of it of diameter $\mathcal{D}_{BL} \ll \mathcal{D}_{CT}$. Tend first the diameter of the Cayley tree $\mathcal{D}_{CT}$ to infinity, and then tend $\mathcal{D}_{BL}$ to infinity, keeping $\mathcal{D}_{BL} \ll \mathcal{D}_{CT}$. After passing to these limits, consider any properties of the (now infinite) central part without cutting it from the infinite Cayley tree. The leaves of the Cayley tree appear to be infinitely far from the internal part (*Bethe lattice*) and do not influence its properties, including a long-range order and a percolation transition. Roughly speaking, a Bethe lattice is an infinite Cayley tree without its leaves, and in this sense, it is not a tree at all. If we look at an infinite random regular graph, keeping in mind its local tree-likeness, we see that it is equivalent to the corresponding Bethe lattice. This allows one to construct Bethe lattices for simulations.

The average branching (the average degree minus one of an edge) equals

$$\langle b \rangle = \frac{\langle q^2 \rangle}{\langle q \rangle} - 1 \tag{4.8}$$

for an arbitrary network, which gives $\langle b \rangle = \langle q \rangle$ for the classical random graphs, since for the Poisson degree distribution, we have $\langle q^2 \rangle - \langle q \rangle = \langle q \rangle^2$. Taking into account the tree-likeness of uncorrelated networks up to long distances, which enables us to estimate the number of the n-th nearest neighbours of a vertex as $\sim \langle b \rangle^n$, we can obtain the mean separation of vertices in the spirit of Section 1.3 as

$$\langle \ell \rangle \cong \frac{\ln N}{\ln \langle b \rangle}, \tag{4.9}$$

that is

$$\langle \ell \rangle \cong \frac{\ln N}{\ln[(\langle q^2 \rangle - \langle q \rangle)/\langle q \rangle]} \tag{4.10}$$

for uncorrelated networks and

$$\langle \ell \rangle \cong \frac{\ln N}{\ln \langle q \rangle} \tag{4.11}$$

for the classical random graphs.

Similarly, one can get the average shortest-path distance from a vertex of degree q to other vertices:

$$\langle \ell \rangle (q) \cong \frac{\ln(N\langle q \rangle / q)}{\ln \langle b \rangle}, \qquad (4.12)$$

and the average shortest-path distance between two vertices degree q and q':

$$\langle \ell \rangle (q, q') \cong \frac{\ln[N\langle q \rangle / (qq')]}{\ln \langle b \rangle}, \qquad (4.13)$$

(Hołyst, Sienkiewicz, Fronczak, Fronczak, and Suchecki, 2005).[7]

According to Eq. (4.11), fat-tailed degree distributions with large $\langle q^2 \rangle$ provide networks with short intervertex distances $\langle \ell \rangle$. This effect is particularly strong for networks having power-law degree distributions $P(q) \sim q^{-\gamma}$ with the exponent $\gamma \leq 3$, where the second moment of the degree distribution diverges. In this situation, the average separation of vertices $\langle \ell \rangle (N)$ grows with N slower than logarithmically, and so such networks are called *ultra-small worlds* (Cohen and Havlin, 2002; Dorogovtsev, Mendes, and Samukhin, 2003b).

The locally tree-like structure of sparse uncorrelated networks has one more direct consequence. It is clear intuitively that a giant connected component can exist in such a network only if the average branching exceeds 1, $\langle b \rangle > 1$. So

$$\langle q^2 \rangle - \langle q \rangle > \langle q \rangle, \qquad (4.14)$$

which is the renowned *Molloy–Reed criterion* (Molloy and Reed, 1995, 1998).[8] This gives $\langle b \rangle_c = \langle q \rangle_c = 1$ for the birth point of a giant connected component in classical random graphs, which is the value already discussed in Chapter 3. Equation (4.14) also shows that if $P(1) = 0$, the network has a giant connected component. Finally, note that the right-hand side of Eq. (4.9) diverges when the average branching approaches 1 at the critical point. This

---

[7]The term $\langle q \rangle$ in the logarithm, in principle, could be neglected. Also, note the relations between $\langle \ell \rangle (q, q')$, $\langle \ell \rangle (q)$, and $\langle \ell \rangle$, valid for arbitrary random networks:

$$\langle \ell \rangle (q) = \sum_{q'} P(q') \langle \ell \rangle (q, q'),$$

$$\langle \ell \rangle = \sum_{q} P(q) \langle \ell \rangle (q) = \sum_{q, q'} P(q) P(q') \langle \ell \rangle (q, q').$$

[8]We shall show later how to get it in a more strict way.

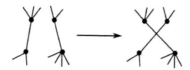

Fig. 4.3 Rewiring operation for the randomization algorithm (Maslov, Sneppen, and Alon, 2003). This rewiring preserves degrees of vertices and does not create self-loops and multiple edges.

indicates that a giant component at this point is not a small world, as is true for general percolation problems (see Appendix D).

4.3   Generating maximally random networks

For simulations, generation of a full statistical ensemble, as in the configuration model is often infeasable. How can we construct an equilibrium network in practice if we want it to satisfy a set of structural constraints? In addition to the degree sequence as in the configuration model, we can, for instance, forbid self-loops and multiple connections.[9] The idea is, as in physics, to start with a graph satisfying these constraints and relax it to an 'equilibrium' by performing a chain of random steps such that all the constraints are fulfilled at each step. Maslov, Sneppen, and Alon (2003) proposed this kind of practical randomization algorithm just for random networks with a fixed degree sequence and without multiple edges and self-loops. Suppose we want to generate a uniformly random network of this sort consisting of a single connected component. To do this, we must perform the following steps:

(i) Create an arbitrary connected graph, with a given degree sequence, and without multiple connections and self-loops.

(ii) Rewire a pair of uniformly randomly chosen edges in a way shown in Figure 4.3. This retains the degrees of all vertices and does not create multiple edges. Check whether this rewiring splits the network into two parts or not. If it splits, then abandon this rewiring and repeat the step.

(iii) Repeat (ii) until the network relaxes to an equilibrium state.

The slowest part of this algorithm is checking the connectedness of the networks. If we do not need our network to be connected, then this algorithm is very fast.

---

[9]This additional constraint can lead to correlations, which we discuss later.

4.4 Hidden variables

After the configuration model using a microcanonical statistical ensemble, the second way to construct a network with a desired degree distribution was found in 2001–2002. This grand canonical ensemble construction (Goh, Kahng, and Kim, 2001; Chung and Lu, 2002; Caldarelli, Capocci, De Los Rios, and Munoz, 2002; Söderberg, 2002), generalizes the $G(N,p)$ model, and it is essentially based on the notion of *hidden variables*. The idea is to ascribe non-negative real numbers f_i (hidden variables) to the vertices, $i = 1, 2, \ldots, N$ and to interconnect each pair of vertices, i and j, with a probability $p_{ij}(f_i, f_j)$ depending on the hidden variables of these two vertices. In different versions of the model, the hidden variables f_i may be a given sequence of numbers, or they may be drawn from some probability distribution $P_h(f)$. As the result, we get a grand canonical ensemble of simple graphs. In particular, when $f_i = c > 0$, $i = 1, 2, \ldots, N$, we arrive at the standard $G(N,p)$ random graph. Various versions of this model are called the *static model* (Goh, Kahng, and Kim, 2001),[10] the *Chung–Lu model* after Chung and Lu (2002), or, more generally, random networks with hidden variables.

First, without specifying the form of $p_{ij}(f_i, f_j)$, let us look at the probability of realization of a member of this ensemble—a simple graph described by an $N \times N$ adjacency matrix A, namely

$$\mathcal{P}(A) = \prod_{i<j} [p_{ij} A_{ij} + (1 - p_{ij})(1 - A_{ij})]. \tag{4.15}$$

Using this probability, we can get all observables for this random network in terms of the probabilities p_{ij}. For example, in the same way as for the $G(N,p)$ model, Eq. (3.10), one can obtain the entropy

$$S = -\sum_{i<j} [p_{ij} \ln p_{ij} + (1 - p_{ij}) \ln(1 - p_{ij})], \tag{4.16}$$

derive an expression for the degree distribution (Lee, Kahng, Cho, Goh, and Lee, 2018), which we do not show here. Of these observables, here we need the average degree of vertex i in this ensemble, which can be easily written:

$$\langle q \rangle_i = \sum_j p_{ij}. \tag{4.17}$$

---

[10] According to Goh, *et al.*, 'the name "static" originates from the fact that the number of nodes is fixed from the beginning.'

We immediately see that if the probability to have an edge between vertices i and j has the factored form

$$p_{ij}(f_i, f_j) = \frac{f_i f_j}{N\langle f \rangle}, \tag{4.18}$$

then $\langle q \rangle_i = f_i$, $\langle q \rangle = \sum_i \langle q \rangle_i / N = \langle f \rangle = \sum_i \langle f \rangle_i / N$, and the hidden variables f_i play the role of the expected degrees of the vertices.[11] (Here we have to assume that $f_i \leq \sqrt{N\langle f \rangle}$ for satisfying $p_{ij} \leq 1$.) This has an important consequence. Clearly, if the probability distribution $\rho(f)$, from which hidden variables are drawn, is fat-tailed (for example, power-law), then the degree distribution of this network, asymptotically approaches $\rho(q)$ for large degrees, $P(q) \cong \rho(q)$, although these two distributions may significantly differ from each other in the region of small degrees. Importantly, the factored probability, Eq. (4.18), produces uncorrelated networks which have a locally tree-like structure. The resulting ensemble, however, appears to be not equivalent to the configuration model with the same degree distribution in the sense that their entropies do not coincide (Anand and Bianconi, 2009, 2010).[12] Nonetheless, it is difficult to indicate an observable which would dramatically differ in these ensembles.

The nuisance that the probability p_{ij} given by Eq. (4.18) can exceed 1 for large values of the hidden variables can be corrected by choosing, for example, the following form of the probability p_{ij}:[13]

$$p_{ij}(f_i, f_j) = 1 - e^{-f_i f_j / (N\langle f \rangle)}. \tag{4.19}$$

This affects only the cut-off region of the degree distribution but also introduces correlations if the distribution $P_h(f)$ has a fat tail.

---

[11] Note the title 'Connected components in random graphs with given expected degree sequences' of the paper of Chung and Lu (2002).

[12] Notably, this happens not only due to the fact that the macrocanonical statistical ensembles in this section are based on simple graphs while the members of the statistical ensemble in the configuration model are multigraphs since Anand and Bianconi (2009, 2010) actually accounted only for simple graphs in the entropy of the configuration model.

[13] We can propose the following heuristic rationale behind this form. Let us introduce the small probabilities $p_i = f_j / (N\langle f \rangle) \ll 1$, $i = 1, 2, \ldots, N$ and try to connect vertices i and j $N\langle f \rangle$ times, each time with the probability $p_i p_j$. This leads to the final probability that vertices i and j are connected just in the desired form:

$$p_{ij} = 1 - \left(1 - \frac{f_i}{N\langle f \rangle} \frac{f_j}{N\langle f \rangle}\right)^{N\langle f \rangle} \cong 1 - e^{-f_i f_j / (N\langle f \rangle)}.$$

4.5 Annealed networks and graphons

At this point, we mention a useful deterministic construction known as the *annealed network approximation* based on relations visually similar to formulas in the previous section. This approximation is actually applicable not only to equilibrium or evolving random networks but even to non-random graphs. Let us consider a large graph whose vertices have degrees q_i, $i = 1, 2, \ldots, N$. (For the sake of brevity, we assume here that this graph is simple, which is actually not necessary.) This graph is approximated by the 'annealed network' that is the fully connected weighted graph of N vertices, whose edges have the weights[14]

$$w_{ij}(q_i, q_j) = \frac{q_i q_j}{N \langle q \rangle}. \tag{4.20}$$

Then the sum of the weights of the edges of a vertex coincides with the degree of the original graph,

$$\sum_j w_{ij} = q_i. \tag{4.21}$$

Compare Eqs. (4.18) and (4.20), (4.17) and (4.21). Even if the original graph is random, the annealed network will be deterministic in the limit $N \to \infty$, which makes it one of the widely used simple instruments in the theory of complex networks.[15]

The annealed network approximation provides a quite cropped image of a graph. In many interesting situations, the infinite size limit, $N \to \infty$, allows a detailed continuous description (Lovász, 2012). Let us roughly describe the idea of this description, focusing on simple graphs. A sequence of graphs of increasing sizes N converges to a well defined limit when the densities of all 'motifs' (graph blocks) in the graphs converge.[16] If this graph limit exists,

---

[14]The term 'annealed' comes from physics. Imagine that the edges of a graph rapidly hop between possible positions, and that the time of observation is much longer than the typical times of these hops. Then the observer will 'see' the annealed network, instead of individual graphs at distinct instances.

[15]Typically the annealed network approximation is used with Eq. (4.20), although one can similarly introduce weights accounting for given degree–degree correlations; see Eq. (7.24).

[16]The density of a motif M of $|M|$ vertices in a graph G of $|G|$ vertices is the ratio

$$\frac{\mathcal{N}(M, G)}{|G|(|G| - 1) \ldots (|G| - |M| + 1)},$$

where $\mathcal{N}(M, G)$ is the number of subgraphs of $|M|$ vertices in G isomorphic to M and the denominator is the number of all subgraphs of $|M|$ vertices in G.

then it can be described by the *graphon* (short for graph function), which is a weighted continuous version of the adjacency matrix rescaled to 1. The rescaling from the limit $N \to \infty$ of the $N \times N$ adjacency matrices of the graph sequence gives the graphon $W(x, y)$, which is the symmetric function defined for $0 \le x, y \le 1$, such that $0 \le W(x, y) \le 1$. Namely,

$$W(x, y)dxdy = \lim_{N \to \infty} \left(\frac{1}{N^2} \sum_{\substack{Nx < i < N(x+dx) \\ Ny < j < N(y+dy)}} A_{ij} \right). \tag{4.22}$$

For example, the graphon of the $G(N, p)$ model is $W(x, y) = p$, and so one could think that graphons are more intended for dense graphs. Nevertheless, the graphons turned out to be also well applicable to the limits of sparse graphs (Borgs, Chayes, Cohn, and Holden, 2017), and numerous models of complex networks, equilibrium and evolving, allow a rescaled continuous description of this sort.

4.6 Clustering, cycles, and cliques

Let us look closer at the local tree-likeness of large sparse uncorrelated networks. To do this, we inspect the average number $\mathcal{N}_{\text{cycles}}(\ell)$ of cycles of a given length ℓ in undirected networks of this kind. Here we consider the configuration model and adapt the derivation from Bianconi and Marsili (2005). The number $\mathcal{N}_{\text{cycles}}(\ell)$ can be written as the following product:

$$\mathcal{N}_{\text{cycles}}(\ell) = R(N, \ell, \{P(q)\}) \, G(\ell) \, W(N, \ell). \tag{4.23}$$

Here the first term $R(N, \ell, \{P(q)\})$, is the number of ways to select pairs of edges connected to ℓ vertices from the edges of these vertices. This number depends on the form of the degree distribution $P(q)$. We easily get

$$R(N, \ell, \{P(q)\}) = \frac{[N \langle q(q-1) \rangle]^\ell}{\ell!}. \tag{4.24}$$

The second term in the product on the right-hand side of Eq. (4.23), $G(\ell)$, is the number of ways to connect ℓ given vertices in a cycle. Equally easily, we have

$$G(\ell) = \frac{\ell!}{2\ell}. \tag{4.25}$$

Finally, the third term of the product, $W(N, \ell)$, is the fraction of graphs in the ensemble containing the cycle. To get it, one should note that the number of graphs in the configuration model is proportional to $(\sum_i q_i - 1)!! =$

$(N\langle q \rangle - 1)!!$ (Section 4.1). This expression is the number of ways to connect $2E = \sum_i q_i$ stubs in the configuration model construction in pairs, that is, in E edges. Indeed, you can match one stub with the remaining $2E - 1$ stubs, then take another stub and match it with the remaining $2E - 3$ stubs, and so on, which results in the double factorial. For the number of graphs containing the ℓ cycle, one should count the number of ways to connect the edges not forming part of the ℓ-cycle, which is, similarly, $(2E - 2\ell - 1)!!$. One can show that in the expressions for the total number of graphs and the number of graphs with this cycle, only the factors with the double factorial differ from each other,[17] so

$$W(N, \ell) = \frac{(N\langle q \rangle - 2\ell - 1)!!}{(N\langle q \rangle - 1)!!} \overset{\ell \ll N}{\cong} \frac{1}{N^\ell \langle q \rangle^\ell}. \tag{4.26}$$

These three terms together give the average number of cycles:

$$\mathcal{N}_{\text{cycles}}(\ell) = \frac{1}{2\ell} \left(\frac{\langle q(q-1) \rangle}{\langle q \rangle} \right)^\ell. \tag{4.27}$$

More thorough calculations show that for this formula, we actually need $\ell \ll N\langle q^2 \rangle^2 / \langle q^4 \rangle$.[18] Thus, if the second moment of the degree distribution is finite, then the number of finite cycles in infinite sparse uncorrelated networks is constant, which validates their tree-like structure.[19] In particular, for the average number of triangles, Eq. (4.27) reduces to

$$\mathcal{N}_3 = \frac{1}{6} \left(\frac{\langle q(q-1) \rangle}{\langle q \rangle} \right)^3, \tag{4.28}$$

and so the clustering coefficient in the configuration model equals

$$C = \frac{1}{N} \frac{\langle q(q-1) \rangle^2}{\langle q \rangle^3}. \tag{4.29}$$

Similarly, for large sparse directed networks, the average number of cycles equals

---

[17]Notice the comment in Section 4.1, concerning the expression for the number of graphs in the configuration model.

[18]For the corresponding hidden variable model, one should only change $\langle q(q-1) \rangle$ in Eq. (4.27) to $\langle q^2 \rangle$ (Bianconi and Marsili, 2005), which demonstrates the non-equivalence of these two ensembles.

[19]Interestingly, for the uncorrelated network with $P(2) = 1$, that is, the equilibrium ensemble of rings, Eq. (4.27) gives $\mathcal{N}_{\text{cycles}}(\ell) = 1/(2\ell)$, showing how the number of rings decays with ℓ.

$$\mathcal{N}_{\text{cycles}}(\ell) = \frac{1}{\ell} \left(\frac{\langle q_i q_o \rangle}{\langle q_i \rangle} \right)^{\ell}, \tag{4.30}$$

when $\ell \ll N \langle q_i q_o \rangle^2 / \langle (q_i q_o)^2 \rangle$ (Bianconi, Gulbahce, and Motter, 2008; Bianconi and Gulbahce, 2008).

In a similar way, one can obtain the average number of k-cliques in large sparse undirected uncorrelated networks, although the resulting expressions turned out to be less compact than for the number of cycles (Bianconi and Marsili, 2006a, 2006b; Janssen, van Leeuwaarden, and Shneer, 2019). The verdict is that when the second moment of the degree distribution is finite, $\langle q^2 \rangle < \infty$, the situation coincides with that for classical random graphs, namely, the $(k{>}3)$-cliques are absent in these networks. On the other hand, when the second moment of the degree distribution diverges, these networks have many cliques even with $k > 3$, and the size of the maximum clique increases with N (although, of course, its relative size approaches zero).

4.7 Correlations

In contrast to the network models described in the preceding sections, the real-world networks are correlated. Furthermore, all growing network models are correlated. The simplest of the structural correlations in networks are the correlations between degrees of the nearest neighbours. We focus here on these *degree–degree correlations*.

4.7.1 Joint degree–degree distribution

The correlations between degrees of the nearest-neighbouring vertices (or, equivalently, between degrees of the end vertices of a uniformly randomly chosen edge) are described by the probability distribution:[20]

$$P(q, q') = \frac{1}{N \langle q \rangle} \sum_{i,j=1}^{N} \langle \delta(q_i, q) A_{ij} \delta(q_j, q') \rangle. \tag{4.31}$$

---

[20]How can we obtain the degree–degree distribution from empirical data? Let the number of edges in a graph be E. Measure the number of edges $E(q, q')$ connecting vertices of degree q and q'.

$$\text{If } q \neq q', \text{ then } P(q, q') = \frac{E(q, q')}{2E},$$

$$\text{if } q = q', \text{ then } P(q, q) = \frac{E(q, q)}{E}.$$

Clearly, $P(q, q') = P(q', q)$ and $\sum_{q,q'} P(q, q') = 1$. When the network is uncorrelated, $P(q, q') = qP(q)q'P(q')/\langle q \rangle^2$, Eq. (4.5).

Similarly to the constructions for uncorrelated networks, considered previously, there are two ways to build equilibrium networks with given degree–degree correlations. The first one is the direct generalization of the configuration model, roughly speaking, the uniformly random (that is, microcanonical) ensemble with a given joint distribution $P(q, q')$. The second ensemble is grand canonical, based on hidden variables. For example, one can construct such an ensemble as in Boguñá and Pastor-Satorras (2003):

- draw the hidden variables f_i of vertices $i = 1, 2, \ldots, N$ from a given distribution $\rho(f)$,
- for each pair i, j, set an edge between the vertices with a given probability $p(f_i, f_j)$.

The probability distribution $\rho(f)$ and the symmetric non-negative function $p(f, f')$ determine the joint distribution $P(q, q')$.[21] Both these constructions provide locally tree-like networks if degree distributions decay sufficiently rapidly and the networks are sparse.

We now discuss these correlations without specifying ensembles. In an arbitrary random network, so

$$\sum_{q'} P(q, q') = \frac{qP(q)}{\langle q \rangle} \tag{4.32}$$

and

$$\sum_{q,q'} q^n P(q, q') = \frac{\langle q^{n+1} \rangle}{\langle q \rangle}. \tag{4.33}$$

Instead of the joint degree–degree distribution, equivalently one can use the conditional probability $P(q|q')$ that if one end vertex of an edge is of degree q', then its second end node is of degree q:

$$P(q|q') = \frac{P(q, q')}{\sum_q P(q, q')} = \langle q \rangle \frac{P(q, q')}{q'P(q')}, \tag{4.34}$$

$\sum_q P(q|q') = 1$. Since $P(q, q') = P(q', q)$, the relation

$$qP(q'|q)P(q) = q'P(q|q')P(q') \tag{4.35}$$

holds (Boguñá and Pastor-Satorras, 2002).

---

[21] Interestingly, an exponential distribution $\rho(f) = e^{-f}$ and $p(f, f') = \theta(f + f' - A)$, where $A > 0$ is a constant, generate scale-free networks (Caldarelli, Capocci, De Los Rios, and Munoz, 2002).

Unfortunately, it is difficult to measure the full joint distribution $P(q, q')$ or $P(q'|q)$ in networks with fat-tailed degree distributions due to insufficient statistics.[22] To get over this difficulty, Pastor-Satorras, Vázquez, and Vespignani (2001) proposed to use a less-informative though convenient characteristic, namely the average degree of the nearest neighbour of a vertex as a function of the degree of this vertex:

$$\langle q \rangle_{nn}(q) = \sum_{q'} q' P(q'|q). \tag{4.36}$$

This dependence relates the degree of a vertex and the average degree of its nearest neighbours and allows one to evaluate how vertices of different degrees are interconnected. If a network is uncorrelated, then $\langle q \rangle_{nn}(q) = \langle q^2 \rangle / \langle q \rangle$. In correlated networks, dependences $\langle q \rangle_{nn}(q)$ can be very diverse in shape. Still, $\langle q \rangle_{nn}(q)$ is constrained by the relation

$$\sum_{q'} \frac{q' P(q')}{\langle q \rangle} \langle q \rangle_{nn}(q') = \frac{\langle q^2 \rangle}{\langle q \rangle}, \tag{4.37}$$

which follows from Eqs. (4.33) and (4.36) (Boguñá, Pastor-Satorras, and Vespignani, 2003). In particular, if the second moment of the degree distribution diverges, $\langle q \rangle_{nn}(q)$ cannot be arbitrary. Its form must guarantee the divergence of the left-hand side of relation Eqs. (4.37).

4.7.2 Pearson correlation coefficient

Sociologists usually characterize degree–degree correlations by an even more rough quantity than $\langle q \rangle_{nn}(q)$. This is the *Pearson correlation coefficient* defined as

$$r = \frac{\langle qq' \rangle_l - \langle q \rangle_l \langle q' \rangle_l}{\langle q^2 \rangle_l - \langle q \rangle_l^2} = \frac{\left\langle \sum_{i,j} [A_{ij} - q_i q_j / (2E)] q_i q_j \right\rangle}{\left\langle \sum_{i,j} [q_i \delta_{ij} - q_i q_j / (2E)] q_i q_j \right\rangle}. \tag{4.38}$$

Here q and q' are the degrees of the end nodes of an edge, and $\langle \ \rangle_l$ denotes the average over all edges in a network, $\langle q \rangle_l = \langle q' \rangle_l$.[23] The Pearson coefficient is a

---

[22]See however Maslov and Sneppen (2002) and Maslov, Sneppen, and Alon (2003).

[23]Equivalently, one can substitute $q - 1$ and $q' - 1$ for q and q' in this definition. $\langle qq' \rangle_l$ is the average product of the degrees of the end vertices of an edge. $\langle q \rangle_l = \langle q' \rangle_l$ is the average degree of an end vertex of an edge. As we explained, $\langle q \rangle_l = \langle q^2 \rangle / \langle q \rangle$ for an arbitrary network. Similarly, the average square of the degree of an end vertex of an edge, $\langle q^2 \rangle_l$, equals $\langle q^3 \rangle / \langle q \rangle$. Equation (4.38) appears to be not very convenient for the computation of r from empirical data. See Newman (2002a) for a more practical form of this definition.

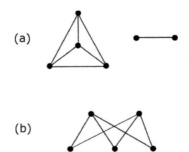

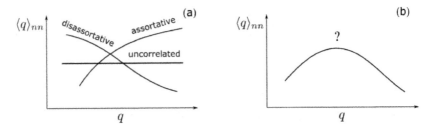

Fig. 4.4 Two small networks with ultimately assortative (a) and disassortative (b) mixing. In network (a), only vertices of equal degree are interlinked. In network (b), only vertices of different degrees are interlinked. Check that r equals 1 and -1 for (a) and (b), respectively.

Fig. 4.5 (a) Typical curves $\langle q \rangle_{nn}(q)$ for various kinds of degree–degree correlations. (b) It is impossible to conclude whether the mixing is assortative or disassortative by inspecting this non-monotonous dependence.

standard pair correlation function $\langle qq' \rangle_l - \langle q \rangle_l \langle q' \rangle_l$ normalized in such a way that r is in the range between -1 and 1. Clearly, if a network is uncorrelated, then the Pearson coefficient is zero. If, on average, strongly connected nodes have strongly connected neighbours, then $r > 0$. These correlations are called *assortative*, and this situation is referred to as *assortative mixing* (Newman, 2002a). In the opposite situation, on average, a weakly connected vertex has a strongly connected neighbour, and vice versa. This is called *disassortative mixing*, and for these correlations, the Pearson coefficient is negative. Figure 4.4 shows examples of networks with ultimately strong assortative and disassortative mixing, $r = 1$ and -1, respectively.

We can distinguish between assortative and disassortative correlations in another way by inspecting the dependence $\langle q \rangle_{nn}(q)$ (Figure 4.5a). Clearly, if this curve is monotonously growing, then the degree–degree correlations are assortative, and if it is monotonously decreasing, then the correlations are disassortative. The problem is that the curve $\langle q \rangle_{nn}(q)$ is often not monotonous as in Figure 4.5b, and then clear distinction between the two kinds of correlations is impossible. Furthermore, sometimes, the Pearson coefficient is itself confusing. To see this, we rewrite Eq. (4.38) using $\langle q \rangle_{nn}(q)$. This gives

$$r = \frac{\langle q \rangle \sum_q q^2 \langle q \rangle_{nn}(q) P(q) - \langle q^2 \rangle^2}{\langle q \rangle \langle q^3 \rangle - \langle q^2 \rangle^2}. \tag{4.39}$$

Note the third moment of the degree distribution, $\langle q^3 \rangle$, in the denominator. In infinite scale-free networks, $\langle q^3 \rangle$ diverges if degree distribution exponent γ is less or equal to 4. So if the numerator is finite, we get zero Pearson's coefficient in an infinite network. It turns out that this is the case only if exponent γ is between 3 and 4. In this situation for finite networks, r strongly depends on network size. So, it is virtually impossible to compare the values of the Pearson coefficient for different networks and, for example, determine which one of them is 'more assortative' (Dorogovtsev, Ferreira, Goltsev, and Mendes, 2010). Only if exponent γ is outside of this (3, 4) interval, is r finite in infinite networks and can it be used for the characterization of correlations.[24] Furthermore, zero Pearson's coefficient does not guarantee the absence of degree–degree correlations. In principle, some non-monotonous $\langle q \rangle_{nn}(q)$ substituted in Eq. (4.39), may give $r = 0$.

These drawbacks of r seriously hinder comparison of empirical data for different networks with fat-tailed degree distributions. Nonetheless, the Pearson correlation coefficient is always listed among main network characteristics, providing information at least about the type of degree–degree correlations. For example, practically all empirical data on social and collaboration networks indicate strong assortative correlations (see a large survey of empirical data on networks by Costa, Rodrigues, Travieso, and Villas Boas, 2007). That is, a sociable person typically has sociable friends, while (very few) friends of an introvert are also unsociable. On the other hand, many technological networks demonstrate disassortative mixing. In particular, the Internet on the Autonomous Systems level has these kinds of correlations (Pastor-Satorras and Vespignani, 2007). However, the network of routers in the Internet does not have such clear degree–degree correlations. The WWW

[24]If $\gamma < 3$, in infinite networks, both the numerator and denominator on the right-hand side of Eq. (4.39) diverge, and these divergences compensate each other resulting in a finite r. The divergence of the numerator is explained by relation (4.39). It was found that $\langle q \rangle_{nn}(q)$ is strongly size-dependent in this range of exponent γ (Boguñá, Pastor-Satorras, and Vespignani, 2003).

and the network of protein interactions have disassortative degree–degree correlations.

4.7.3   Branching matrix

One can express observables of random networks with degree–degree correlations in terms of $P(q, q')$ or $P(q|q')$, but in many cases it is more reasonable and elucidating to use a non-symmetric *branching matrix* (Vázquez and Moreno, 2003; Boguñá, Pastor-Satorras, and Vespignani, 2003) with entries

$$B_{qq'} \equiv (q' - 1)P(q'|q). \tag{4.40}$$

The entry is equal to the branching $(q' - 1)$ of an edge adjacent to a vertex of degree q and having a vertex of degree q' at its second end, multiplied by the probability $P(q'|q)$ that the second end has degree q'. With this matrix, the network observables can be expressed in terms of matrix products and, consequently, in terms of eigenvalues and eigenfunctions of the branching matrix. One can see that[25]

$$q(q - 1)P(q)B_{qq'} = q'(q' - 1)P(q')B_{q'q}. \tag{4.41}$$

We can introduce the average branching $b(q)$ of an edge adjacent to a vertex with degree q,

$$b(q) = \sum_{q'} B_{qq'} = \sum_{q'} (q' - 1)P(q'|q), \tag{4.42}$$

which gives

$$b(q) = \langle q \rangle_{\mathrm{nn}}(q) - 1. \tag{4.43}$$

We also reproduce the formula for the average branching

$$\langle b \rangle = \sum_q b(q) \frac{qP(q)}{\langle q \rangle} = \frac{\langle q^2 \rangle}{\langle q \rangle} - 1 \tag{4.44}$$

---

[25] Due to this relation, the eigenvalues of the branching matrix coincide with those of the symmetric matrix

$$\tilde{B}_{qq'} = \sqrt{\frac{\langle q \rangle(q - 1)}{qP(q)}} P(q, q') \sqrt{\frac{\langle q \rangle(q' - 1)}{q'P(q')}},$$

and their eigenvectors are related. See Goltsev, Dorogovtsev, and Mendes (2008) for the detailed description of the spectral properties and applications of the branching matrix. In the particular case of uncorrelated networks, the first (largest) eigenvalue of the branching matrix equals the average branching, $\lambda_1 = \langle b \rangle$, and the remaining eigenvalues are all zero.

(notice that the average branching of a correlated network coincides with the average branching of the uncorrelated network with the same degree distribution) and express Pearson's coefficient in terms of $b(q)$,

$$r = \frac{\frac{1}{\langle q \rangle} \sum_q [b(q) - \langle b \rangle] q(q-1) P(q)}{\frac{\langle q^3 \rangle}{\langle q \rangle} - \frac{\langle q^2 \rangle^2}{\langle q \rangle^2}}. \tag{4.45}$$

This expression shows that the condition

$$\max_q \frac{|b(q) - \langle b \rangle|}{\langle b \rangle} \ll 1 \tag{4.46}$$

guarantees that the network is weakly correlated.

One can express the average number of cycles in such correlated networks, the average number of cycles passing through a vertex of degree q, the average number of vertices of degree q' at a distance ℓ from a vertex of a given degree, and so on in terms of matrix powers (and traces) of the branching matrix. For example, for the average clustering of a vertex of degree q, we get

$$C(q) = \frac{t(q)}{q(q-1)/2} = \frac{(B^3)_{qq}}{Nq(q-1)P(q)} \tag{4.47}$$

(Dorogovtsev, 2004; Goltsev, Dorogovtsev, and Mendes, 2008), where $t(q)$ is the average number of triangles adjacent to a vertex of degree q. This enables us to get the clustering coefficient

$$C = \frac{\sum_q q(q-1) P(q) C(q)}{\langle q^2 \rangle - \langle q \rangle} \tag{4.48}$$

and the average clustering of a vertex

$$\langle C \rangle = \sum_q P(q) C(q), \tag{4.49}$$

where for uncorrelated networks, $C = \langle C \rangle = C(q)$.[26] Note that if the second moment of the degree distribution, $\langle q^2 \rangle$, diverges and $C(q)$ decays sufficiently rapidly, then $C \to 0$ due to the term $\langle q^2 \rangle$ in the denominator on the right-hand side of Eq. (4.48), while $\langle C \rangle$ is a constant.

---

[26]Using the joint degree–degree distribution $P(q, q')$, one can relate two local characteristics of clustering, namely, $C(q)$ and the average multiplicity $m(q, q')$ of the edges connecting vertices of degrees q and q'. Recall that the so-called edge multiplicity m_{ij} for edge (ij) is defined as the number of triangles to which this edge belongs; see Section 2.5.

Finally, when there is a finite gap between the largest eigenvalue λ_1 of the branching matrix of a locally tree-like network and the second largest one, which is not universally true, the average intervertex distance equals

$$\langle \ell \rangle \cong \frac{\ln N}{\ln \lambda_1} \tag{4.50}$$

(Goltsev, Dorogovtsev, and Mendes, 2008), directly generalizing Eq. (4.9) for uncorrelated networks. In this situation, λ_1 coincides with the 'global branching' of a network, which is the limiting value of the ratio between the numbers of edges in two consecutive spherical layers, ℓ-th and $(\ell - 1)$-th, around an arbitrary vertex, with $\ell \to \infty$. This value coincides with the average branching $\langle b \rangle = \langle q^2 \rangle / \langle q \rangle - 1$ only in uncorrelated networks.

If the condition guaranteeing weak correlations, Eq. (4.46), is fulfilled, then the components v_q of the principal eigenvector of the branching matrix are

$$v_q \approx c\, b(q) = c \sum_{q'} (q' - 1)P(q'|q), \tag{4.51}$$

where c is a constant.[27] Then we get the largest eigenvalue, λ_1,

$$\lambda_1 \approx \frac{\sum_{q,q'} v_{q'}(1 - q')P(q, q')(q - 1)v_q}{\sum_q q(q - 1)P(q)v_q^2} \approx \frac{1}{\langle b \rangle} \sum_q b(q)q(q - 1)P(q). \tag{4.52}$$

Taking into account this definition, we can write

$$m(q, q') = \left\langle \frac{\sum_{i \in \mathcal{V}(q), j \in \mathcal{V}(q')} m_{ij} A_{ij}}{E(q, q')(1 + \delta_{q,q'})} \right\rangle,$$

where A_{ij} are entries of an adjacency matrix, $\mathcal{V}(q)$ is the full set of vertices of degree q in a graph, and $E(q, q')$ is the number of edges connecting vertices of degrees q and q'. For an arbitrary random network, the following identity is valid:

$$\sum_{q'} P(q, q')m(q, q') = \frac{q(q - 1)P(q)}{\langle q \rangle} C(q)$$

(Serrano and Boguñá, 2006a), and the average multiplicity equals

$$\langle m \rangle = \sum_{q,q'} P(q, q')m(q, q') = \frac{\langle q(q - 1)C(q) \rangle}{\langle q \rangle}.$$

The condition $C(q) \leq 1/(q - 1)$ guarantees that the average number of triangles to which a randomly chosen edge belongs is finite for any degree distribution. In this case, triangles, on average, do not overlap, which enables one to analytically treat various problems for such networks (Serrano and Boguñá, 2006b).

[27]To get Eq. (4.51), substitute the principal eigenvector of the matrix B of an uncorrelated network, which is a constant, into the right-hand side of the eigenvector equation $\lambda_1 \mathbf{v} = B\mathbf{v}$.

This can be expressed in terms of the Pierson coefficient r, Eq. (4.38),

$$\lambda_1 \approx \langle b \rangle + r \frac{1}{\langle b \rangle} \left(\frac{\langle q^3 \rangle}{\langle q \rangle} - \frac{\langle q^2 \rangle^2}{\langle q \rangle^2} \right), \tag{4.53}$$

and hence the largest eigenvalues for networks with a given degree distribution satisfy the following inequalities:

$$\lambda_1^{(\mathrm{dis})} < \lambda_1^{(\mathrm{un})} = \langle b \rangle = \frac{\langle q(q-1) \rangle}{\langle q \rangle} < \lambda_1^{(\mathrm{as})}, \tag{4.54}$$

where the superscripts (dis), (un), and (as) indicate disassortative, uncorrelated, and assortative networks, respectively. The same is all the more true for strongly correlated networks. This leads to the corresponding inequalities for an average intervertex distance,

$$\langle \ell \rangle^{(\mathrm{as})} < \langle \ell \rangle^{(\mathrm{un})} < \langle \ell \rangle^{(\mathrm{dis})}, \tag{4.55}$$

and for a clustering coefficient,

$$C^{(\mathrm{dis})} < C^{(\mathrm{un})} < C^{(\mathrm{as})}. \tag{4.56}$$

4.8   Cut-offs and rich club phenomenon

Discussing finite scale-free networks, one cannot miss the problem of the *cut-off* of the degree distribution. Let us begin this discussion with the case of the configuration model or any other model, in which all the members of the statistical ensemble have the same sequence of vertex degrees q_i, where $i = 1, 2, \ldots, N$, drawn from a given distribution $P(q)$. The extremal value statistics readily gives the distribution of the largest degree in this sequence:

$$Q(q) = [1 - P_{\mathrm{cum}}(q+1)]^N - [1 - P_{\mathrm{cum}}(q)]^N, \tag{4.57}$$

where $P_{\mathrm{cum}}(q) \equiv \sum_{k \geq q} P(q)$ is the *cumulative degree distribution*. If, for example, a degree distribution has a power-law tail, $P(q) \cong A q^{-\gamma}$, $\gamma > 1$, that is, $1 - P_{\mathrm{cum}}(q) \cong 1 - \frac{A}{\gamma - 1} q^{1-\gamma}$, then for large N, we have

$$Q(q) \cong \frac{d}{dq} \left(1 - \frac{A}{\gamma - 1} q^{1-\gamma} \right)^N \cong A N q^{-\gamma} \exp\left[-\frac{A N q^{-(\gamma-1)}}{\gamma - 1} \right] \tag{4.58}$$

91

(Gumbel, 2004), and hence the probability that the most connected vertex in this sequence has degree q rapidly decays with q after the 'cut-off degree':[28]

$$q_{nat}(N) \sim N^{1/(\gamma-1)}. \tag{4.59}$$

This cut-off is often called *natural*, as it is the largest possible value for a single realization of a random network (or of a random sequence of degrees) of a given size N. Then for estimations for the configuration model ensemble, one can approximately use a power-law degree distribution, cut to zero above $q_{cut}(N)$.[29]

In general, the tails of the degree distributions $P(q)$ of various models of scale-free random networks (full ensembles), may have rapidly decaying ends in the region of large degrees, $q \gtrsim q_{cut}(N)$. These cut-offs of the degree distributions differ from model to model. In some scale-free networks, $q_{cut}(N) \sim N^{1/(\gamma-1)}$ similarly to $q_{nat}(N)$, Eq. (4.59),[30] while in other

---

[28]More naively, one can estimate q_{cut} in the following way

$$N \int_{q_{nat}(N)}^{\infty} dq\, P(q) \sim 1,$$

i.e., conditioning that there is about one vertex of degree greater than q_{nat} in the network, which leads to the same formula.

[29]On the other hand, the probability of the degree sequence drawn from a given distribution $P(q)$, in which N_1 vertices have degree 1, N_2 vertices have degree 2, etc., $N = \sum_q N_q$, is

$$p(N_1, ..., N_{q_{max}}) = \frac{N!}{\prod_q N_q!} \prod_q [P(q)]^{N_q}.$$

With this distribution, one can check that for an arbitrary $N \geq 1$ and not only for $N \to \infty$, the observable—the average relative number of vertices of degree q,

$$\left\langle \frac{N_q}{N} \right\rangle = \sum_{N_1,...,N_{q_{max}}:\sum_q N_q = N} \frac{N_q}{N} \frac{N!}{\prod_q N_q!} \prod_q [P(q)]^{N_q}$$

$$= \sum_{N_q} \frac{N_q}{N N_q!} [P(q)]^{N_q} \sum_{N_k \neq N_q : \sum_{k \neq q} N_k = N - N_q} \frac{N!}{\prod_{k \neq q} N_k!} \prod_{k \neq q} [P(k)]^{N_k}$$

$$= \sum_{N_q} \frac{N_q}{N N_q!} \frac{N!}{(N - N_q)!} [P(q)]^{N_q} [1 - P(q)]^{N - N_q} = P(q),$$

coincides with the degree distribution as is should be, and so it has no any cut-off. The averaging here is, in fact, performed over the infinite number of the N-sequences drawn from $P(q)$. Note that for a single sequence (or for a finite number of them), the term 'degree distribution', strictly speaking, is not applicable. So $q_{cut}(N)$ in Eq. (4.59) is actually not about the shape of the degree distribution but rather about the largest degree in a realization of a random network.

[30]For example, this is the case for recursive networks with linear preferential attachment; see Section 5.2.

networks the asymptotic dependence on N is very different. For a single realization of a random network, the largest vertex degree is estimated as

$$q_{max}(N) \approx \min[q_{cut}(N), q_{nat}(N)]. \tag{4.60}$$

When deriving the number of self-loops in the configuration model in Section 4.2, we obtained the average number of self-loops of a vertex of degree q, $\langle N_{self\text{-}loops} \rangle(q) \cong q(q-1)/(2N\langle q \rangle)$. This shows that if the degree of a vertex exceeds some threshold, $q \gtrsim \sqrt{N}$, then the average number of self-loops attached to a vertex exceeds 1. In a similar way, one can show that if two vertices have degrees larger than this threshold, then the average number of connections is of the order of 1 or greater. Then the set of vertices with $q \gtrsim \sqrt{N}$ in uncorrelated network turns out to be strongly interconnected forming a dense cluster in which most of vertices are linked to each other. The presence of this dense club in networks with fat-tailed degree distributions was called the *rich club phenomenon* (Zhou and Mondragón, 2004), and it was observed in numerous real networks (Colizza, Flammini, Serrano, and Vespignani, 2006).

Clearly, for the existence of this cluster in a scale-free uncorrelated network, the equality $\sqrt{N} < N^{1/(\gamma-1)}$ must be fulfilled. So, in uncorrelated networks, it exists only if $\gamma < 3$.[31] We can easily estimate the size of this rich club:

$$N_{rich\ club} \sim N \int_{\sqrt{N}} dq\, q^{-\gamma} \sim N^{(3-\gamma)/2}. \tag{4.61}$$

Thus this club contains a vanishingly small fraction of vertices, although the effect becomes stronger as γ decreases.

In the standard configuration model, the self-loops and multi-edges are allowed, that is the members of the ensemble are multigraphs. Let us ban the self-loops and multi-edges in this ensemble. Then the members with such connections, mostly occurring between vertices of large degrees, disappear from the ensemble. This effectively produces redistribution of connections in a random network and the emergence of degree–degree correlations, noticeable when the second moment of the degree distribution diverges, that is, when $\gamma < 3$. One can easily escape these correlations in a version of the configuration model without self-loops and multi-edges by introducing the

---

[31]We already mentioned that uncorrelated networks with the divergent second moment of a degree distribution (i.e. with $\gamma \leq 3$) have large cliques; see Section 4.6. This agrees with the rich-club phenomenon.

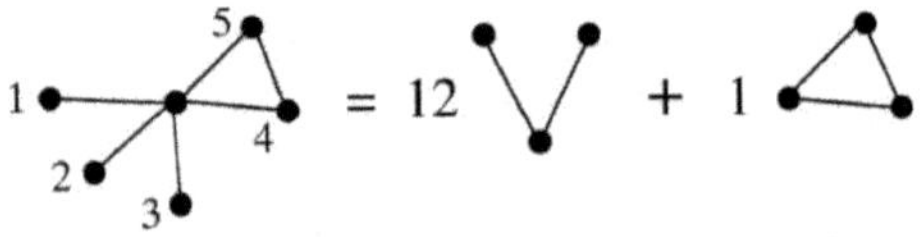

Fig. 4.6 Two 3-vertex motifs of the graph on the left-hand side occur 12 and 1 times. The 2-vertex motif (two vertices and an edge between them) occurs 6 times. The 4-vertex motif (triangle and a vertex attached to it) occurs 3 times. The 5-vertex motif (triangle and two vertices attached to it) occurs 3 times.

cut-off $q_{cut}(N) \lesssim \sqrt{N}$ in the degree distributions (Catanzaro, Boguñá, and Pastor-Satorras, 2005b).

4.9 Motifs in networks

A *network motif* is an unlabelled copy of repeating connected isomorphic subgraphs in a network.[32] Motifs can be treated as building blocks of networks, and it is reasonable to characterize networks using a set of occurrence frequencies for motifs (Figure 4.6). Clearly, the number of different n-vertex motifs is significantly larger in directed networks than in undirected ones. Figure 4.7 shows some of 3-vertex motifs for directed networks. Milo, Shen-Orr, Itzkovitz, Kashtan, Chklovskii, and Alon (2002) compared the occurrence frequencies of the motifs in a number of real-world situations. It turned out that the occurrence frequencies of a given motif are very different in different networks and their randomized models. Moreover, the observed number of motif copies in a real network typically exceeds that for its randomized counterpart. A tempting idea stimulating the numerous studies of motifs in networks was that each specific motif is responsible for some function of a network, and then the statistics of motifs should essentially describe and even determine the function of a network.

Most studies focus on small motifs. What about the frequency of occurrence of different motifs with fixed numbers of vertices, n, and edges, m, in a network? Vazquez, Dobrin, Sergi, Eckmann, Oltvai, and Barabási (2004) explored this occurrence frequency, and the number of the n, m-motifs in networks was estimated in terms of their degree-dependent clustering $C(q)$,

---

[32]Recall that a subgraph on a set of vertices can include part of edges connecting these vertices in the original network, not necessarily all these edges. Another, nonequivalent approach to network characterization is based on so-called *graphlets* (Pržulj, Corneil, and Jurisica, 2004). A *network graphlet* is an unlabelled copy of repeating connected isomorphic *induced subgraphs* in a network. By definition, induced subgraphs include all edges connecting their vertices in the original network.

Fig. 4.7 A few of the thirteen 3-vertex motifs for directed networks. For the full list, see Milo, Shen-Orr, Itzkovitz, Kashtan, Chklovskii, and Alon (2002).

Fig. 4.8 Newman's generalization of the configuration model (Newman, 2009). For each vertex i in this network, two numbers are given: the number t_i of triangles to which this vertex belongs and the number s_i of its rest connections. Note that the triangles have no joint edges.

see also Itzkovitz, Milo, Kashtan, Ziv, and Alon (2003). It is clear that in a sparse network, densely connected motifs of a given size are less frequent than motifs with a small number of edges, and cliques are the least frequent motifs.

It is hard to directly construct networks with a given set and number of motifs. Mark Newman (2009), however, proposed a way to build a 'uniform' random network of standard blocks (connected subgraphs) overlapping by single vertices. This is a direct generalization of the configuration model, which has a finite clustering coefficient even if the network is infinite. The original version of his network is made of two simple blocks, 2- and 3-cliques: triangles having no joint sides and single edges which do not belong to any of the triangles. Instead of a given degree sequence in the configuration model, consider a sequence of pairs of numbers. For each vertex i two numbers are given: the number of triangles to which this vertex belongs, t_i, and the number of the remaining connections (single edges), s_i (Figure 4.8). So the degree of the vertex i is $q_i = s_i + 2t_i$. Newman's network is defined as a uniformly random graph with a given sequence of the pairs (s_i, t_i), $i = 1, 2, \ldots, N$. Therefore this network is a superposition of two 'configuration models': for single links and for triangles.[33] By construction, this network is

---

[33]The local clustering in this model cannot be too strong:

$$C_i \leq \frac{q_i/2}{q_i(q_i-1)/2} = \frac{1}{q_i-1}.$$

'locally tree-like' in the sparse regime in the following sense. When $N \to \infty$, the network has very few finite cycles other than 3-cycles (triangles). This construction can be easily generalized allowing an arbitrary set of different building blocks, including complicated ones.

Notably, this network has degree–degree correlations. If we need to prepare a clustered network of this kind for numerical simulations, it is more convenient to use a hidden variable construction. For example, similarly to the Chung–Lu model, let $((s_1, t_1), (s_2, t_2), \ldots, (s_N, t_N))$ be a given sequence of desired numbers of single edges and triangles. Using these numbers, connect pairs of vertices (i, j) with probability $p_{ij} \propto s_i s_j$, and interconnect triples of vertices (i, j, k) with probability $p_{ijk} \propto t_i t_j t_k$.

4.10   Equilibrium network ensembles

The configuration model and models based on hidden variables are actually only special cases of far more general equilibrium statistical constructions allowing one to get a random network with a desired set of structural features. Interestingly, these equilibrium statistical ensembles, known as *exponential random graph models* (Strauss, 1975; Strauss, 1986; Harris, 2013), were originally developed for sociological applications.[34] The structure of these random networks (ensembles) is completely determined by a set of *structural constraints*. These constraints imposed on a set of arbitrary structural features, local or global,[35] can be of two kinds:

(i) *rigid (hard)*, like in the configuration model leading to the product of the corresponding Kronecker symbols in the probability $\mathcal{P}(G)$ of realization of a graph G in the ensemble $\mathcal{G}$;

(ii) *soft*, that are fixed averages of structural features $\mathcal{X}_k(G)$, $k = 1, 2, \ldots$:

$$\sum_{G \in \mathcal{G}} \mathcal{P}(G) \mathcal{X}_k(G) = \langle \mathcal{X}_k \rangle. \qquad (4.62)$$

We also fix the number of vertices, N, and, of course, we have the condition $\sum_{G \in \mathcal{G}} \mathcal{P}(G) = 1$. The idea is to get a statistical ensemble (in fact, the probabilities $\mathcal{P}(G)$ of its members) with all these constraints, maximizing the entropy, $S = \langle \ln \mathcal{P}(G) \rangle$, Eq. (3.9).

---

[34] As far as we know, for scale-free networks, a construction of this kind was first used by Burda, Correia, and Krzywicki (2001). Here we follow Park and Newman (2004).

[35] These, for example, may be vertex degrees, numbers of triangles, cycles, motifs, and so on—any structural characteristics.

The hard constraints lead to microcanonical statistical ensembles, the soft constraints lead to grand canonical or canonical ones, but, in principle, a combination of distinct constraints—hard and soft—for different quantities is also possible, leading to mixed ensembles.

As is usual in statistical mechanics, to find the maximum of the entropy under these constraints, one should apply the method of Lagrange multipliers (Vasiliev, 2019), that is, to maximize the expression:

$$-\sum_{G\in\mathcal{G}} \mathcal{P}(G)\ln\mathcal{P}(G) - \nu\Big[1 - \sum_{G\in\mathcal{G}}\mathcal{P}(G)\Big] - \sum_{k}\mu_k\Big[\langle\mathcal{X}_k\rangle - \sum_{G\in\mathcal{G}}\mathcal{P}(G)\mathcal{X}_k(G)\Big].$$

(4.63)

Here ν and μ_k, $k = 1, 2, \dots$ are Lagrange multipliers. Equating the derivative of this expression with respect to $\mathcal{P}(G)$ gives the probabilities of the members,

$$\mathcal{P}(G) = e^{-(1-\nu)}\exp\Big[\sum_{k}\mu_k\mathcal{X}_k(G)\Big] = \frac{1}{Z}e^{-\mathcal{H}(G)} \qquad (4.64)$$

with the normalization denoninator—the partition function $Z = e^{1-\nu}$ and the *network Hamiltonian*

$$\mathcal{H}(G) = -\sum_{k}\mu_k\mathcal{X}_k(G). \qquad (4.65)$$

Comparing Eqs. (3.9) and (4.64), we arrive at the relation

$$S = \ln Z + \langle\mathcal{H}(G)\rangle, \qquad (4.66)$$

in which $-\ln Z$ can be interpreted as the *free energy* and $\langle\mathcal{H}(G)\rangle$ as the *internal energy* of a random network. Note that the three quantities entering this relation are expressed in terms of the Lagrange multipliers μ_k, $k = 1, 2, \dots$. Comparing the constraints, Eq. (4.62), with Eq. (4.64) one expresses the averages entering the constraints in terms of the set of the Lagrange multipliers,

$$\langle\mathcal{X}_k\rangle = \frac{\partial\ln Z}{\partial\mu_k}. \qquad (4.67)$$

Vice versa, from these relations one can express all μ_k in terms of the set of the averages $\langle\mathcal{X}_k\rangle$.

To obtain the average of any quantity $\widetilde{\mathcal{X}}(G)$ which differs from the members of the constrained set in Eq. (4.62), that is, $\widetilde{\mathcal{X}}(G) \neq \mathcal{X}_k(G)$, derive the partition function Z with the Hamiltonian

$$\mathcal{H}(G) = -\sum_{k}\mu_k\mathcal{X}_k(G) - \tilde{\mu}\,\widetilde{\mathcal{X}}(G) \qquad (4.68)$$

and take the derivative

$$\langle \tilde{X}_k \rangle = \left. \frac{\partial \ln Z}{\partial \tilde{\mu}} \right|_{\tilde{\mu}=0}. \tag{4.69}$$

As an example, one can consider the ensemble with the soft constraints for the vertex degrees $X_i(G) = q_i = \sum_{j=1}^{N} A_{ij}$, $i = 1, 2, \dots, N$ (Anand and Bianconi, 2009):

$$\sum_{G \in \mathcal{G}} P(G) \sum_{j=1}^{N} A_{ij} = \langle q_i \rangle, \tag{4.70}$$

which leads to the probabilities of the members of the ensemble

$$P(G) = \frac{1}{Z} \exp\left[\sum_{i<j} (\mu_i + \mu_j) A_{ij}\right] = \prod_{i<j} \left[p_{ij} A_{ij} + (1 - p_{ij})(1 - A_{ij})\right] \tag{4.71}$$

(see Eq. (4.15) for the last equality), which gives the probabilities

$$p_{ij} = \frac{1}{1 + e^{\mu_i + \mu_j}}, \tag{4.72}$$

$\sum_j p_{ij} = \langle q_i \rangle$. Thus this ensemble is equivalent to the Chung–Lu model discussed in Section 4.4.

Park and Newman (2004) explored ensembles of this kind with various constraints, including essentially more complex constraints than shown here. In particular, the additional term

$$\mathcal{H}_2(G) = -\sum_{i<j} \mu_{ij} g\left(\sum_l A_{il}, \sum_l A_{jl}\right) \tag{4.73}$$

in the network Hamiltonian generates random networks with degree–degree correlations, where a given function $g(q_i, q_j)$ determines the resulting joint degree–degree distribution $P(q, q')$ (Berg and Lässig, 2002).[36]

There is another, less formal way to build equilibrium ensembles of networks by launching an evolutionary process for a network with a fixed number of vertices and waiting till the system will approach an equilibrium (Dorogovtsev, Mendes, and Samukhin, 2003c). This can be progressive rewiring of edges or their random hops between different positions, etc. Such evolutionary designed random networks can have a condensation phase, and at the critical point these networks are scale-free.

---

[36]In contrast to the strongly correlated networks with this Hamiltonian, Eq. (4.73), the correlations in the Chung–Lu model showing up in the expression for p_{ij}, Eq. (4.71), which does not factors, turn out to be weak when the degree distribution has no long tail.

4.11   The case of triangles

Networks with large number of triangles are particularly interesting. The *Strauss model* of clustering is an exponential model with two soft constraints, for the number of edges and for the number of triangles (Strauss, 1975, 1986). The grand canonical ensemble of the Strauss model is described by the network Hamiltonian

$$\mathcal{H}(G) = -\mu_2 \frac{1}{2} \mathrm{Tr} A^2 - \mu_3 \frac{1}{6} \mathrm{Tr} A^3, \qquad (4.74)$$

whose second term favours triangles (high clustering). This Hamiltonian can be processed similarly to standard spin models, if we treat the elements of the adjacency matrix as 'spin variables', like in matrix models. Then a usual mean-field theory can be applied to this 'spin system'. The mean-field solution of the problem reveals two phases: normal phase, in which triangles are spread uniformly over the network, and the condensation phase, in which a fraction of vertices have large degrees and clustering about 1, stealing triangles and connections from the remaining vertices.[37] The phase transition between them is of the first order, and so there is a large region on the phase diagram of this system, where these two phases co-exist.[38] In simulations, in this region, networks rapidly fall into the condensation phase. At first sight, this is a drawback of the Strauss model originally aimed at real-world network systems with strong though uniform clustering. In other words, it seems that the Strauss model can be either in a uniform state with low clustering or in the condensation state with many triangles but not in a uniform state with high clustering. Intrigued by this controversy, Burda, Jurkiewicz, and Krzywicki (2004) explored the dynamics of the transition of this system from the metastable normal state to the stable condensation one in this model, asking, how long does it take to pass the barrier separating the states? It turned out that the height of the barrier rapidly diverges with N. Therefore in the metastability region of large networks, it is actually impossible to approach the condensation state, starting from the homogeneous configuration, even with high clustering, and they stay homogeneous virtually for ever.

In some networks, this condensation phase looks rather special. Let the ensemble be mixed, for example, a random q-regular graph with the sole

---

[37]This condensation phenomenon was confirmed by graph theory methods (Radin, Ren, and Sadun, 2014; Kenyon, Radin, Ren, and Sadun, 2017*b*, 2017*a*).

[38]This is similar to the liquid–gas phase transition.

term in the Hamiltonian, favouring clustering. Then, the network in the condensation phase looks as a set of isolated $(q+1)$-cliques (Avetisov, Hovhannisyan, Gorsky, Nechaev, Tamm, and Valba, 2016; Avetisov, Gorsky, Nechaev, and Valba, 2020).

Furthermore, rigid constraints for triangles may result in quite regular structures. Consider a maximal entropy ensemble with two fixed numbers: a fixed degree q of every vertex, and, for example, the fixed number $(q - 1)(q - 2)/2$ of triangles to which the vertex belongs. That is, of the nearest neighbours of every vertex, all but one are the nearest neighbours of each other. This constraint fixes the local clustering of every vertex, $C_i = 1 - 2/q$. Then the resulting network is a set of q-cliques in which each vertex has a single connection to a vertex from another clique. The network has a 'locally tree-like' structure in the sense that it has no finite cycles longer than 3.

4.12   Weighted networks

Here we only briefly touch upon random weighted networks, focusing on edge weighted networks and assuming that weights are non-negative real numbers. In this situation, the entries of the adjacency matrix of a weighted graph are weights, $A_{ij} = w_{ij}$ with zeros on the diagonal. The unweighted projection of the adjacency matrix of a weighted graph is the matrix $A_{\text{unweighted}} = \theta(A)$, where the θ-function, $\theta(x \leq 0) = 0$, $\theta(x > 0) = 1$, is applied to all entries of the matrix A. The simplest local characteristics of a vertex are the *strength*

$$s_i = \sum_{j=1}^{N} A_{ij} = \sum_{j \in \partial_i} A_{ij} \tag{4.75}$$

and the degree

$$q_i = \sum_j \theta(A_{ij}). \tag{4.76}$$

Note that, in contrast to the degree of a vertex in an unweighted graph, the pair of numbers, namely the strength and degree, s_i and q_i, provide only a small piece of information about the edges adjacent to vertex i. The complete information (ignoring the neighbours) should be given by the set of the weights of the adjacent edges. An additional, yet non-complete, information in given by the *disparity* of a vertex (Barthélemy, Barrat, Pastor-Satorras, and Vespignani, 2005a),

$$Y_i = \frac{1}{s_i^2} \sum_j A_{ij}^2, \tag{4.77}$$

where $Y_i = 1/q_i$ when all weights of the adjacent edges are equal, and Y_i is close to 1 when one of the adjacent weights is much greater than the weights of the remaining adjacent edges.[39]

A number of phenomena, discussed earlier for unweighted networks, also occur in some form in weighted networks, for instance, the reach club effect (Zlatić, Bianconi, Díaz-Guilera, Garlaschelli, Rao, and Caldarelli, 2009). Numerous issues related to the characterization of weighted networks, particularly important to applications, have been extensively explored (Barrat, Barthelemy, Pastor-Satorras, and Vespignani, 2004; Newman, 2001b, 2004a; Almaas, Kovacs, Vicsek, Oltvai, and Barabási, 2004; Bhattacharya, Mukherjee, Saramäki, Kaski, and Manna, 2008).[40] The ensembles of random weighted networks is a more difficult and less studied topic. Let us introduce the probability $p_{ij}(A_{ij})$ that the edge (ij) has weight $A_{ij} = w_{ij}$, where $p_{ij}(0)$ is the probability that the edge (ij) is absent. Then in the spirit of the theory of random matrices, one can define the ensemble with the following probabilities for its members (Garlaschelli and Loffredo, 2009):

$$\mathcal{P}(G) = \prod_{i<j} p_{ij}(A_{ij}). \qquad (4.78)$$

In the particular case of unweighted networks, this general formula is reduced to Eq. (4.15). The form of the function $p_{ij}(A_{ij})$ shapes the ensemble.[41]

To grasp the difficulty of random weighted networks, the reader can try to imagine an uncorrelated weighted network. In uncorrelated networks, the local characteristics of the nearest neighbours, for example, their triples q, s, Y and q', s', Y', must be independent in the sense that their joint distribution must factor into two parts. Let vertex i have degree q_i and strength s_i. Then the sum of the strengths of its nearest neighbours must be greater than s_i, $\sum_{j \in \partial i} s_j \geq s_i$. This restriction means the presence of correlations, which are particularly strong when the degree q_i is small, especially, when $q_i = 1$. Thus, uncorrelated weighted networks do not exist (Serrano, Boguñá, and Pastor-Satorras, 2006). One can construct a maximally random weighted network with, say, a given joint degree–strength distribution, $P(q, s)$,[42] but

[39]This disparity is a direct analogy of the inverse participation ratio, see Section 9.3.

[40]Note that it turned out difficult to generalize some measures, including the clustering coefficient, to weighted networks (Opsahl and Panzarasa, 2009).

[41]It may occur that this probability itself is expressed in terms of some hidden variables.

[42]Here q and s are for the same vertex.

it will have correlations. It is the incidence of the two nearest neighbouring vertices to the same weight that makes correlations in weighted networks inevitable.[43]

4.12.1 The strength of weak ties

In 1973 American sociologist Mark Granovetter published his landmark paper 'The strength of weak ties'. According to Granovetter, individuals in social networks are connected by 'ties'—weighted edges. The 'tie strength' (actually, the edge weight in a one-partite weighted network) is defined to be proportional to the frequency (intensity) of social interaction between two individuals. The tie strength enables one to distinguish 'strong' and 'weak' social ties. In the view of Granovetter,

> 'our acquaintances (weak ties) are less likely to be socially involved with one another than are our close friends (strong ties). Thus the set of people made up of any individual and his or her acquaintances comprises a low-density network ... whereas the set consisting of the same individual and his or her close friends will be densely knit'.

As a result, a social network looks as shown in Figure 4.9: dense communities of strongly tied close friends are connected together by weak acquaintance ties. In this scheme,

> 'The weak tie between Ego and his acquaintance ... becomes not merely a trivial acquaintance tie but rather a crucial bridge between the two densely knit clumps of close friends.'

This organization of social networks has an important consequence also indicated by Granovetter. Ego receives information from his or her close friends through strong ties, while information from other, outer parts of a social network reaches Ego through weak ties. So a lack of weak ties would significantly delay the receipt of information coming from the outer social world. In this case, Ego can hear all news only after his or her close friends.

---

[43]For demonstrating the contrast between the ensembles of weighted and unweighted networks, let us formulate an analogy of the configuration model for weighted networks in a situation where all edge weights are different. Let a network have N vertices and E edges with different given weights. The sequence of degrees is given, $\{q_i\}$, $i = 1, 2, \ldots, N$, $\sum_i q_i = 2E$. Create a list of length $2E$ in which each weight appears twice. To each vertex i ascribe q_i weights— q_i weighted stubs—taken from this list. Join the stubs in all possible ways, demanding that two stubs can merge together only if they have equal weights. This defines the ensemble. Clearly, there is only one possible way to connect vertices, and so there is only one member in this ensemble. Thus this model with different weights is not random.

Fig. 4.9 Organization of social networks according to Granovetter. The width (weight) of each edge represents the strength of the corresponding social tie. Weak ties connect together dense communities of strongly tied individuals. An individual in the centre of the right community (an open dot) has no weak ties and receives any 'external' information only after his close friends.

The assumption of Granovetter is often called the strength-of-weak-ties hypothesis. Granovetter tested his idea by surveying workers to find who did tell them about their current job. In most cases, the information came from 'not a friend, an acquaintance' (Granovetter, 1973). A new possibility of verifying Granovetter's ideas more thoroughly was found by Onnela, Saramäki, Hyvönen, Szabó, Lazer, Kaski, Kertész, and Barabási (2007*b*) and Onnela, Saramäki, Hyvönen, Szabó, De Menezes, Kaski, Barabási, and Kertész (2007*a*). Large log files stored by cellular network operators contain traces of all mobile phone calls made within these networks. From these records, in particular, one can get the total duration of calls between each pair of customers during a given period. This number characterizes the intensity of social interaction—the strength of a given social tie, or the weight of an edge, treated in these works as undirected. In this way, these authors constructed a large weighted undirected network of 4.6×10^6 vertices and 7.0×10^6 edges. The question was what the relation was between the weight of an edge and information flow though this edge. It was possible to answer the question by analysing the structure of this weighted network. Accepting a minimal model for the information transfer in a network, assume that each vertex produces the flow of news at equal rate, and that this information is sent to all other vertices through the shortest paths. Then the information flow through an edge is proportional to the number of shortest paths between all pairs of vertices, which pass through this edge. Instead of the numbers of shortest paths, Onnela and coauthors studied the closely related betweenness centralities of edges. In the phone call network, the edge betweenness centrality turned out to be, on average, high for weak ties and low for strong ties, varifying Granovetter's hypothesis.

4.13 Random geometric graphs

Section 4.4 considered networks with vertices connected according to some hidden variables of the vertices. Imagine a network whose vertices are uniformly randomly located in some area of a D-dimensional Euclidian space. Then the geographic coordinates of the vertices can play the role of their 'hidden variables', and the connection between two vertices can be established according to their geographic closeness. These networks are called the *random geometric graphs* (Gilbert, 1961; Penrose, 2003), and in many respects they strongly differ from the networks discussed earlier.[44]

Let N vertices be uniformly distributed within the hypercube $[0,1]^D$, and each two vertices, i and j, are connected by an edge if the geographic distance r_{ij} between them is shorter than some number $r > 0$.[45] For $D = 2$, this random geometric graph is distinct from planars,, since its edges can cross each other. Consider a D-dimensional ball of radius r with one of the vertices in its centre. Since the remaining vertices are uniformly distributed in space, the number of those of them occurring in the ball should follow the binomial distribution.[46] For large N, this leads to the Poisson degree distribution, like in the classical random graphs, with the average degree equal to the product of N and the volume of the ball,

$$\langle q \rangle = N \frac{\pi^{D/2} r^D}{\Gamma(1 + D/2)},$$ (4.79)

where $\Gamma(x)$ is the Gamma function. The geometry of these graphs, however, dramatically differs from the classical random graphs. As one could guess, the average shortest path distance of a random geometric graph is

$$\langle \ell \rangle \sim \frac{1}{r},$$ (4.80)

assuming that $r \ll 1$ (Friedrich, Sauerwald, and Stauffer, 2013). Thus the random geometric graphs are D-dimensional large worlds.

The point of the birth of a giant connected component (percolation threshold in terms of continuum percolation) in the random geometric graphs

[44]These networks were intensively investigated in continuum percolation theory in the 1970s while studying hopping conductivity in doped semiconductors (Shklovskii and Efros, 1984; Böttger and Bryksin, 1985; Meester and Roy, 1996).

[45]One can introduce a more general linking rule, interconnecting vertices with probability given by a decaying function $P(r_{ij})$ (Waxman, 1988; Dettmann and Georgiou, 2016). We discuss this possibility in the following sections.

[46]We ignore border effects.

sits higher than in the classical random graphs: $\langle q \rangle_c(D{=}2) = 4.512...$ (Quintanilla, Torquato, and Ziff, 2000), $\langle q \rangle_c(D{=}3) \approx 2.73$ (Rintoul and Torquato, 1997), monotonously decreasing with D to $\langle q \rangle_c(D{\to}\infty) = 1$ (Dall and Christensen, 2002). This limit looks natural, since at $D \to \infty$ these graphs become small worlds approaching the classical random graph regime.

In stark contrast to the classical random graphs, the random geometric graphs have high clustering. The clustering coefficient was found for an arbitrary D,

$$C(D) = \frac{3}{\sqrt{\pi}} \left(\frac{3}{4}\right)^{(D+1)/2} \frac{{}_2F_1(\frac{1}{2}, \frac{D+1}{2}, \frac{D+3}{2}, \frac{3}{4})}{D+1} \overset{D \gg 1}{\cong} 3\sqrt{\frac{2}{\pi}}\sqrt{\frac{1}{D}}\left(\frac{3}{4}\right)^{(D+1)/2},$$
$$(4.81)$$

(Dall and Christensen, 2002), where ${}_2F_1(a, b, c, z)$ is the hypergeometric function. In particular, this gives $C(1) = 3/4$, $C(2) = 1 - 3\sqrt{3}/(4\pi) = 0.5865...$, $C(3) = 15/32 = 0.4687...$, and so on. Remarkably, while the local clustering is independent of degree, and so the clustering coefficient (transitivity) coincides with the average clustering, $C = \langle C \rangle$, the random geometric graphs show an assortative mixing. In particular, for $D = 2$, in a range of degrees (excluding small and large degrees),

$$\langle q \rangle_{nn}(q) \cong (1 - C)\langle q \rangle + Cq \qquad (4.82)$$

(Antonioni and Tomassini, 2012).

4.14 Small-world networks

In early empirical studies of networks, the main focus was on degree distributions and clustering and on the average separation of vertices. It may appear strange nowadays that one of the biggest surprises was a large clustering coefficient in the great majority of real-world networks. This combination— the small-world phenomenon and the strong clustering—seemed to be completely incomprehensible in the late 1990s, when the only reference model widely used for comparison was a classical random graph with a very weak clustering. Today it is clear that there exist plenty of easy ways to get a strongly clustered small world. The nature of strong clustering (and, in general, of numerous cycles) is not considered a serious problem now. Nonetheless, it was the desire to understand the high values of clustering coefficient in the empirical data that inspired sociologist Duncan Watts and applied mathematician Steven Strogatz to propose an original model, interconnecting themes of networks and lattices (Watts and Strogatz, 1998). This very popular model significantly influenced the development of the field.

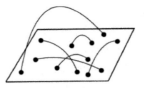

Fig. 4.10 The idea of a small-world network. Long-range shortcuts—edges connecting randomly chosen vertices—are added to a D-dimensional lattice. Together with the vertices and connections within the mother lattice (which are not shown) these edges form a small-world network.

The basic idea of Watts and Strogatz is as follows. Suppose somebody, who knows only lattices and the Erdős–Rényi model, wants to construct a network with the small-world feature and numerous triangles. The classical random graphs demonstrate the small-world phenomenon but have few triangles. On the other hand, many lattices have numerous triangles (if their unit cells include triangles of bonds) but the lattices have no small-world feature. We combine a lattice with many triangles and a classical random graph with the small-world feature. Technically, Watts and Strogatz connected pairs of randomly chosen vertices in a lattice by edges—'shortcuts' (Figure 4.10). Thanks to these long-range shortcuts, even when widely separated geographically within the mother lattice, vertices have a chance to become nearest neighbours. Clearly, the shortcuts make the resulting networks more compact than the original lattice. Watts and Strogatz called networks of this kind *small-world networks*. This term is used even if a mother lattice is disordered, or it has no triangles.

Figure 4.11a shows the original Watts–Strogatz network where randomly chosen edges of a mother lattice were rewired to randomly selected vertices. The rewiring produces the same effect as the added long-range shortcuts, shown in Figure 4.11b, which also explains why we have superposition of a lattice and a classical random graph. Indeed, if we let the connections within the mother lattice be absent, then the shortcuts and the vertices form a classical random graph. The problem is: what happens when the number of shortcuts increases? It is easy to see that even a single long-range shortcut sharply diminishes the average shortest-path distance $\langle \ell \rangle$ between vertices. For example, the first shortcut added to a one-dimensional lattice (as in Figure 4.11) may reduce $\langle \ell \rangle$ by half! So, the influence of shortcuts on the shortest path lengths in a small-world network is dramatic. Let p be the relative number of shortcuts in a small-world network (with respect to the

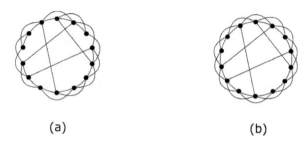

(a)   (b)

Fig. 4.11 The original Watts–Strogatz model (a) with rewiring of links (Watts and Strogatz, 1998), and its variation (b) with addition of shortcuts (b) (Newman, 2000). Notice numerous triangles within the mother lattice.

overall number of connections. Watts and Strogatz observed that even at a very low p, where the clustering is still nearly the same as in the mother lattice, the average intervertex distance is tremendously reduced. In fact, instead of clustering for the sake of comparison, any other local characteristics could be used with the same result. Thus even at small (but non-zero) p the network shows the small-world phenomenon. In other words, a smooth crossover between two regimes—from large to small worlds—occurs at very small p. In particular, in the limit of the infinite number of vertices N, we obtain a large world if the number of shortcuts N_s is finite, and a small world if the relative number of shortcuts p is finite.

Let us estimate the average distance between vertices $\langle \ell \rangle (p)$. We assume that $N \to \infty$ and p is finite, that is, the number $N_s \sim pN$ of shortcuts is large enough—the small-world regime. In this case the shortcuts determine the global architecture of the network, and the network resembles a classical random graph with N_s edges and of the order of N_s 'supernodes'. These 'supernodes' are regions of the mother lattice, which contain neighbouring ends of shortcuts. To this graph, we can apply the expression $\langle \ell \rangle (N)$ for the classical random graphs, Eq. (4.11), that is, $\langle \ell \rangle \sim \ln N$, with two changes. In our estimation we ignore constant factors and for the sake of simplicity suppose that the network is based on a one-dimensional mother lattice—a chain or a ring. In the resulting network, the average distance between ends of different shortcuts is $\langle d \rangle \sim 1/p \sim N/N_s$. Then, (i) substitute N for N_s, and (ii) take into account that the shortest paths in the network pass the mother lattice between ends of different shortcuts, so multiply the classical expression by $\langle d \rangle$. This gives the result:

$$\langle \ell \rangle (p) \sim \langle d \rangle \ln N_s \sim (1/p) \ln(Np). \tag{4.83}$$

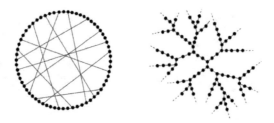

Fig. 4.12 This small-world network locally has the structure of a tree.

Thus the diameter of this small-world network is about $N_s/\ln(N_s) \gg 1$ times smaller than that of the mother lattice. Note that when the mother lattice is a chain or ring, the resulting small-world network is locally tree-like (Figure 4.12).

Crossover phenomena produced by a small fraction of shortcuts can be found in a range of networks, much broader than the Watts–Strogatz model. For example, consider a network consisting of two modules—small worlds— and explore how the distribution of shortest path lengths in this network varies when we interconnect these blocks by random shortcuts. Section 1.3 explained that the vertices of the small worlds are quasi-equidistant in the limit $N \to \infty$, which is a consequence of the exponential increase of the number of the n-th nearest neighbours of a vertex, $z_n \sim \langle b \rangle^n$, with n. That is, the width $\delta\ell$ of the distribution of intervertex distances $\mathcal{P}(\ell)$ in these networks is finite, even if these nets are infinitely large. This is typical for small worlds but it may fail in networks with a block structure. Let the numbers of vertices in the blocks in our example, $N_1, N_2 \gg 1$, be of the same order, and we assume the same for the numbers of edges in the blocks. When the blocks are isolated, the distribution has two narrow peaks (Figure 4.13a). When there is a single shortcut between the blocks, the distribution has three peaks (Figure 4.13b). Finally, when the number of edges interconnecting the blocks is any finite fraction of the number of edges in the modules, the three peaks merge into one (Figure 4.13c).

For another example of a crossover of this kind between distinct network architectures, let us add a vertex to a large network and attach it to a finite fraction p of the vertices, like in Figure 3.3. In this situation, $\langle \ell \rangle (N)$ will be finite as $N \to \infty$. If the original network was a small world, then the crossover in the region of small p is from a small to an ultra-small world.

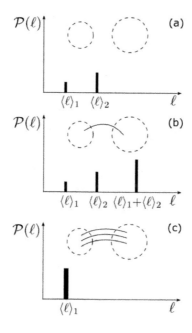

Fig. 4.13 The evolution of an intervertex distance distribution with the increasing number of shortcuts between two large subnetworks—small worlds: (a) two separate blocks; (b) two blocks with a single shortcut between them; (c) two interconnected blocks, when the number of shortcuts is a finite fraction of the number of edges in the network. $\langle \ell \rangle_1$ and $\langle \ell \rangle_2$ are the average intervertex distances in the first and in the second subnetworks, respectively (Dorogovtsev, Mendes, Samukhin, and Zyuzin, 2008b).

4.15 Networks embedded in metric spaces

A large number of real networks attracting a particular attention (Internet, WWW, etc.) share the following five features: (i) they are small worlds, (ii) they are scale-free, (iii) they are highly clustered, (iv) they are correlated, and (v) they are sparse. The models with hidden variables considered provide features (i), (ii), (v), and even (iv), but their clustering is vanishingly low. So these models are far from the real networks. On the other hand, the random geometric graphs lack features (i) and (ii) but they are highly clustered. So they also are far from the important real networks. A good way exists to combine these two classes of network models and obtain all these features together (Serrano, Boguñá, and Díaz-Guilera, 2005; Serrano, Krioukov, and Boguñá, 2008). In this section we briefly outline the idea of this construction.

Let us assume that N vertices, $i = 1, 2, \ldots, N$, with standard hidden variables f_i (expected degrees) are uniformly distributed with unit density in some metric space, where the hidden variables are drawn from a power-law distribution $\rho(f) \sim f^{-\gamma}$, $2 < \gamma < 3$. First, let this metric space be 1D Euclidian, a circle of length $2\pi R = N$, and the position of a vertex be described by its angle coordinate θ_i, $0 \leq \theta < 2\pi$. So for each vertex i we have the pair (f_i, θ_i), the hidden variable (expected degree), and the angle coordinate. Let the probability that two vertices are connected be a function of the metric distance between two vertices, that is, the geodesic distance over the circle, $d(\theta, \theta')$, like it was in general random geometric graphs, but now $d(\theta, \theta')$ is rescaled by the distance scale $d_s(f, f')$ depending on the hidden variables of these vertices,

$$p(f, \theta; f', \theta') = \mathcal{F}\left(\frac{d(\theta, \theta')}{d_s(f, f')}\right). \tag{4.84}$$

If the function $\mathcal{F}$ is chosen in the following form:

$$p(f, \theta; f', \theta') = \left\{1 + \left[\frac{d(\theta, \theta')}{ff'/(\langle q\rangle/2)}\right]^{\beta}\right\}^{-1}, \quad \beta > 1, \tag{4.85}$$

where the form of the distance scale $d_s(f, f') = ff'/(\langle q\rangle/2)$ ensures $\langle q_i\rangle = \langle f_i\rangle$ and proper normalization,[47] then for any $\beta > 1$: (i) the network is a small world; (ii) with a power-law degree distribution $P(q) \sim q^{-\gamma}$ having the same exponent γ as $\rho(f)$; (iii) with a large clustering coefficient controlled by the parameter β; and, clearly, (iv) this network is correlated and (v) sparse (Serrano, Krioukov, and Boguñá, 2008).[48] It's easy to see the origin of the small-world phenomenon in these networks. Due to $d_s(f, f') \propto ff'$, remote vertices with large expected degrees have a chance to be interlinked even when they are far from each other geographically. These long-range links produce the same effect as the long-range shortcuts in the small-world networks of Strogatz and Watts.

With properly chosen distribution $\rho(f)$ and parameter β, the resulting networks get the local clustering $C(q)$ and the average degree of the nearest neighbours of a vertex with degree q, $\langle q\rangle_{nn}(q)$, well agreeing with empirical

---

[47]In this construction, one can equally use a D-dimensional Euclidean space with an arbitrary $D \geq 1$. In this situation, the distance scale in Eqs. (4.84) and (4.85) should be $d_s(f, f') \propto (ff')^{1/D}$ (Serrano, Boguñá, and Díaz-Guilera, 2005). One should also note that the chosen form of the connection probability, Eq. (4.85), is not unique.

[48]In fact, for $1 < \beta < 2$, these networks are small worlds with any γ.

data for a number of real networks, including the Internet. Namely, $C(q)$ decays with q, and the shape of $\langle q \rangle_{\mathrm{nn}}(q)$ indicates the presence of disassortative correlations.

Let us now think about all versions and variations of random geometric graphs in more general terms, assuming that the underlying space, in which vertices of a network sit, is an arbitrary metric space. Recall that a *metric space* is a set of elements (e.g., points) with a metric (distance between two elements) $d(x, y)$, satisfying the triangle inequality: $d(x, z) \leq d(x, y) + d(y, z)$. Due to the triangle inequality, if three vertices are close to each other in the underlying metric space, then there is a good chance that all these vertices will be connected in the observed network, which leads to high clustering of such networks. Moreover, one can even say that the architecture of the observed network is essentially the consequence of the underlying metric space (Boguñá, Bonamassa, De Domenico, Havlin, Krioukov, and Serrano, 2021).[49]

The model on a circle, described earlier can be mapped to a hyperbolic (Poincaré) disk (Krioukov, Papadopoulos, Kitsak, Vahdat, and Boguñá, 2010; Papadopoulos, Kitsak, Serrano, Boguñá, and Krioukov, 2012).[50] For each vertex, the mapping is $(f_i, \theta_i) \rightarrow (r_i, \theta_i) = (R - 2 \ln f_i, \theta_i)$, where (r_i, θ_i) are the polar coordinates of vertex i on the hyperbolic disc of the radius $R \sim 2 \ln N$. With this mapping, the vertices are distributed over the disk, depending on the form of the probability distribution $\rho(f)$. If, in particular, the exponent of this distribution $\gamma = 3$, then the vertices are distributed uniformly over the disk. After the mapping, the probability of connection between two vertices appears to be the Fermi–Dirac function of the hyperbolic distance d_{ij} between the vertices:

$$ p_{ij} = \left[1 + e^{\beta(d_{ij} - R)/2} \right]^{-1}. \tag{4.86} $$

This allows a natural interpretation of this random network as a grand canonical ensemble of edges—fermions with energies ϵ_{ij} equal to halves of the hyperbolic distances d_{ij} between their vertices, $\epsilon_{ij} = d_{ij}/2$.

One can move in the opposite direction and try to reveal the metric structures underlying an observed network. Uncovering hyperbolic and other

---

[49]The role of such metric spaces can be played even by small worlds (binary trees, etc.).

[50]*Hyperbolic space* has a constant negative curvature at every point. The Poincaré hyperbolic disk models hyperbolic geometry in a two-dimensional space. In this model, the hyperbolically straight lines (geodesics) are represented by arcs with ends perpendicular to the disk's boundary.

metric spaces hidden beneath real networks has immediate practical applications, including detection of communities, efficient routing protocols, and
many others (Kleinberg, 2007; Boguná, Papadopoulos, and Krioukov, 2010;
Serrano, Krioukov, and Boguñá, 2008).

5
Evolving Networks

Most real networks are evolving. Here we discuss how basic models of these networks provide a spectrum of complex architectures.

5.1 Random recursive trees

We begin with the simplest growing networks, the random recursive trees introduced in Section 1.5, namely the labelled recursive trees where each added vertex is attached to a vertex chosen uniformly at random. Recall that the total number of arbitrary labelled recursive trees of N vertices equals $(N-1)!$, Eq. (1.22). The reader can easily derive this number. It is equally easy to check that in random recursive trees, for a given N, each labelled tree (the member of this ensemble) occurs with equal probability $1/(N-1)!$, which readily follows from the uniform selection of existing vertices for the attachment of the new one. This is a specific feature of random recursive trees.

We already explained that these trees are small worlds. Their average shortest-path distance is

$$\langle \ell \rangle \sim 2 \ln N \tag{5.1}$$

for large N.[1] Let us find the degree distribution of these trees. For the sake

---

[1]To get this expression, compare the total lengths of the shortest paths between all pairs of vertices in this tree at 'times' N and $N+1$. These are $\langle \ell \rangle(N)\, N(N-1)/2$ and $\langle \ell \rangle(N+1)\,(N+1)N/2$, respectively. The difference due to the new attached vertex is $1 + [\langle \ell \rangle(N)+1](N-1)$. Together these three terms give the following equation for $\langle \ell \rangle$:

$$\langle \ell \rangle(N+1)\frac{(N+1)N}{2} = \langle \ell \rangle(N)\frac{N(N-1)}{2} + 1 + [\langle \ell \rangle(N)+1](N-1),$$

directly leading to Eq. (5.1).

On the other hand, one can estimate the largest distance of a vertex in this tree from its root (the depth of a recursive tree, $d_{\max}$) (Krapivsky and Redner, 2001). Let us explore the average number of vertices at distance d from the root, $\langle W \rangle (d, N)$, so that $\langle W \rangle (d, N)/N$ is the distribution of vertices over their distance from the root. One can write an exact master equation describing evolution of this average

$$\langle W \rangle (d, N+1) = \langle W \rangle (d, N) + \frac{1}{N} \langle W \rangle (d-1, N),$$

of convenience, we introduce a 'discrete time' variable $t \equiv N$ for growing networks. For the evolution of the average number of vertices of degree q in this random tree, $\langle N(q,t) \rangle$, one can write the rate equation:

$$\langle N(q, t+1) \rangle = \langle N(q,t) \rangle + \frac{1}{t} \langle N(q-1, t) \rangle - \frac{1}{t} \langle N(q,t) \rangle + \delta_{q,1}. \qquad (5.2)$$

Since $P(q,t) = \langle N(q,t) \rangle / t$, we arrive at the master equation for the degree distribution:

$$(t+1)P(q, t+1) - tP(q,t) = P(q-1, t) - P(q,t) + \delta_{q,1}. \qquad (5.3)$$

The initial condition is $P(q, 1) = \delta_{q,0}$, that is, the growth starts with an isolated vertex. This equation can be equally derived from the master equation for the probability $p(q, s, t)$, where $P(q,t) = \sum_{s=1}^{t} p(q, s, t)/t$, that a vertex added at time s has degree q at time t:

$$p(q, s, t+1) = \frac{1}{t} p(q-1, s, t) + \left(1 - \frac{1}{t}\right) p(q, s, t) \qquad (5.4)$$

with the initial condition $p(q, 1, 1) = \delta_{q,0}$. Assuming the existence of the stationary solution $P(q) = P(q, t \to \infty)$ of Eq. (5.3), we get the equation

$$P(q) = P(q-1) - P(q) + \delta_{q,1}, \qquad (5.5)$$

whose solution is an exponential degree distribution

$$P(q) = 2^{-q}, \qquad (5.6)$$

which decays markedly more slowly than the Poisson degree distribution of the classical recursive graphs. From Eq. (5.4), one can derive the degree

$\langle W \rangle (-1, N) \equiv 0$. In the N-continuum approximation, this equation takes the form

$$\frac{d \langle W \rangle (d, N)}{dN} = \frac{1}{N} \langle W \rangle (d-1, N),$$

leading to the Poissonian form:

$$\langle W \rangle (d, N) \cong \frac{(\ln N)^d}{d!},$$

which properly satisfies the relation $\sum_{d=0}^{\infty} \langle W \rangle (d, N) = N$. The depth of the tree is estimated from the condition $\langle W \rangle (d_{\max}, N) \sim 1$, which gives

$$d_{\max}(N) \sim e \ln N.$$

114

distributions of individual vertices, $p(q, s, t)$. For large s and t and fixed $s/t < 1$, this distribution has a Poisson form,

$$p(q, s, t) = \frac{s}{t} \frac{1}{(q+1)!} \left[\ln \left(\frac{t}{s} \right) \right]^{q+1},$$ (5.7)

so this distribution is much narrower than $P(q)$. The average degree of vertex s at time t,

$$\langle q \rangle (s, t) = 1 + \sum_{u=1}^{t-s} \frac{1}{s+u} \cong 1 + \ln(t/s),$$ (5.8)

logarithmically diverges at small s indicating that this network is non-uniform—as is natural, older vertices have, on average, more connections than young ones.

Interestingly, the distribution $P_{\text{descendants}}(w)$ of the number of descendants of a vertex in this tree decays very slowly, as w^{-2} (Krapivsky and Redner, 2001). Moreover, it was shown that this law is practically general for recursive trees, and the uniform attachment is not necessary. For branches of the root (the oldest vertex), even more slowly decaying distribution of branch sizes was found (Feng, Su, and Hu, 2005). The mean number of branches with w vertices is exactly $1/w$. This means that in the random recursive tree, on average, the root has just one leaf ($w = 1$).

If we attach each new vertex, not to one of the existing vertices, but to $m > 1$ vertices, then this random recursive graph will have cycles. The number of these cycles is small, and the graph has a locally tree-like structure. Let the attachment be uniform as in the random recursive trees. Then, for example, the average number of 3-cycles (triangles) in this network grows with t very slowly, as $(\ln t)^2$ (see Section 5.2).[2] This growth should be compared to a size-independent average number of triangles in uncorrelated networks. The statistics of degrees in these 'loopy' recursive networks is similar to that in the random recursive trees. Equations (5.2)–(5.8) can be easily generalized to an arbitrary $m \geq 1$. Still, there are nuances concerning the degree–degree correlations. All growing networks are correlated, but the amount of correlations and their type depend on a model. The Pearson correlation coefficient, Eqs. (4.38) and (4.39), of these recursive networks approaches the value

$$r = \frac{(m-1)(3m+1)}{(m+1)(7m+1)}$$ (5.9)

---

[2]Note that the number of $(k>m+1)$-cliques in such recursive networks is zero.

as their size tends to infinity (Dorogovtsev, Ferreira, Goltsev, and Mendes, 2010). Thus the Pearson coefficient is zero in the special case of the random recursive trees, $m = 1$.

5.2 Preferential attachment

The *preferential attachment* process is probably the simplest way to get growing scale-free networks. Vertices for attachment are selected with probability proportional to some function of their degrees:

$$\text{Prob}(i) \propto \mathcal{F}(q_i), \tag{5.10}$$

where $f(q)$ is called the *preference function*. When the preference function is monotonously increasing, 'popular' vertices with many connections have a better chance to get new connections than poorly connected vertices— 'popularity is attractive'. Specifically, a linear preference function

$$\mathcal{F}(q) = q + A, \tag{5.11}$$

where the constant A is an *additional (or initial) attractiveness*, generates scale-free networks.

5.2.1 Linear preferential attachment

Recursive trees growing with the proportional preference, $A = 0$, were first considered in graph theory by Szymański (1987). These recursive trees actually coincide with the particular case $m = 1$ of the *Barabási–Albert model* (Barabási and Albert, 1999). Assuming $A > -1$, the realization probability $\mathcal{P}(\mathcal{T})$ for a labelled tree $\mathcal{T}$ with a possible sequence of degrees $\{q_i\}$, $i = 1, 2, \ldots, N$, $\sum_i q_i = 2(N - 1)$, in this ensemble is, explicitly,

$$\mathcal{P}(\mathcal{T}) = \frac{\displaystyle\prod_{i=1}^{N}\prod_{j=1}^{q_i-1}(A+j)}{(A+2)^{N-2}\displaystyle\prod_{u=3}^{N}\left(u - \frac{A+4}{A+2}\right)}$$

$$= (A+2)^{2-N}\frac{\Gamma\left(3 - \dfrac{A+4}{A+2}\right)}{\Gamma\left(N+1 - \dfrac{A+4}{A+2}\right)}\prod_{q=2}^{q_{\max}}\prod_{k=1}^{q-1}(k+A)^{N(q)} \tag{5.12}$$

(Timár, da Costa, Dorogovtsev, and Mendes, 2020), where $N(q)$ is the number of vertices with degree q in this tree, $\sum_q N(q) = N$. Thus this probability is determined only by the set of the numbers $N(q)$ despite correlations in this tree, which is somewhat unique. When $A \to \infty$, this probability is reduced to $1/(N-1)!$ corresponding to the case of random recursive trees.

For $m \geq 1$ and an arbitrary additional attractiveness $A > -m$,[3] the probability of attachment becomes

$$\mathrm{Prob}(i) = \frac{q_i + A}{(2m + A)t}. \tag{5.13}$$

For this preference function, one can easily reproduce derivations in Section 5.1 and obtain the discrete difference equation for the degree distribution of the infinite network:

$$P(q) = \frac{m}{2m + A}[(q - 1 + A)P(q-1) - (q + A)P(q)] + \delta_{q,m}, \tag{5.14}$$

(Dorogovtsev, Mendes, and Samukhin, 2000; Krapivsky, Redner, and Leyvraz, 2000). The solution of this equation is

$$P(q) = (2 + A/m)\frac{\Gamma[m + 2 + A + A/m]}{\Gamma(m + A)} \frac{\Gamma(q + A)}{\Gamma[q + A + 3 + A/m]} \overset{q \gg m}{\propto} q^{-\gamma}, \tag{5.15}$$

where the exponent γ of the degree distribution is

$$\gamma = 3 + A/m. \tag{5.16}$$

Here the asymptotics $\Gamma(z+c)/\Gamma(z) \overset{z \gg 1}{\cong} z^c$ was used.[4] This solution provides the range of power laws with exponent $2 < \gamma < 3$. In the special case of the

---

[3]One can obtain the same result as with a positive additional attractiveness, $A > 0$, by mixing proportional ($A = 0$) and uniform ($A \to \infty$) preferential attachment. With some probability, make connections as in the Barabási–Albert model and with the complementary probability attach a new vertex to uniformly randomly selected vertices.

[4]The formulae in this section are given for undirected networks. One can easily modify them and apply to recursive networks with preferential attachment determined by, for example, in-degrees. (The recursive graphs, each of whose new vertices has either all outgoing edges or all incoming edges, belong to the class of directed acyclic graphs—the graphs without directed cycles.) Let each added vertex have m outgoing edges that become attached to vertices selected with probability $\mathrm{Prob}(i) \propto q_i^{(\mathrm{in})} + \tilde{A}$, where $\tilde{A} > 0$. Hence all but the first vertices have equal out-degree $q_i^{(\mathrm{out})} = m$, and their full degree is $q = q_i^{(\mathrm{in})} + m$. Then one can easily see that this rule is equivalent to Eq. (5.11) for undirected networks if $\tilde{A} = A + m$, and the exponent of the in-degree distribution $P(q^{(\mathrm{in})}) \propto [q^{(\mathrm{in})}]^{-\gamma}$ is $\gamma = 2 + \tilde{A}/m$.

Barabási–Albert model, the preference is proportional, $A = 0$, which results in the degree distribution

$$P(q) = \frac{2m(m+1)}{q(q+1)(q+2)} \overset{q \gg m}{\propto} q^{-3}. \tag{5.17}$$

In contrast to the recursive networks with uniform attachment discussed in Section 5.1, the degree distribution for vertices born at time s and observed at time t, $p(q, s, t)$, is not Poisson but rater exponential, having the following scaling form:

$$p(q, s, t) = \left(\frac{s}{t}\right)^{\beta} f\left(q\left(\frac{s}{t}\right)^{\beta}\right), \tag{5.18}$$

where the scaling function $f(\xi)$ is

$$f(\xi) = \frac{1}{\Gamma(A+m)} \xi^{A+m-1} e^{-\xi} \tag{5.19}$$

and the exponent $\beta = 1/(2 + A/m)$ is expressed in terms of exponent γ:

$$\beta = \frac{1}{\gamma - 1}, \tag{5.20}$$

where $0 < \beta < 1$ and $\gamma > 2$. Still, $p(q, s, t)$ has a peak and decays with q much faster than the power-law degree distribution $P(q)$. Then the average degree of vertices born at time s and observed at time t has a power-law singularity,

$$\langle q \rangle(s, t) \cong (A + m) \left(\frac{s}{t}\right)^{-\beta} \tag{5.21}$$

reaching the maximum value $(A+m)t^{\beta}$ for the first vertex. As a consequence of the scaling form of $p(q, s, t)$, Eq. (5.18), the degree distribution of a finite network depends on the network size $t \equiv N$,

$$P(q, t) = P(q) g\left(q t^{-1/(\gamma-1)}\right) \tag{5.22}$$

(Dorogovtsev, Mendes, and Samukhin, 2001 c; Krapivsky and Redner, 2002), where the function $g(\eta)$ depends on an initial condition for the growth and, importantly, it is a rapidly decreasing function. This function shapes the size-dependent cut-off of the degree distribution,

$$q_{\text{cut}} \sim t^{1/(\gamma-1)}, \tag{5.23}$$

coinciding in this case with the natural cut-off, Eq. (4.59). For example, for the Barabási–Albert model, $g(\eta) = (1 + \frac{1}{4}\eta^2 + \frac{1}{8}\eta^4)e^{-\eta^2/4}$, where the first multiplier depends on the initial condition, producing a hump in the degree distribution.

5.2.2   Continuum approach

The treatment of master equations like Eq. (5.14) can be greatly simplified
by applying the continuum (in terms of q) approach, which is often called
the mean-field theory for these problems. For the probability $p(q, s, t)$ that
a vertex born at moment s, has degree q at time t, the continuum approach
uses the approximation:

$$p(q, s, t) = \delta(q - \langle q \rangle (s, t)), \tag{5.24}$$

where $\langle q \rangle (s, t)$ is the average degree of vertices born at s and observed at
time t, and the degree and time variables are continuous.[5] Then the degree
distribution equals

$$P(q, t) = \frac{1}{t} \int_0^t ds\, \delta(q - \langle q \rangle (s, t)) = -\frac{1}{t} \left(\frac{\partial \langle q \rangle (s, t)}{\partial s} \right)^{-1} [s = s(q, t)], \tag{5.25}$$

where $s(q, t)$ is the solution of the equation $q = \langle q \rangle (s, t)$.

For the sake of demonstration, let us apply this ansatz to the following directed recursive network whose growth is determined by in-degrees,
denoted here by q:

- At each time step, a new vertex with m outgoing edges is added to the
  network. The target ends of these edges become preferentially attached
  to vertices with probability proportional to a linear function $\mathcal{F}(q) =
  q + A$ of their in-degree q, where $A = 0$.

Our δ-ansatz, Eq. (5.24), immediately leads to the equation for the average
degrees of vertices:

$$\frac{\partial \langle q \rangle (s, t)}{\partial t} = m \frac{\langle q \rangle (s, t) + A}{\int_0^t du\, [\langle q \rangle (u, t) + A]}. \tag{5.26}$$

The initial condition is $\langle q \rangle (0, 0) = 0$ and the boundary one is $\langle q \rangle (t, t) = 0$,
since new vertices have no incoming edges. One may check that

---

[5]Equation (5.24) is the solution of the evolution equations for $p(q, s, t)$ with derivatives
over q substituted for the discrete differences (continuum approximation to the discrete
evolution equations), see details in Dorogovtsev and Mendes (2001b, 2003). The exact
solution $p(q, s, t)$ of the discrete evolution equation typically decays rapidly with q, as in
Eq. (5.18), and so this δ-function, Eq. (5.24), appears to be a good approximation. The
term 'mean-field theory' here means that all vertices born at a given moment are assumed
to have the same degree. This approximation works well when degree distributions $P(q)$
decay sufficiently slowly. In particular, for a power-law degree distributions, it gives an
exact exponent γ, and even for exponentially decaying degree distributions, it provides
qualitatively correct results.

$$\frac{\partial}{\partial t}\int_0^t ds\, \langle q\rangle(s,t) = \int_0^t ds\, \frac{\partial}{\partial t}\langle q\rangle(s,t) + \langle q\rangle(t,t) = m, \qquad (5.27)$$

so $\int_0^t ds\, [\langle q\rangle(s,t) + A] = mt$, as is must be, and we arrive at the equation

$$\frac{\partial \langle q\rangle(s,t) + A}{\partial t} = \frac{1}{1 + A/m}\frac{\langle q\rangle(s,t)}{t}, \qquad (5.28)$$

where

$$\beta = \frac{1}{1 + A/m}. \qquad (5.29)$$

Its general solution is $\langle q\rangle(s,t) + A = C(s)t^\beta$, where $C(s)$ is an arbitrary function of s. Taking into account the boundary condition, $\langle q\rangle(t,t) = 0$, we have

$$\langle q\rangle(s,t) + A = A\left(\frac{s}{t}\right)^{-\beta}. \qquad (5.30)$$

A slight difference between this expression and Eq. (5.21) is due to the undirected network in the latter case. Substituting Eq. (5.30) into Eq. (5.25) leads to

$$P(q,t\to\infty) \propto q^{-(1-1/\beta)}, \qquad (5.31)$$

and we readily obtain the exponent of the degree distribution, $\gamma = 2 + A/m$, for this directed network, compare with Eq. (5.16).

5.2.3 Diameter

The average shortest-path length (and, also, diameter), characterizing the compactness of the networks with linear preferential attachment, is markedly different in the trees, $m = 1$, and in the recursive networks with cycles, $m \geq 2$. It is convenient to use the exponent γ as the parameter instead of A. For $m = 1$ and any $\gamma > 2$, the average shortest-path length increases logarithmically with network size, $t \equiv N$,

$$\langle \ell\rangle \cong C(\gamma)\ln t \qquad (5.32)$$

(Pittel, 1994), where the constant $C(\gamma)$ monotonously decreases with decreasing γ, approaching zero as $C(\gamma) \sim 1/\ln(\gamma-2)^{-1}$ when γ approaches 2. In contrast to this, for the recursive scale-free networks with cycles, $m \geq 2$,

$$\langle \ell\rangle \sim \begin{cases} \ln t & \gamma > 3, \\ \dfrac{\ln t}{\ln\ln t} & \gamma = 3, \\ \ln\ln t & 2 < \gamma < 3 \end{cases} \qquad (5.33)$$

(Bollobás and Riordan, 2003; Dommers, Van Der Hofstad, and Hooghiemstra, 2010) indicating ultra-small-world architectures for $\gamma \leq 3$.

5.2.4 Number of cycles

The recursive networks with $m \geq 2$ are locally tree-like. For example, the
average number of triangles in them varies with t as

$$\langle N_3 \rangle \sim \begin{cases} (\ln t)^2 & \gamma > 3, \\ (\ln t)^3 & \gamma = 3, \\ t^{(3-\gamma)/(\gamma-1)} \ln t & 2 < \gamma < 3 \end{cases} \tag{5.34}$$

(Gleiss, Stadler, Wagner, and Fell, 2001; Bianconi and Capocci, 2003; Bollobás and Riordan, 2003).[6] Thus $\langle N_3 \rangle (t)$ grows much slower than t. Notice
that as $\gamma \downarrow 2$, the term $t^{(3-\gamma)/(\gamma-1)} \to t$, and the local tree-likeness disappears. In particular, for the Barabási–Albert model, $\gamma = 3$,

$$\langle N_3 \rangle \cong \frac{(m-1)m(m+1)}{48}(\ln t)^3 \tag{5.35}$$

(Bollobás and Riordan, 2003), and the average number of cycles of length
ℓ,

$$\langle N_\ell \rangle \sim \left[\frac{m}{2} \ln t \right]^\ell, \tag{5.36}$$

generalizing Eq. (5.35) (Bianconi and Capocci, 2003).[7] Equation (5.35) provides the following clustering coefficient of the Barabási–Albert model,

$$C \cong \frac{m-1}{8} \frac{(\ln t)^2}{t}. \tag{5.37}$$

---

[6]These dependencies can be obtained in the following way. For the emergence of a new
triangle, an added vertex should be attached to a connected pair of vertices, that is,

$$\frac{\partial \langle N_3 \rangle (t)}{\partial t} = \frac{1}{2} \sum_{u=1}^{t} \sum_{v=1}^{t} p_{ut} p_{vt} p_{uv},$$

where p_{uv} is the probability that vertices u and v are connected (Bianconi and
Capocci, 2003). From Eqs. (5.13) and (5.21), we get the connection probability $p_{uv} \propto$
$u^{-1/(\gamma-1)} v^{-1+1/(\gamma-1)}$ which, after substituting into the preceding equation, directly leads
to Eq. (5.34).

[7]It is clear from these expressions that the clustering coefficient vanishes in these networks as they grow. For details on clustering and local clustering, and on degree dependent clustering, see Fronczak, Fronczak, and Hołyst (2003) and Krot and Prokhorenkova
(2015).

5.2.5 Degree–degree correlations

In Section 4.7, we have already mentioned that the Pierson correlation coefficient r, Eq. (4.39), is a widely used but not very informative integral characteristics of degree–degree correlations in networks. It can falsely signal the absence of these correlations even when they are actually present, and even strong. In the scale-free recursive graphs, this happens not only when a network is infinite and $\gamma \leq 4$, but also in all recursive trees, $m = 1$, of this sort with an arbitrary γ. The derivations of the perfectly informative joint degree–degree distribution $P(q, q')$ are straightforward but rather cumbersome compared to $P(q)$ (Krapivsky and Redner, 2001; Dorogovtsev, Ferreira, Goltsev, and Mendes, 2010). One can analyse the form of the dependence of the average degree of the nearest neighbours of a vertex on its degree, $\langle q \rangle_{nn}(q)$, in the continuum approximation, neglecting the discrete nature of degrees, (Dorogovtsev and Mendes, 2001b). The conclusion is that the correlations are assortative when $\gamma > 3$, disassortative, when $\gamma < 3$, and they are weak when $\gamma = 3$, which is the case for the Barabási–Albert model.

5.2.6 Nonlinear preferential attachment

Only linear preference functions generate scale-free networks. Krapivsky and Redner (2001) considered non-linear, power-law preference $\mathcal{F}(q) = q^y$, where y is some non-negative constant. The case with $y = 1$ is exactly the Barabási–Albert model, $P(q) \propto q^{-3}$, while $y = 0$ corresponds to uniform attachment, $P(q) \propto e^{-\text{const}\, q}$. When $0 \leq y < 1$ (sublinear preference), the degree distribution is

$$P(q) \propto q^{-y} \exp\left(-\frac{\mu(y)}{1-y} q^{1-y}\right),\qquad(5.38)$$

where $\mu(y)$ monotonously increases from 1 to 2 as y increases from 0 to 1.

This degree distribution can be obtained in the following way. For an arbitrary preference function $\mathcal{F}(q)$, Eq. (5.14) should be replaced by the equation

$$P(q) = \frac{1}{\mu}[\mathcal{F}(q-1)P(q-1) - \mathcal{F}(q)P(q)] + \delta_{q,m},\qquad(5.39)$$

where

$$\mu = \sum_{q \geq m} \mathcal{F}(q)P(q).\qquad(5.40)$$

Consequently

$$P(q) = \frac{\mathcal{F}(q-1)P(q-1) + \mu\delta_{q,m}}{\mathcal{F}(q) + \mu}, \tag{5.41}$$

which leads to the expression

$$P(q) = \frac{\mu}{\mathcal{F}(q)} \prod_{k=m}^{q} \left(1 + \frac{\mu}{\mathcal{F}(k)}\right)^{-1}. \tag{5.42}$$

An equation for μ is obtained by substituting this expression into Eq. (5.40), which gives

$$1 = \sum_{q \geq m} \prod_{k=m}^{q} \left(1 + \frac{\mu}{\mathcal{F}(k)}\right)^{-1}. \tag{5.43}$$

For $\mathcal{F}(q) = q^y$, where $0 < y < 1$, Eq. (5.42) leads to the asymptotics of the degree distribution in Eq. (5.38), in which $\mu = \mu(y)$ is the solution of Eq. (5.43).[8]

Finally, if $y > 1$ (superlinear preference), then almost all vertices in the large network are leaves, assuming $m = 1$. Moreover, if $y > 2$, then there is a finite probability that all attachments occur to the first vertex, and the resulting network is a proper star.

5.3 Origin of preferential attachment

For numerical simulations, researchers usually need to generate very large networks, say of 10^7 vertices or more. This can be done easily if a graph is recursive and preference is linear. For generating these networks, it is inefficient to select vertices by using their degrees. Indeed, for each attachment, we would have to examine all the degrees. Instead, we form a list where each vertex is repeated as many times as its degree (or, if you wish, degree minus m). Selecting uniformly an entry from this list, we get proportional preference. To get linear preference, we must also form the relevantly weighted (with weight A/m) list of all vertices, and choose uniformly from the combination of these two lists. After every attachment, add new entries to these lists.

Still, we naturally ask what the origin is of the proportional or linear attachment. Indeed, preferential attachment explains power laws, but what can explain preferential attachment? In other words, could one get linear preference from uniform selection? Unfortunately, there is no completely

---

[8]To obtain the asymptotics of Eq. (5.42), the reader should rewrite the product as the exponential of a sum and approximate this sum by an integral.

satisfactory, general answer up to now, and this is the weakest point of the self-organization (or self-organized criticality) concept for networks. The first explanation that comes to mind is the so-called *link copying* mechanism (Kumar, Raghavan, Rajagopalan, Sivakumar, Tomkins, and Upfal, 2000). In the link copying models, attachment is to one of randomly chosen nearest neighbours of a uniformly randomly chosen vertex.[9] In other models of this sort, the added vertex attaches to both a uniformly randomly chosen vertex and to one of its randomly chosen nearest neighbours (Bianconi, Darst, Iacovacci, and Fortunato, 2014) or to both a preferentially chosen vertex and to one of its randomly chosen nearest neighbours (Holme and Kim, 2002). This *triadic closure* is a way to get many triangles in these networks.[10] It

---

[9]In some versions of this model, attachment can be to one of the nearest neighbours of a preferentially selected vertex, which does not produce great difference.

[10]Another natural way, for non-recursive networks, is linking a selected vertex to one of its second nearest neighbours. In addition to generating triangles, this rule leads to a preferential emergence of new edges between vertices with a large number m of common nearest neighbours.

new edge

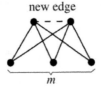

m

This phenomenon was detected in the time evolution of scientific collaboration networks (Newman, 2001a). The probability of scientists collaborating at a given moment increased with the number of other co-authors they had in common. The figure for the evolving net of co-authorships in the arXiv, adapted from Newman (2001a), shows the ratio $R(m)$ between (a) the probability $P(m;t)$ that the two scientists connected by an edge added at a given moment t already had m mutual co-authors and (b) the relative number $N(m;t)/[N(t)(N(t)-1)/2]$ of pairs with m mutual co-authors, where $N(t)$ is the current size of the network.

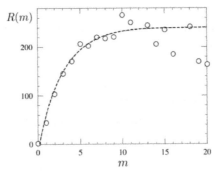

Note that R_m is nearly independent of time.

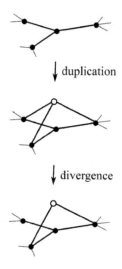

Fig. 5.1 A duplication–divergence event. The white vertex is a new protein.

is widely believed that link copying effectively produces proportional preference, and so it should lead to scale-free networks. This common belief is actually wrong, as was quantitatively demonstrated in Bianconi, Darst, Iacovacci, and Fortunato (2014). Indeed, as explained in Sections 4.2 and 4.7, proportional preference for such attachment can be only in uncorrelated networks, while all growing networks are correlated.[11] Thus link copying cannot explain proportional preference.

A simple way to get proportional attachment was used by Dorogovtsev, Mendes, and Samukhin (2001c). At each step, an added vertex is attached to both end vertices of a uniformly randomly chosen edge. It is still unclear how such uniform selection of edges (instead of vertices) can occur in real networks and what the rationale is behind it.

Another way is to select for attachment *all* nearest neighbours of a uniformly randomly chosen vertex (or each of these nearest neighbours with equal probability p). Section 4.2 explained that the degree distribution of these neighbours equals $qP(q)/\langle q \rangle$ for arbitrary networks which results in proportional preference. When p is small, these networks grow sparse (see Section 5.7 for more detail). A similar process drives the evolution of genetic networks (Pastor-Satorras, Smith, and Solé, 2003; Ohno, 2013). In cellular biology, this process is called *duplication–divergence*. In protein–protein in-

---

[11]We mentioned in Section 5.2 that just proportional preference produce minimal correlations. Still these correlations exist.

teraction networks,[12] new proteins are born as copies of the original ones (duplication), copying all connections of the original proteins, but afterwards proteins lose some of their functions (divergence) (Figure 5.1). (In addition, they acquire new functions but this can be treated separately.) In network terms, duplication–divergence means precisely the connection of a new vertex to a fraction of the nearest neighbours of a selected vertex. Thus we again arrive at a proportional preferential attachment.

5.4 Condensation phenomenon

5.4.1 Heterogeneous preference function

In the preferential attachment models of the previous sections, the preference function had only one variable, a vertex degree q. However, one can introduce however a more general probability of attachment allowing different vertices with the same degree to differently attract new connections—the *Bianconi–Barabási model* (Bianconi and Barabási, 2001*b*). For the recursive networks with the linear preferential attachment, where vertices are labelled in order of their addition, one can write

$$\text{Prob}(q, s) \propto f(s)q + A(s), \qquad (5.44)$$

where s is the label ('birth time') of a vertex. Traditionally, the heterogeneous non-negative coefficient $f(s)$ is called the *fitness*, while $A(s)$ is the heterogeneous additional attractiveness. If only $A(s)$ is heterogeneous, the result is quite simple. Instead of $\gamma = (2\,\text{or}\,3) + A$ for homogeneous preferential attachment, Eq. (5.16), the exponent of the resulting degree distribution becomes $\gamma = (2\,\text{or}\,3) + \bar{A}$, where the average of A is over all vertices and the first term, $(2\,\text{or}\,3)$, depends on the model (Ergün and Rodgers, 2002).[13]

The case of fluctuating fitness described by Bianconi and Barabási (2001*b*) is more interesting. Let us discuss this case considering the following network for the sake of demonstration:[14]

- At each time step, a new vertex is added to the network. This vertex has its own value of fitness f, which is a random variable drawn from

---

[12]Vertices in these networks are proteins, while edges indicate functional and other associations between protein pairs.

[13]Here 2 is for directed recursive networks, if q in the preference function is for the vertex in-degree, and 3 is for undirected recursive networks.

[14]Ergün and Rodgers (2002) described a more general model in which fitness and additional attractiveness were both heterogeneous.

a given probability distribution $\rho(f)$. The fitness of a vertex does not change during the evolution of the network.

- This added vertex becomes attached to one of existing vertices, say, vertex s, with probability proportional to the product of the fitness and degree of vertex s, namely $f(s)q(s)$.

With fitness equal to a fixed number, this growing network is reduced to the tree version of the Barabási–Albert model having the degree distribution exponent $\gamma = 3$. The idea is to introduce the average number $\langle N(q, t; f) \rangle$ of vertices with fitness f having degree q at time t. Then the degree distribution of the network is

$$P(q, t) = \frac{1}{t} \int df \rho(f) \langle N(q, t; f) \rangle \equiv \frac{1}{t} \sum_f \langle N(q, t; f) \rangle. \qquad (5.45)$$

The evolution of $\langle N(q, t; f) \rangle$ is described by the equation:

$$\frac{\partial \langle N(q, t; f) \rangle}{\partial t} = \frac{1}{M(t)} [f(q-1)\langle N(q-1, t; f) \rangle - fq \langle N(q, t; f) \rangle] + \delta_{q,1} \rho(f),$$
$$(5.46)$$

where

$$M(t) = \sum_{q,f} fq \langle N(q, t; f) \rangle. \qquad (5.47)$$

For the derivative of $M(t)$, we have

$$\frac{\partial M(t)}{\partial t} = \frac{1}{M(t)} \sum_f f^2 \sum_q q \langle N(q, t; f) \rangle + \bar{f}, \qquad (5.48)$$

where $\bar{f} \equiv \int df f \rho(f)$ is the average fitness of vertices. For large t, one can expect that $M(t) = \mu t$ and $\langle N(q, t; f) \rangle = n(q, f)t$, where μ and $n(q, f)$ are time-independent. Then the limiting stationary degree distribution $P(q) = \lim_{t \to \infty} P(q, t)$ can be written as

$$P(q) = \int df \rho(f) n(q, f), \qquad (5.49)$$

Eq. (5.46) leads to the equation for $n(q, f)$,

$$n(q, f) = \frac{1}{\mu} [f(q-1)n(q-1, f) - fq\, n(q, f)] + \delta_{q,1}\rho(f), \qquad (5.50)$$

and Eq. (5.48) leads to the equality

$$\mu = \frac{1}{\mu} \sum_f f^2 \sum_q q\, n(q, f) + \bar{f}. \tag{5.51}$$

Let us first treat μ in Eq. (5.50) as a given number. Then the solution of this equation has the form

$$n(q, f) = \frac{\Gamma(q)}{\Gamma(q + 1 + \mu/f)} \frac{\mu}{f} \Gamma(1 + \mu/f)\rho(f), \tag{5.52}$$

and hence the asymptotics of $n(q, f)$ is

$$n(q, f) \sim q^{-(1+\mu/f)}. \tag{5.53}$$

Substituting Eq. (5.52) into Eq. (5.51), we get the equation for μ,

$$\mu = \int df \rho(f) \frac{f^2}{\mu - f} + \bar{f}. \tag{5.54}$$

To obtain the degree distribution, one should solve this equation for a given fitness distribution, substitute the solution into the expression for $n(q, f)$, Eq. (5.52), and integrate the result over f according to Eq. (5.49). In particular, for the uniform fitness distribution $\rho(f) = f_0^{-1}\theta(f)\theta(f_0 - f)$, and so $\bar{f} = f_0/2$, the solution of Eq. (5.54) is $\mu/f_0 = 1.255\ldots$ for any $f_0 > 0$. This leads to the following asymptotics of the degree distribution:

$$P(q) \sim \int_0^{f_0} df\, q^{-(1+\mu/f)} \frac{\mu}{f} \sim \frac{q^{-(1+\mu)}}{\ln q}. \tag{5.55}$$

Hence in this case,

$$P(q) \sim \frac{q^{-2.255}}{\ln q} \tag{5.56}$$

(Bianconi and Barabási, 2001b).

5.4.2   Condensation

It turns out, however, that for strongly heterogeneous fitness distributions $\rho(f)$, such that a number of vertices in the network have high fitness, Eq. (5.54) has no solution, and something interesting happens—a small set of vertices attract a finite fraction of connections (Bianconi and Barabási, 2001a). For demonstration purposes, we consider a simple model showing this effect (Dorogovtsev and Mendes, 2001b). Suppose that one vertex has a higher fitness f than the unit fitness of all other vertices. Now the growth rules for our network are:

- At each time step, a new vertex is added. The fitness of all vertices, except vertex added at instant w, is unit. The fitness of vertex w equals $f \geq 1$.
- The new vertex becomes attached to one of existing vertices with the probability $q(s)/[fq(w) + \sum_{s \neq w} q(s)]$ to vertex $s \neq w$ and the probability $fq(w)/[fq(w) + \sum_{s \neq w} q(s)]$ to vertex w.

The evolution of this network is described by two equations, for vertex w and for the remaining vertices,

$$\frac{\partial \langle q(w,t) \rangle}{\partial t} = \frac{1}{M(t)} f \langle q(w,t) \rangle, \tag{5.57}$$

$$\frac{\partial \langle N(q,t) \rangle}{\partial t} = \frac{1}{M(t)} \left[(q-1)\langle N(q-1,t;f) \rangle - q \langle N(q,t) \rangle \right] + \delta_{q,1}, \tag{5.58}$$

where $\langle N(q,t) \rangle$ is the average number of vertices other than w at time t, $\langle q(w,t) \rangle$ is the average degree of vertex w at time t,[15] and

$$M(t) = f \langle q(w,t) \rangle + \sum_q q \langle N(q,t) \rangle. \tag{5.59}$$

Clearly, at long times, the total degree of vertices in the network is $\langle q(w,t) \rangle + \sum_q q \langle N(q,t) \rangle \cong 2t$.

At long times, two distinct situations are possible.

(i) The average degree of the vertex with high fitness grows slower than t, the denominators on the right-hand sides of Eqs. (5.57) and (5.58) approach $2t$, and the contribution of vertex w to the degree distribution and other macroscopic characteristics of the network is negligible. In this situation, Eq. (5.57) takes the form

$$\frac{\partial \langle q(w,t) \rangle}{\partial t} = \frac{f}{2t} \langle q(w,t) \rangle, \tag{5.60}$$

and its solution at long times is $\langle q(w,t) \rangle = \text{const } t^{f/2}$. Therefore the slow growth is realized only when f is below some threshold:

$$f < f_c = 2. \tag{5.61}$$

(ii) Now, let the fitness f of vertex w be greater than the critical value f_c. The average degree of this vertex $\langle q(w,t) \rangle$ cannot grow more rapidly than

---

[15]For the sake of brevity, we assume that $\langle q(w,t) \rangle$ is sufficiently large, which is reasonable in the interesting for us region of fitness f when time t is large.

t, since the maximum number of connections which a vertex can receive is t. Then this growth should be asymptotically $\langle q(w,t) \rangle \cong dt$, where $d < 1$.

This means that, for $f > f_c$, a finite fraction of all edges is captured by vertex w, which Bianconi and Barabási (2001a) interpreted as the Bose–Einstein condensation with the condensation point f_c. When $f > f_c$, Eqs. (5.57) and (5.58) take the form

$$1 = \frac{f}{(f-1)d+2},\tag{5.62}$$

$$\frac{\partial \langle N(q,t) \rangle}{\partial t} = \frac{(q-1)\langle N(q-1,t;f) \rangle - q\langle N(q,t) \rangle}{[(f-1)d+2]t} + \delta_{q,1},\tag{5.63}$$

and so for $f > f_c = 2$, the following fraction of all edges in the network is captured by one vertex with high fitness:

$$d = \frac{f-2}{f-1}.\tag{5.64}$$

Then the denominator on the right-hand side of Eq. (5.64) equals ft, which indicates that the tail of the degree distribution is a power law with exponent $\gamma = 1 + f$; see Eqs. (5.14) and (5.16). Thus

$$\gamma = \begin{cases} 3 & \text{if } f < 2, \\ 3 + (f-2) & \text{if } f > 2. \end{cases}\tag{5.65}$$

These results for $t \to \infty$ do not depend on the initial conditions, but we can find how the network relaxes to this state by inspecting the asymptotic dependence $\langle q(w,t) \rangle$. It turns out that, interestingly, in the entire condensation phase, $f > f_c$, relaxation to the final state is of a power-law kind:

$$\frac{\langle q(w,t) \rangle - dt}{t} \propto t^{-\mathrm{const}(f-f_c)},\tag{5.66}$$

and at the critical point, the relaxation proceeds slower than any power law.

For continuous fitness distributions $\rho(f)$ with a sufficiently fat tail, the observed phenomenon is similar to our demonstrative example with the only difference being that the condensation occurs on a small number of vertices instead of a single one (Bianconi and Barabási, 2001b).

5.5 Berezinskii–Kosterlitz–Thouless (BKT) transition

The models of growing networks of the preceding sections generated connected random graphs. In general, growing networks can contain a set of connected components. How does a giant connected component emerge in such growing networks? Callaway, Hopcroft, Kleinberg, Newman, and Strogatz (2001) studied the birth of a giant connected component in growing networks and discovered a phase transition, quite distinct from the one in equilibrium networks. They considered the following model:

- At each time step, a new vertex is added to a network.
- Simultaneously, b new undirected edges emerge between pairs of uniformly randomly chosen vertices (b may be non-integer).

The control parameter b allows us to vary the structure of the network, change the size of the giant component, and approach the point of its birth.

Since we focus on the statistics of sizes of connected components, this network growth can be treated as an aggregation process like the Erdős–Rényi random graph process in Sections 3.4 and 3.5. In a similar way to the Erdős–Rényi process, we can derive evolution equations for the probability $\mathcal{P}(s,t)$ that a uniformly randomly selected vertex belongs to a connected component of size s,

$$t\frac{\partial \mathcal{P}(s,t)}{\partial t} + \mathcal{P}(s,t) = \delta_{s,1} + bs \sum_{u=1}^{s-1} \mathcal{P}(u,t)\mathcal{P}(s-u,t) - 2bs\mathcal{P}(s,t), \quad (5.67)$$

compare to Eq. (3.23).[16] This bilinear equation has a stationary solution in the limit $t \to \infty$, which can be find by applying generating functions. The result dramatically differs from equilibrium networks. The giant connected component is present above the critical point, $b > b_c = 1/8$, and its relative size has the critical singularity

$$S(b) \cong 0.590...\exp\left(-\frac{\pi}{2\sqrt{2}}\frac{1}{\sqrt{b-b_c}}\right) \quad (5.68)$$

earlier observed in the renown Berezinskii–Kosterlitz–Thouless (BKT) phase transition in the two-dimensional XY-model (Berezinskii, 1971; Kosterlitz and Thouless, 1973). All the derivatives of $S(b)$ are zero at the critical point. Thus, this transition is of infinite order. Moreover, Dorogovtsev, Mendes, and Samukhin (2001a) found the same critical singularity at the birth point of a giant connected component in a wide range of growing networks with

---

[16]The term $\partial[t\mathcal{P}(s,t)]/\partial t$ in Eq. (5.67) is substituted for $\partial\mathcal{P}(s,t)/\partial t$ in Eq. (3.23), since the number of vertices now grows. The new term $\delta_{s,1}$ added on the right-hand side of Eq. (5.67), indicates that an extra component of size 1 emerges at each time step.

linear preferential attachment. In the phase without a giant connected component in all these networks the distribution $\mathcal{P}(s)$ turned out to be power law

$$\mathcal{P}(s, b < b_c) \sim s^{-[\tau(b)-1]}, \tag{5.69}$$

where the exponent $\tau(b)$ increases from 3 to ∞ as b decreases from the critical value b_c to 0. In the phase with a giant connected component, it is exponential,

$$\mathcal{P}(s, b > b_c) \propto \exp[-s/(\mathrm{const}(b - b_c)], \tag{5.70}$$

and at the critical point,

$$\mathcal{P}(s, b_c) \sim \frac{1}{s^2(\ln s)^2}. \tag{5.71}$$

This singularity provides the finite first moment of the distribution $\langle s \rangle_{\mathcal{P}} = \sum_s s\mathcal{P}(s)$ (the average size of a finite component to which a uniformly randomly chosen vertex belongs, that is, the susceptibility) at the critical point.[17] So we have a power-law distribution of components in the normal phase and an exponential distribution in the phase with a giant component. This is opposite to the original BKT phase transition where correlations decay in a power-law way below the critical temperature and decay exponentially in the normal phase. This is why this transition in growing networks is frequently called the *inverted BKT phase transition*.

5.6 Accelerated growth and densification

In many real networks the number of edges grows faster than the number of vertices, so that the average degree of a vertex is a growing function of the network size—connections become more dense (Leskovec, Kleinberg, and Faloutsos, 2007b).[18] This non-linear, or, one can say, 'accelerated', growth has numerous consequences (Dorogovtsev and Mendes, 2001a, 2003). For example, the diameter of a growing network of this kind may be constant or even shrink as the number of vertices increases. The models of networks that we have considered up to now do not show this accelerated growth. It is not very difficult, however, to incorporate this feature into network models

---

[17]The reader will find the details of these calculations in Dorogovtsev, Mendes, and Samukhin (2001a).

[18]Broder, Kumar, Maghoul, Raghavan, Rajagopalan, Stata, Tomkins, and Wiener (2000) first measured the density of connections in the WWW in May 1999 and found that the average in- and out-degrees of a vertex are equal at 7.22. When they repeated the measurements in October 1999, the average in- and out-degrees were already 7.85.

with preferential attachment. Suppose, for example, that in an undirected random recursive network, the number of connections of a new vertex $m(t)$ is a growing function of the network size $t \equiv N$, and the average degree of a vertex, $\langle q \rangle(t) = 2m(t)$, is also growing. With the acceleration, a network can still be scale-free, if the preferential attachment is linear. There are some differences compared to previously discussed degree distributions. Without the acceleration, a scale-free degree distribution is stationary in the infinite network limit, and its exponent γ cannot be smaller than 2, otherwise the first moment of a degree distribution would diverge. This is the case in most of the studied models with preferential attachment. If the growth is accelerated, both these restrictions are removed. Let us discuss how it can happen, without focusing on specific models.

Let us assume that the average degree grows as a power law, $\langle q \rangle(t) \sim t^a$, $a > 0$, so the number of edges, $E(t) \sim t^{1+a}$, grows faster than t. Let the resulting degree distribution be non-stationary, power law,

$$P(q, t) \sim t^z q^{-\gamma}. \tag{5.72}$$

What are the possible values of exponents z and γ? In fact, the power-law form, Eq. (5.72), is valid only in the limited range of degrees: $q_0(t) \lesssim q \lesssim q_{\text{cut}}(t)$. It's easy to estimate them. For $\gamma > 1$,

$$\int_{q_0(t)}^{\infty} dq\, t^z q^{-\gamma} \sim 1, \tag{5.73}$$

which leads to

$$q_0(t) \sim t^{z/(\gamma-1)}. \tag{5.74}$$

Hence $z \geq 0$ and $P(q_0(t)) \sim t^{-z/(\gamma-1)}$. On the other hand, if we assume that the cut-off of the degree distribution of this network coincides with the natural one, like in recursive graphs, discussed previously,[19] then

$$t \int_{q_{\text{cut}}(t)}^{\infty} dq\, t^z q^{-\gamma} \sim 1, \tag{5.75}$$

and so

$$q_{\text{cut}}(t) \sim t^{(z+1)/(\gamma-1)}. \tag{5.76}$$

[19]This is a strong assumption, since, in general, such cut-offs are model-dependent. We made it here only for the sake of brevity. The reader can check that for any form of $q_{\text{cut}}(t)$, satisfying $q_0(t) \ll q_{\text{cut}}(t) \lesssim q_{\text{nat}}(t)$, where the natural cut-off $q_{\text{nat}}(t)$ is given by Eq. (5.76), the conclusions are the same. Consider, for example, $q_{\text{cut}}(t) \sim t^{c(z+1)/(\gamma-1)}$ with a constant c, where $z/(z+1) < c \leq 1$.

Consequently, when $1 < \gamma < 2$, the average degree is

$$t^a \sim \int^{t^{(z+1)/(\gamma-1)}} dq\, q\, t^z q^{-\gamma} \sim t^{-1+(z-1)/(\gamma-1)}, \tag{5.77}$$

which leads to

$$\gamma = 1 + \frac{z+1}{a+1} \tag{5.78}$$

implying $0 \leq z < a$. In this case, a stationary degree distribution, $z = 0$, has the smallest exponent $\gamma = 1 + 1/(a+1)$.

When $\gamma > 2$, we estimate the average degree as

$$t^a \sim \int_{t^{z/(\gamma-1)}} dq\, q\, t^z q^{-\gamma} \sim t^{z-z(\gamma-2)/(\gamma-1)}. \tag{5.79}$$

Then the exponent of the degree distribution equals

$$\gamma = 1 + \frac{z}{a} \tag{5.80}$$

implying $z > a$. Thus a stationary degree distribution is impossible in such networks if $\gamma > 2$.[20]

5.7 Transitions in dense networks

Even the simplest model of a growing dense recursive network shows curious phenomena (Lambiotte, Krapivsky, Bhat, and Redner, 2016; Bhat, Krapivsky, Lambiotte, and Redner, 2016). These authors explored the following growth model:

- At each step, a new vertex is added to the network.
- This vertex attaches to a uniformly randomly chosen vertex and independently to each of its nearest neighbours with probability p.

For $p = 1$, this rule produces a complete graph at each step. With each new vertex added, the average number of edges $\langle E \rangle (t)$ increases by $1 + p \langle q \rangle (t) = 1 + p2 \langle E \rangle (t)/t$. This leads to the equation

$$\langle E \rangle (t+1) = \langle E \rangle (t) + 1 + \frac{2p}{t} \langle E \rangle (t) \tag{5.81}$$

with the initial condition $\langle E \rangle (1) = 0$. The asymptotic solution of this equation shows the sparse and dense regimes:

---

[20] For preferential attachment models evolving in this way, see Dorogovtsev and Mendes (2001*a*).

$$\langle E \rangle (t) \cong \begin{cases} \dfrac{1}{1 - 2p} t & p < \tfrac{1}{2}, \\[2ex] \dfrac{1}{(2p - 1)\Gamma(1 + 2p)} t^{2p} & \tfrac{1}{2} < p \le 1, \end{cases} \tag{5.82}$$

with the transition between them at $p = \tfrac{1}{2}$. In the sparse regime, the degree distribution is a power law with exponent γ approaching infinity as $p \to 0$. The variance of the number of edges, $\langle E^2 \rangle - \langle E \rangle^2$, grows as

$$\langle E^2 \rangle - \langle E \rangle^2 \sim \begin{cases} t \sim \langle E \rangle & p < \tfrac{1}{4}, \\[2ex] t^{4p} & \tfrac{1}{4} < p \le 1. \end{cases} \tag{5.83}$$

This indicates that the distribution of the number of edges is Gaussian only when $p < 1/4$, and that the dense network, $p > 1/2$, has surprisingly strong fluctuations $\sqrt{\langle E^2 \rangle - \langle E \rangle^2} \sim \langle E \rangle$, implying a wide diversity between the realizations of a network and lack of self-averaging.

Clearly, this growth produces numerous triangles and larger cliques, and it is the evolution of the average number of k-cliques, $\langle N_{k\text{-clique}} \rangle (t)$, that shows a bouquet of distinct regimes and transitions between them. For $k \ge 2$, the average number of cliques demonstrates k different asymptotic behaviours,

$$\langle N_{k\text{-clique}} \rangle (t) \sim \begin{cases} t & p < \dfrac{1}{2}, \\[2ex] t^{2p} & \dfrac{1}{2} < p < \dfrac{2}{3}, \\[2ex] t^{3p^2} & \dfrac{2}{3} < p < \dfrac{3}{4}, \\[1ex] \quad \cdots \\[1ex] t^{kp^{k-1}} & \dfrac{k-1}{k} < p \le 1. \end{cases} \tag{5.84}$$

In particular, $\langle N_{k\text{-clique}} \rangle \sim t$ in the sparse regime of this network, and $\langle N_{k\text{-clique}} \rangle \sim t^k$ in the fully connected graph.

5.8   Growth and decay

The preferential attachment mechanism and its variations can be used not only to generate growing networks but also for equilibrium and even decaying or shrinking networks, in which edges and vertices can disappear with time. Although such networks are not recursive, they can be solved. For

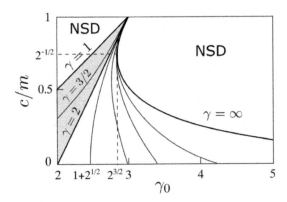

Fig. 5.2 Phase diagram of the network with preferential attachment of new vertices and removal of uniformly selected edges in the plane $c/m-\gamma_0$. Degree distribution is non-stationary in the NSD regions. Thin lines connect points with equal values of exponent γ. Thick lines are for $\gamma = 1$, 2, and ∞. Adapted from Dorogovtsev and Mendes (2001*b*).

example, let a network evolve due to two processes: preferential attachment of new vertices and removal of edges; see Dorogovtsev and Mendes (2000, 2001*b*). The rules of this model are as follows:

- At each step, add a vertex with m edges attached to existing vertices selected with linear preference $\mathrm{Prob}(i) \propto q_i + A$.
- Simultaneously, remove c uniformly randomly chosen edges, where $c \geq 0$ can be non-integer.

When $c > m$, the network losses all its edges with time. The end vertices of a randomly chosen edge in an arbitrary network have degrees distributed according to the probability distribution $qP(q)/\langle q \rangle$. So this model can be analysed by implementing techniques for linear preferential attachment described in the previous sections. Without the second channel, that is, when $c = 0$, the degree distribution exponent $\gamma_0 = 3 + A/m$. For $c \geq 0$, the continuum approximation provides a range of degree distributions. Figure 5.2 shows the phase diagram: each point of the plane $(\gamma_0, c/m)$ indicates the value of the exponent γ of the degree distribution of the resulting network. Notice the regions with non-stationary degree distributions (NSD). Also notice the region, in which exponent γ takes values between 1 and 2, impossible in recursive networks with a fixed average degree considered in Section 5.2.[21]

---

[21]Interestingly, one can get degree distributions with exponent $\gamma = 1$ in models of equilibrium networks (Timár, Dorogovtsev, and Mendes, 2016). For example, at each

Finally, note that when c is sufficiently close to m, this network can consist of many connected components.

Similarly, one can consider networks in which, apart from the progressive addition of vertices, preferentially attaching to linearly selected ones, a number of uniformly selected vertices permanently vanish (Chung and Lu, 2004; Cooper, Frieze, and Vera, 2004; Moore, Ghoshal, and Newman, 2006). Let r vertices be removed in this way per each vertex added. The resulting degree distribution is a power law with exponent γ increasing from 3 to ∞ as r increases from 0 to 1.

5.9 Power of choice

In fact, there are many ways to select vertices with large or small degrees apart from straightforward preferential attachment rules described in the preceding sections. An interesting option for preferential selection was explored by D'Souza, Krapivsky, and Moore (2007). For the sake of brevity, here we touch upon only one of their models. Consider a recursive tree.

- At each step, compare the degrees of k uniformly randomly chosen vertices in the tree, find the vertex with the largest degree among them, and attach a new vertex to it.

For $k = 1$, the model is reduced to the ordinary random recursive tree. The reader remembers that its degree distribution is $P(q) = 2^{-q}$ (Section 5.1). For $k = 2$, straightforward calculations show that the degree distribution remains exponential, $P(q) \sim (3/2)^{-q}$, and, at first sight, the effect of this attachment is not dramatic. If, however, $k \gg 1$, then

$$
P(q) \sim \begin{cases} \dfrac{1}{k} q^{-1} & 1 \ll q \ll k, \\[2mm] \dfrac{1}{k} e^{-q/k} & q \gg k. \end{cases} \tag{5.85}
$$

Thus the attachment exploiting the power of choice can produce degree distributions with exponent γ equal to 1, which is markedly smaller than the values $\gamma > 2$ for recursive graphs with linear preference.

5.10 Evolving weighted networks

Preferential selection mechanisms are applicable to evolving weighted networks as well (Yook, Jeong, Barabási, and Tu, 2001). Two distinct types

time step, (i) choose uniformly at random an edge, (ii) choose one of its end vertices, and (iii) relink the edge from this vertex to one of its nearest neighbours.

of relations between the strength (Section 4.12) and degree of individual vertices were observed in a number of real networks,

$$s(i) \sim \begin{cases} q(i), & \text{networks of coauthorships,} \\ [q(i)]^\theta, & \text{Internet, airport and shareholder networks,} \end{cases} \tag{5.86}$$

where exponent θ is noticeably greater than 1 (Serrano, Boguñá, and Díaz-Guilera, 2005; Barrat, Barthelemy, Pastor-Satorras, and Vespignani, 2004; Garlaschelli, Battiston, Castri, Servedio, and Caldarelli, 2005). Bianconi (2005) proposed and analyzed a model of recursive weighted network demonstrating how these two distinct dependencies can arise. At each step:

- A new vertex attaches to m existing vertices selected with probability proportional to their degrees. Each of these m edges has weight w_o.
- Simultaneously, m_w edges are chosen with probability proportional to their weights,[22] and each of their weights is increased by w_0.

For evolution of connections, this is exactly the Barabási–Albert model. So in the continuum approximation, the average degree of a vertex added at 'time' u and observed at time t equals

$$\langle q \rangle (u, t) = m\, u^{-1/2} t^{1/2}. \tag{5.87}$$

Therefore, its degree distribution is $P(q) \sim q^{-3}$. The continuum approximation equation for the average weight of the edge between vertices added at times u and v and observed at time t has the following form

$$\frac{\partial \langle w \rangle (u, v, t)}{\partial t} = w_0 m_w \frac{\langle w \rangle (u, v, t)}{w_0 (m + m_w) t}. \tag{5.88}$$

The solution of this equation is

$$\langle w \rangle (u, v, t) = w_0 v^{-\beta_w} t^{\beta_w}, \tag{5.89}$$

if $u < v$. Here exponent $\beta_w = m_w / (m + m_w)$. This leads to the distribution of weights

$$P_w(w) \sim w^{-(1+1/\beta_w)} = w^{-(2+m/m_w)}. \tag{5.90}$$

---

[22]Note that selecting edges in this way is equivalent to first selecting a vertex with probability proportional to its strength and then choosing one of its edges with probability proportional to its weight. Indeed,

$$\frac{s(i)}{\sum_k s(k)} \frac{w(i,j)}{s(i)} = \frac{w(i,j)}{\sum_{k,k'} w(k,k')}.$$

The average strength of a vertex added at time i and observed at time t is given by

$$\langle s \rangle (i, t) = \int_0^t dk \, \langle w \rangle (i, k, t) \, p(i, k), \qquad (5.91)$$

where the connection probability $p(i, k) \propto i^{-1/2} k^{-1/2}$. Substituting Eq. (5.89) into Eq. (5.91) and analysing the contributions to the integral on the right-hand side, one can see that $\langle s \rangle (i, t) \propto i^{-1/2}$ if $\beta_w < 1/2$, and $\langle s \rangle (i, t) \propto i^{-\beta_w}$ if $\beta_w > 1/2$. Consequently,

$$\langle s \rangle (i, t) \propto \begin{cases} \langle q \rangle (i, t) & \text{if } \beta_w < 1/2, \text{ i.e. } m_w < m, \\ [\langle q \rangle (i, t)]^{2\beta_w} & \text{if } \beta_w > 1/2, \text{ i.e. } m_w > m, \end{cases} \qquad (5.92)$$

demonstrating the same combination as in the real-world networks; see Eq. (5.86).

Equally easily one can get the disparity, Eq. (4.77), for the vertices of this network:

$$Y(q) \sim \begin{cases} \dfrac{1}{q} & \text{if} \quad \beta_w < 1/4, \quad \text{i.e.} \quad m_w < m/3, \\ q^{-2(1-2\beta_w)} & \text{if } 1/4 < \beta_w < 1/2, \text{ i.e. } m/3 < m_w < m, \\ O(1) & \text{if} \quad \beta_w > 1/2, \quad \text{i.e.} \quad m_w > m, \end{cases} \qquad (5.93)$$

which shows that when $\beta_w > 1/2$, one of the edges of a vertex takes the weight much greater than the weights of its remaining edges.

5.11   Evolution preserving degrees

The random graph with a given degree sequence—the configuration model—is one of the basic models of equilibrium networks. Imagine how a growing network model with a fixed degree sequence could be organized. Let in a given sequence of degrees, $\{q_i\}$, $i = 1, 2, \ldots, t \equiv N$, each element q_i is the degree of a vertex added at step i. This value stays fixed forever. Clearly, any subsequence $\{q_i\}$, $i = 1, 2, \ldots, t' \leq t$ should be graphical, which is a strong constraint. Due to this restriction, the growth should start not from a single vertex but rather from a given graph, say, of n vertices. Note that a new vertex of degree q_i can be inserted into the network only by cutting at least q_i edges, which produces at least $2q_i$ stubs. Of them, q_i stubs attach to vertex i, while the remaining stubs should join each other in accordance to the rules of a specific model. Thus these networks are not recursive, and the initial graph can strongly influence the entire evolution of the network.

One can easily grasp the structure of the simplest network governed by such rules. Let the growth start with an arbitrary graph and at each step a new vertex of degree 2 be added to the network in such a way that the degrees of the previous remain unchanged. If, in addition, we demand the minimum possible redistribution of connections, then each of the edges in the original graph turns into a chain of vertices, elongating with time, which is the only outcome of this growth process.

Kharel, Mezei, Chung, Erdős, and Toroczkai (2022) proposed a set of models of this sort, ranging from growing random regular graphs to growing scale-free networks. This construction is not only of academic interest, chemical complexes provide an example of real-world growing networks with fixed vertex degrees.

5.12 Evolution of simplicial complexes

Let us introduce necessary notions. Consider $d + 1$ vertices—points in an d-dimensional space, such that they cannot be enclosed into lower-dimensional spaces. The smallest convex hull for these vertices is called the d-*dimensional simplex*. A vertex is a 0-dimensional simplex, an edge is a 1-dimensional simplex, a triangle is a 2-dimensional simplex, and so on. These $d + 1$ vertices completely represent the simplex. On any subset of the vertices of a given d-dimensional simplex, one can base a lower-dimensional simplex. These d'-dimensional simplexes ($d' = 0, 1, \ldots, d-1$) are the *faces* of the d-dimensional simplex. Thus a d-dimensional simplex actually contains $2^{d+1} - 1$ simplexes (the simplex itself and all its faces). These $2^{d+1} - 1$ simplexes can be used in description of group interactions within all possible subsets of the $d + 1$ vertices. If we interconnect all the vertices of a d-dimensional simplex by edges, then we get a clique, and the faces are smaller cliques within this maximal clique. One can construct a complex of simplexes—the *simplicial complex*—such that: (i) all faces of the simplexes of the simplicial complex also belong to this complex, and (ii) each two simplexes in the simplicial complex either overlap by some of their faces or any overlap is absent (Matousek, 2013). Simplexes of different dimensions in a simplicial complex can serve as substrates for different processes and interactions. Similarly to a maximal clique, one can introduce a *maximal simplex* that is a simplex not contained in a larger simplex. Interconnecting all vertices within each of the simplexes of a simplicial complex by edges we get a simple graph called the *skeleton* (or 1-skeleton) of the simplicial complex—the set of its 0- and 1-dimensional simplexes, which provides the complete description of this simplicial complex in terms of graphs.[23] Thus for every simple graph we can indicate the corresponding simplicial complex. Every clique in the skeleton of a simplicial complex represents a simplex in this simplicial complex, and every maximal clique in the skeleton represents a maximal simplex in the simplicial complex.

The simplicial complexes are directly related to hypergraphs. The hypergraph whose vertices are the vertices of a given simplicial complex and the hyper-edges correspond to all simplexes completely represents the simplicial complex. This representation however may be not economical, since the total number of simplexes in a simplicial complex can be very large. For a more compact representation, one can use the hypergraph whose hyper-edges correspond to the maximal simplexes in a simplicial complex. The skeleton of the simplicial complex is the clique (or primal) graph of this hypergraph (Section 2.1), which guarantees one-to-one correspondence between a simplicial complex and the hypergraph.

An important class of simplicial complexes render the discrete construction of *manifolds*, topological spaces locally homeomorphic to Euclidean spaces. Simple examples of manifolds are the circle, the sphere, the torus, etc. In these constructions, a simplex is a building block, and a simplicial complex is homeomorphic to a manifold. The simplicial complex construction is possible for, in particular, any smooth (in other words, differentiable) manifold. Dimension d of maximal simplexes in such simplicial complexes equals the dimension of the corresponding manifolds. As an example, recall triangulations, homeomorphic to various surfaces. Every edge in a triangulation, if it is not on a boundary of the surface, belongs to two triangles. If it is on a boundary, it has a single adjacent triangle.[24]

As a first simple example of a growing simplicial complex, we indicate the array of triangles (2-dimensional simplexes) evolving by the following rules.

- Start the growth with a single triangle.
- At each step, select one of triangles uniformly at random and attach a new triangle to one of the edges (faces of 2-dimensional simplex) of the selected triangle.

---

[23]Consequently, there are two alternative complete representations of a simplicial complex, allowing to get all its simplexes: (i) the complex of maximal simplexes, that is the array of subsets—maximal simplexes—of the full set of vertices in the simplicial complex; (ii) the skeleton graph.

[24]A boundary can be external or it can be the boundary of a hole in the surface.

In terms of evolving graphs, the model of this kind was considered and solved in Dorogovtsev, Mendes, and Samukhin (2001c).[25] The uniform selection of triangles means the proportional selection of edges that results to the Barabási–Albert model, and so, in particular, to a power-law degree distribution, $P(q) \sim q^{-3}$, of the skeleton graph.

The growing simplicial complex described above has a rather primitive tree-like structure in the sense that its skeleton has no cycles longer than of length 3. Wu, Menichetti, Rahmede, and Bianconi (2015) proposed a significantly richer class of growing simplicial complexes, which, in the particular case of triangles, range from the simplicial complex above to a triangulation growing through the attachment of new triangles to its boundary. For the sake of brevity, here we focus on the case of triangles. The idea is to introduce the maximum number m_{max} of triangles that can be incident to an edge (face of triangle—2-dimensional simplex). Then the number m of triangles incident to an edge must satisfy $1 \leq m \leq m_{max}$. The growth starts from a single triangle. At each step:

- attach a new triangle to an edge selected uniformly at random from the set of all edges with $m < m_{max}$;
- with probability p, select a uniformly random edge from the set of all edges with $m < m_{max}$, select one of its adjacent edges with $m < m_{max}$, such that the two selected edges do not belong to the same triangle (if this is not possible, repeat the attempt), and add a new edge between the free ends of these adjacent edges, creating a new triangle.

Figure 5.3 explains this process in the case of $m_{max} = 2$. In the particular case of $m_{max} = \infty$ and $p = 0$, this simplicial complex is reduced to the previous one, having a scale-free skeleton. In this case, it is a small world. In the other limit, $m_{max} = 2$ (Figure 5.3) for any p, this growing simplicial complex is a planar—a large world,, and the degree distribution of its skeleton is exponential. For $2 < m_{max} < \infty$ and various p, the resulting simplicial complexes are small worlds with finite spectral dimensions, having a wide set of architectures.[26]

---

[25]In fact, in Dorogovtsev, Mendes, and Samukhin (2001c), at each step a uniformly random edge was selected and a new triangle was attached to it. This does not produce much difference.

[26]For $p = 0$, the skeletons of these simplicial complexes have no k-cores with $k > 1$, see Section 6.7, that is, progressive pruning their vertices of degree 2 eliminates them completely. On the other hand, if $p > 0$, then the k-cores of the skeletons are nontrivial.

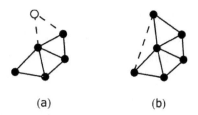

(a)   (b)

Fig. 5.3 Growth of simplicial complex with $m_{max} = 2$. (a) A new triangle is attached to one of the edges with $m = 1$. (b) With probability p, the free ends of two adjacent edges with $m = 1$ are connected by a new edge, creating a new triangle. The new edges are shown by dashed lines. Adapted from Wu, Menichetti, Rahmede, and Bianconi (2015).

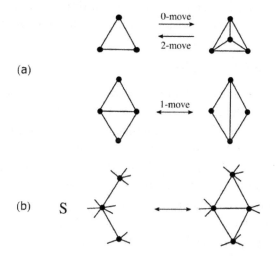

Fig. 5.4 Elementary operations that keep triangulations within the complete set of triangulations: (a) Pachner moves (Pachner, 1991); (b) splitting and merging of adjacent edges and their joint vertex. Pachner moves can be reduced to a chain of the S-operations.

For $m_{max} = 2$, these models generate triangulations with expanding boundaries. Da Silva, Bianconi, da Costa, Dorogovtsev, and Mendes (2018) explored the alternative process of the evolution of triangulations and other simplicial complexes representating manifolds without boundaries. Let us focus on triangulations homeomorphic to topological surfaces. The set of local transformations of triangulations that keep them within the complete set of triangulations (triangular mesh operations) can be reduced to a few elementary ones (Figure 5.4). These evolution models are organized in the following way. At each time step:

(i) An element or neighbouring elements of a simplicial complex are chosen with some preference or, in the simplest particular case, uniformly at random. For triangulations, such elements are vertices, edges, and triangles.

(ii) Then, a specific transformation from the set of operations that keep the simplicial complex intact is applied to this element.

Depending on specific (i) and (ii), one can get a zoo of evolution scenarios, including, in general, growing, decaying, and equilibrium simplicial complexes with diverse structures, space dimensions, and even topologies.[27] Some of the features of these simplicial complexes are surprising, in particular, particularly the Hausdorff dimension, which can differ from the value 4 typical for random planar graphs (Ambjørn, Durhuus, Jonsson, and Jonsson, 1997).

5.13 Deterministic graphs

This book mostly concerns random systems, but, interestingly, a number of deterministic graphs demonstrate many features resembling those of random growing networks. The resemblance can be so close that deterministic graphs were even used for interpretation of empirical observations (Ravasz, Somera, Mongru, Oltvai, and Barabási, 2002; Ravasz and Barabási, 2003). As examples, we indicate a few deterministic graphs that can be naturally treated as simplicial complexes: *Farey-sequence based graphs* (Hardy and Wright, 1979), *pseudo-fractals* (Dorogovtsev, Goltsev, and Mendes, 2002b), and *Apollonian graphs* (Andrade Jr, Herrmann, Andrade, and Da Silva, 2005) (Figure 5.5).

These three graphs are small worlds with the 'average' separation between their vertices $\langle \ell \rangle \sim N$. Their Hausdorff dimension is infinite. In particular, considering the relative number of vertices $N(q)$ of degree q, one can easily get $N(q) \propto 2^{-q/2}$ for the Farey graph, where $q = 2, 4, 6, \ldots$, which resembles the exponential degree distribution of random recursive graphs. In the spectra of degrees of the pseudofractals and Apollonian graphs, the gaps between consecutive degrees exponentially increase with q, so once should consider the cumulative number of vertices of degree equal or larger than q. In both cases, this number has the asymptotics

---

[27] Imagine a triangulation of a sphere. Choose two of its sufficiently remote triangular faces, merge them together, and annihilate. Then the triangulation of a sphere transforms into the triangulation of a torus.

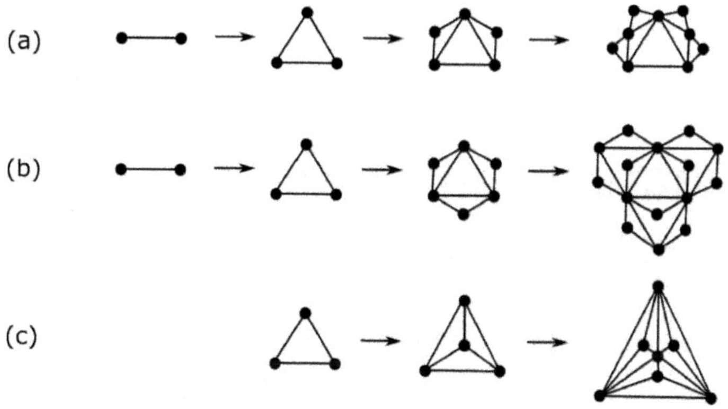

(a)

(b)

(c)

Fig. 5.5 (a) Farey-sequence based graphs, (b) the pseudo-fractals, and (c) Apollonian deterministic graphs.

$\sum_{k \geq q} N(k) \sim q^{-\gamma+1}$ corresponding to the degree distribution with exponent $\gamma = 1 + \ln 3 / \ln 2 = 2.584 \ldots$ In this sense, these two deterministic graphs are 'scale-free'.

In contrast to the infinite Hausdorff dimension of these networks, their spectral dimensions are finite. Bianconi and Dorogovstev (2020) applied the renormalization group techniques of Hwang, Yun, Lee, Kahng, and Kim (2010) to obtain the spectral dimensions [28]

$$D_s = \begin{cases} 2, & \text{Farey graph,} \\ 2\dfrac{\ln 3}{\ln 2} = 3.169 \ldots, & \text{pseudofractal,} \\ 2\dfrac{\ln 3}{\ln 9/5} = 3.738 \ldots, & \text{Apollonian graph.} \end{cases} \tag{5.94}$$

[28]The reader will find the spectral dimensions of a wide set of simplicial complexes of various dimensions in Bianconi and Dorogovstev (2020).

6

Connected Components

6.1 Giant connected component

Let us return to the configuration model of uncorrelated networks with a given degree distribution $P(q)$ and exploit its local tree-likeness to explore basic structural features of these networks. The elements of the techniques that we use here were first implemented in graph theory (Pittel, 1990; Molloy and Reed, 1995, 1998), and physicists developed it into a convenient and powerful mathematical apparatus (Newman, Strogatz, and Watts, 2001) applicable to various locally tree-like networks, including directed, multipartite, and correlated ones, and many others. For the sake of simplicity, we first consider uncorrelated undirected networks. The generating functions techniques is ideally suited for random trees and tree-like structures (Appendix C). The generating function for the degree distribution is defined as

$$G(z) \equiv \sum_{q=0}^{\infty} P(q)z^q, \tag{6.1}$$

hence its first moment $\langle q \rangle = G'(1)$. Then

$$G_1(z) \equiv \frac{G'(z)}{G'(1)} = \sum_{q=0}^{\infty} \frac{qP(q)}{\langle q \rangle} z^{q-1} = \sum_{k=0}^{\infty} \frac{(k+1)P(k+1)}{\langle q \rangle} z^k \tag{6.2}$$

is the generating function of the branching of a randomly chosen edge in a locally tree-like network. In other words, if you choose a vertex uniformly at random and follow one of its edges to the second its end, then the distribution of the number of outgoing edges of the end vertex (degree of the vertex minus 1) has the generating function $G_1(z)$. Thus $G_1(1) = G(1) = 1$, and $G_1'(1)$ equals the average branching,

$$G_1'(1) = \frac{\langle q^2 \rangle - \langle q \rangle}{\langle q \rangle} = \langle b \rangle. \tag{6.3}$$

Using Eqs. (C.5) and (C.6), one can get the generating function of the number of the second nearest neighbours of a vertex

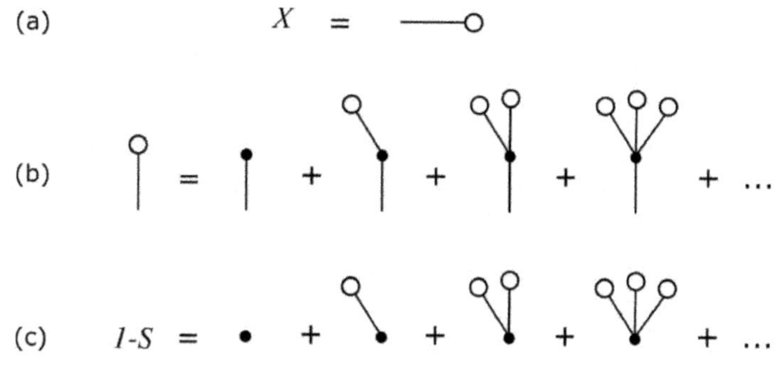

(a) X =

(b)

(c) $1\text{-}S$ =

Fig. 6.1 (a) Graphical notation for the probability X that, following a randomly chosen edge to one of its end vertices, we further reach a finite cluster of vertices. (b) Equation for X, Eq. (6.8), in graphical form. (c) Expression for the probability that a vertex is in one of the finite components, which is equal to $1 - S$, Eq. (6.9), where S is the relative size of the giant connected component.

$$G^{[2]}(z) = \sum_{q} P(q)[G_1(z)]^q = G(G_1(z)), \qquad (6.4)$$

of the third-nearest neighbours

$$G^{[3]}(z) = G(G_1(G_1(z))), \qquad (6.5)$$

and so on, where we used the locally tree-likeness of the network. The average number of the second nearest neighbours of a vertex is

$$z_2 = \left[\frac{d}{dz}G^{[2]}(z)\right]_{z=1} = G'(1)G_1'(1) = \langle q \rangle \frac{\langle q^2 \rangle - \langle q \rangle}{\langle q \rangle} = \langle q \rangle \langle b \rangle, \qquad (6.6)$$

and, in general, of the n-th nearest neighbours is

$$z_n = \langle q \rangle \left(\frac{\langle q^2 \rangle - \langle q \rangle}{\langle q \rangle}\right)^{n-1} = \langle q \rangle \langle b \rangle^{n-1} = \langle q \rangle \left(\frac{z_2}{\langle q \rangle}\right)^{n-1}. \qquad (6.7)$$

When $\langle q^2 \rangle < \infty$, this leads to Eqs. (4.9) and (4.10) for the average shortest path distance, $\langle \ell \rangle \cong \ln N / \ln \langle b \rangle$.

The next step is to use this technique to find the relative size of the giant connected component in the infinite uncorrelated network. The key idea is to introduce the probability X that following an edge chosen uniformly at random to one of its ends we can reach only a finite number of vertices. Note that these vertices are not necessarily within a finite component. Indeed, if we follow the edge in the opposite direction, the number of reachable vertices may be infinite. In other words, X is the probability that if we cut a uniformly randomly chosen edge, then its end vertex is within a finite component. It is convenient to introduce the graphical notation for the probability X, shown in Figure 6.1a. With this notation, one can easily derive the equation for X and the expression, in terms of X, for the probability that a randomly chosen vertex is within a finite component (Figure 6.1b, c). In fact, the graphical forms in this figure explain the following exact equation and expression:[1]

$$X = \sum_q \frac{qP(q)}{\langle q \rangle} X^{q-1},$$ (6.8)

$$1 - S = \sum_q P(q)X^q.$$ (6.9)

Finding the probability X from Eq. (6.8), one can then obtain S and other quantities for this network, for example, the probability that a uniformly randomly chosen edge belongs to a finite component:

$$\text{O}\!\!-\!\!-\!\!\text{O} = X^2.$$ (6.10)

Hence it belongs to the giant connected component with the probability $1-X^2 = 2X(1-X)$, which enables one to easily measure X. With generating functions, Eqs. (6.8) and (6.9) take the compact form:

$$X = G_1(X),$$ (6.11)

$$1 - S = G(X).$$ (6.12)

Notice that only finite components, which are proper trees, were involved in the derivation of Eqs. (6.8) and (6.9) (and in the definition of the probability X). So the numerous infinite cycles in uncorrelated networks were naturally ignored. Nonetheless, one can derive equivalent equations by considering the infinite component and boldly neglecting the infinite cycles. We briefly show this derivation here, since close approaches and notations are applied to a few other problems in later sections. Let us introduce the probability Z that following a uniformly randomly chosen edge to one of its end vertices, we further reach an infinite number of vertices. In other words, this

---

[1]These equations can also be obtained from the equations for the distribution of the number of vertices reachable by following a randomly chosen edge, which we discuss in Section 6.6.

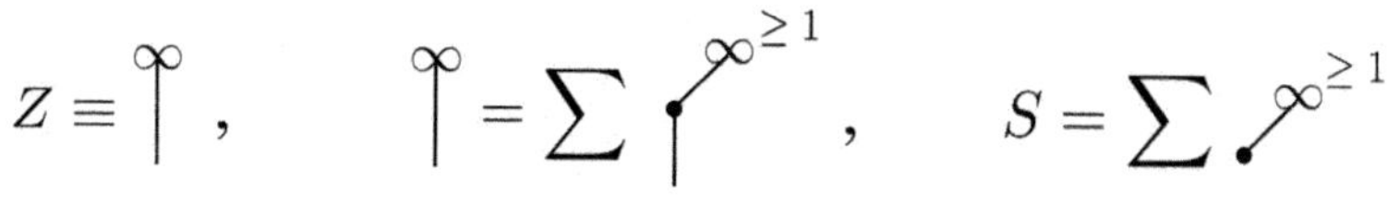

(a)   (b)   (c)

Fig. 6.2 (a) Graphical notation for the probability Z that, following a randomly chosen edge to one of its end vertices, we further reach an infinite set of vertices. (b) Equation for Z, Eq. (6.13), in graphical form. (c) Expression for the relative size of the giant connected component, Eq. (6.14). Here the sums are over degree q and over all configurations with at least one branch leading to infinity. For the sake of compactness, we do not show other branches leading to finite components, whose number is arbitrary.

is the probability that, after we cut the selected edge, this end vertex is the root of an infinite tree. Clearly, $Z = 1 - X$. Introducing the graphical notation for the probability Z shown in Figure 6.2a, one can derive the equation for Z and the expression of S in terms of X explained in Figure 6.2b, c. It is convenient to use the compact graphical representation showing only the infinite branches when the number of finite branches is arbitrary.

The equation for Z and the expression of S in terms of Z shown in graphical form in Figure 6.2b, c can be written as[2]

$$Z = \sum_q \frac{qP(q)}{\langle q \rangle}[1 - (1 - Z)^{q-1}],\tag{6.13}$$

$$S = \sum_q P(q)[1 - (1 - Z)^q]\tag{6.14}$$

and

$$Z = 1 - G_1(1 - Z),\tag{6.15}$$

$$S = 1 - G(1 - Z),\tag{6.16}$$

which are equivalent to Eqs. (6.8), (6.9) and Eqs. (6.11), (6.12), respectively, as is natural, given $Z = 1 - X$.[3] With a Poisson degree distribution, $G(x) =$

[2]Note that

$$1 - (1 - Z)^q = \sum_{k \geq 1} \binom{q}{k} Z^k (1 - Z)^{q-k}.$$

[3]The equivalence of these two sets of equations is not that trivial, as one would think. The point is that the giant connected component of an uncorrelated network has different

$e^{\langle q \rangle (x-1)}$ (see Appendix C), Eqs. (6.11), (6.12) or Eqs. (6.15), (6.16) lead to Eq. (3.30) for the classical random graphs, namely, $S = 1 - e^{-S\langle q \rangle}$.

The degree distribution for the vertices within the giant connected component is

$$P_{\text{Giant}}(q) = \frac{1}{S}P(q)[1 - (1 - Z)^q]. \tag{6.17}$$

In particular, at the critical point, we have

$$P_{\text{Giant}}(q) \xrightarrow{Z \to 0} \frac{qP(q)}{\langle q \rangle}. \tag{6.18}$$

That is, if for the entire network, $P(q) \sim q^{-\gamma}$, then at the critical point, $P_{\text{Giant}}(q) \sim q^{-(\gamma-1)}$. From Eq. (6.17) we can easily find the average degree of vertices within the giant connected component. In the critical region, the fraction of edges belonging to the giant connected component equals $1 - X^2 = 1 - (1 - Z)^2 \cong 2Z$, and $S \cong \langle q \rangle Z$. This immediately leads to the following average degree $\langle q \rangle_{\text{Giant}}$ of a vertex within a giant connected component at the critical point:

$$\langle q \rangle_{\text{Giant}} \xrightarrow{Z \to 0} 2. \tag{6.19}$$

This shows that the giant connected component becomes a tree at the point of its disappearance, which also follows from Eq. (6.18).

Analysing Eq. (6.13) for Z graphically (Figure 6.3), one can see that the phase transition associated with the birth of the giant connected component is continuous, that Z can be treated as the order parameter of this transition, and that the giant connected component is present in the network, when the derivative of the right-hand side of this equation with respect to Z exceeds 1 at $Z = 0$, that is, when

$$\langle q^2 \rangle - \langle q \rangle > \langle q \rangle, \tag{6.20}$$

which is the Molloy–Reed criterion (Molloy and Reed, 1995), which we derived in Section 4.2 by heuristic reasoning. The reader can also easily check that if $P(1) = 0$, then the giant connected component includes all the vertices in the network except isolated vertices (Figure 6.3c).

structure compared to the network itself. It has a different degree distribution, and it is actually correlated (Bialas and Oleś, 2008; Mizutaka and Hasegawa, 2018). These specific disassortative correlations are related to the presence of leaves. The equivalence of these equations demonstrates that, in calculations of this kind, the correlations in the giant component should not be taken into account.

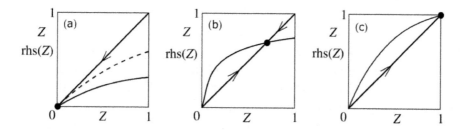

Fig. 6.3 Graphical solution of Eq. (6.13) for Z. For the sake of brevity, we assume that $P(0) = 0$, that is, isolated vertices are absent. (a) $P(1) > 0$. The solid curve shows the right-hand side of the equation when the giant connected component is absent, out of the critical point. The dashed curve shows the right-hand side at the critical point. (b) $P(1) > 0$. The right-hand side of the equation when the giant connected component is present. (c) $P(1) = 0$. The giant connected component includes all vertices. Looking at the arrows, one can grasp which of the fixed points (solutions) is stable.

Substituting the solution of Eq. (6.13) or (6.15) into Eq. (6.14) or (6.16) we obtain the critical singularity of S. In the critical region,

$$S \cong \langle q \rangle Z. \tag{6.21}$$

Consequently, if $\langle q^3 \rangle < \infty$, that is, $\gamma > 4$, then

$$S \cong \frac{2\langle q \rangle \langle q(q-2) \rangle}{\langle q(q-1)(q-2) \rangle}, \tag{6.22}$$

where the deviation from the critical point, $(\langle q^2 \rangle - \langle q \rangle) - \langle q \rangle$, plays the role of the control parameter. Consequently, this phase transition is of the second order. For a Poisson degree distribution, Eq. (6.22) leads to Eq. (3.31) for the classical random graphs, $S \cong 2(\langle q \rangle - 1)$.

Let $\langle q^2 \rangle < \infty$ while $\langle q^3 \rangle$ diverge, that is, $3 < \gamma < 4$ for scale-free networks. Introducing the amplitude A of the asymptotics of the degree distribution $P(q) \cong Aq^{-\gamma}$ and substituting the leading terms of the generating function $G(1 - Z)$ for small Z, Eq. (C.19), into Eq. (6.15), we get

$$S \cong \langle q \rangle \left[\frac{\langle q(q-2) \rangle}{-A(\gamma-1)\Gamma(1-\gamma)} \right]^{1/(\gamma-3)}, \tag{6.23}$$

and so the critical exponent β of the order parameter is

$$\beta = \frac{1}{\gamma - 3}. \tag{6.24}$$

Thus the order of the continuous phase transition increases with decreasing γ, and it becomes an infinite order phase transition when γ reaches 3. Here the deviation of β from the value 1, standard for percolation in sufficiently uniform systems above the upper critical dimension (6 for percolation), actually does not mean any violation of the mean-field theory.[4] Indeed, the exact self-consistent equations shown (for Z or for X) are, in essence, the mean-field theory. The reason for why this exact mean-field theory produces the exponents, non-standard for mean-field theories, is the strong heterogeneity of these networks, that is, the presence of hubs in them.

When $\langle q \rangle < \infty$ while $\langle q^2 \rangle$ diverges, that is, $2 < \gamma < 3$ for scale-free networks, proceeding in a similar way, we get the following singularity of S:

$$S \cong \langle q \rangle \left[\frac{\langle q \rangle}{A(\gamma - 1)\Gamma(1 - \gamma)} \right]^{1/(3-\gamma)}. \tag{6.25}$$

Thus, remarkably, the giant connected component is present for any small average degree $\langle q \rangle$ of a network having infinite $\langle q^2 \rangle$, and the transition actually disappears. Clearly, this also occurs due to the presence of strongly connected hubs in this range of networks.

6.2 Site and bond percolation

Using the terms of condensed matter theory, there are two percolation problems, site and bond percolation, concerned with removal of vertices (sites) and edges (bonds), respectively. Although these problems belong to the same universality class of critical behaviour, the difference is notable in some networks. For bond percolation, one can easily obtain the degree distribution of a network after the removal of a fraction $1 - p$ of vertices (see Appendix E). In particular, if the degree distribution of the undamaged network is power law, $P(q) \cong A q^{-\gamma}$, then the tail of the degree distribution of the damaged network has the same exponent but the amplitude is different, $P_p(q) \cong A p^{\gamma-1} q^{-\gamma}$. Substituting this distribution into the formulas in Section 6.1 leads to the following set of singularities of S:

---

[4]In the standard theory of percolation, $\beta = 1$ when the dimension of a lattice $D \geq D_{uc} = 6$, and $0 < \beta < 1$ when $D_{lc} = 1 < D < D_{uc}$. Here D_{uc} and D_{lc} are the upper critical and lower critical dimensions, respectively.

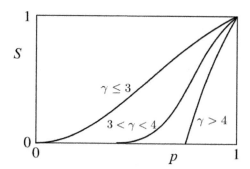

Fig. 6.4 Relative size of a giant connected component versus the fraction of vertices retained after random damage. The three curves show the typical dependences at various values of the exponent of a power-law degree distribution.

$$
S_{bond} \sim
\begin{cases}
p - p_c & \gamma > 4, \\
\dfrac{p - p_c}{\ln[1/(p - p_c)]} & \gamma = 4, \\
(p - p_c)^{1/(\gamma - 3)} & 3 < \gamma < 4, \\
e^{-2/(p\langle q \rangle)} & \gamma = 3, \\
p^{1/(3-\gamma)} & 2 < \gamma < 3
\end{cases}
\tag{6.26}
$$

(Cohen, Erez, Ben-Avraham, and Havlin, 2000; Cohen, Ben-Avraham, and Havlin, 2002), hence $p_c = 0$ if $2 < \gamma \leq 3$ (Figure 6.4). This shows that the uncorrelated networks with $\gamma \leq 3$, that is with diverging $\langle q^2 \rangle$, are hyper-resilient against random damage. The giant connected component is still present in a network for any small (though finite) fraction p of undeleted edges.

One can approach these percolation problems in a different way (Callaway, Newman, Strogatz, and Watts, 2000) and straightforwardly derive equations similar to Eqs. (6.8), (6.9) or (6.13), (6.14) from Section 6.1 for site percolation,

$$
Z = p \sum_q \frac{qP(q)}{\langle q \rangle} [1 - (1 - Z)^{q-1}],
\tag{6.27}
$$

$$
S_{site} = p \sum_q P(q)[1 - (1 - Z)^q],
\tag{6.28}
$$

and for bond percolation,

$$Z = \sum_q \frac{qP(q)}{\langle q \rangle}[1 - (1 - pZ)^{q-1}], \tag{6.29}$$

$$S_{\text{bond}} = \sum_q P(q)[1 - (1 - pZ)^q]. \tag{6.30}$$

Introducing $\tilde{Z} = pZ$ for the bond percolation problem, we get

$$\tilde{Z} = p\sum_q \frac{qP(q)}{\langle q \rangle}[1 - (1 - \tilde{Z})^{q-1}], \tag{6.31}$$

$$S_{\text{bond}} = \sum_q P(q)[1 - (1 - \tilde{Z})^q], \tag{6.32}$$

which leads to the relation[5]

$$S_{\text{site}} = pS_{\text{bond}}, \tag{6.33}$$

and to the percolation threshold

$$p_c = \frac{\langle q \rangle}{\langle q(q - 1) \rangle} \tag{6.34}$$

for both site and bond percolation. So for site percolation, we have

$$S_{\text{site}} \sim \begin{cases} p - p_c & \gamma > 4, \\ \dfrac{p - p_c}{\ln[1/(p - p_c)]} & \gamma = 4, \\ (p - p_c)^{1/(\gamma-3)} & 3 < \gamma < 4, \\ pe^{-2/(p\langle q \rangle)} & \gamma = 3, \\ p^{1+1/(3-\gamma)} & 2 < \gamma < 3. \end{cases} \tag{6.35}$$

Instead of uniform removal of vertices, one can apply intentional damage to networks, that is, retain vertices with probability $p(q)$ depending on their degrees (Callaway, Newman, Strogatz, and Watts, 2000) This modifies Eqs. (6.27) and (6.28) in the following way:

$$Z = \sum_q p(q)\frac{qP(q)}{\langle q \rangle}[1 - (1 - Z)^{q-1}], \tag{6.36}$$

$$S_{\text{site}} = \sum_q p(q)P(q)[1 - (1 - Z)^q]. \tag{6.37}$$

---

[5] This simple relation can be violated in more difficult situations, including multiplex networks (Baxter, da Costa, Dorogovtsev, and Mendes, 2020).

Linearizing Eq. (6.36), one can easily get the condition

$$\langle p(q)q(q-1)\rangle > \langle q\rangle \tag{6.38}$$

that the network has a giant connected component, which is a version of the Molloy–Reed criterion for this problem. Analysing Eqs. (6.36) and (6.37) or Eq. (6.38), one can see that the removal of a fraction of vertices with largest degrees is an efficient method for eliminating the giant connected component.

Knowing the exponent β of the size of the giant connected component, we can immediately obtain the Hausdorff dimension D_H of the giant component at the point of its birth by using the relation $D_H = 1 + \beta$ taken from percolation theory (Appendix D),

$$D_H = \begin{cases} 2 & \gamma \ge 4, \\ \dfrac{\gamma - 2}{\gamma - 3} & 3 < \gamma \le 4. \end{cases} \tag{6.39}$$

As we explained, at the critical point, the giant connected component turns out to be an equilibrium tree with a degree distribution having exponent $\gamma - 1$. Hence from Eq. (6.39) we readily get the Hausdorff dimension of equilibrium scale-free trees with exponent γ, namely, (i) $D_H = 2$, if $\gamma \ge 3$, and (ii) $D_H = (\gamma-1)/(\gamma-2)$ if $2 < \gamma \le 3$, which we have already mentioned in Section 1.5.

6.3 Message passing

The technique described in the two preceding sections is applicable to locally tree-like networks. We use it extensively in this book. There exists a more powerful approach applicable also to a range of networks with cycles and even to finite networks, namely *message-passing algorithms* (Mezard and Montanari, 2009). In these algorithms, a given graph is approximated by an infinite tree, and hence the message-passing techniques provide exact results for infinite locally tree-like random networks. In particular, for uncorrelated networks, they lead to the same exact equations as in the preceding sections. We first consider the version of the message passing allowing such reduction most directly and focus on the finding of a giant connected component.

Let the problem be site percolation on a given (non-random) undirected graph. Each vertex i has two states: $v_i = 0$ (vertex i is removed) and $v_i = 1$ (vertex i is present). For each edge (ij), there are two messages: the message $\sigma_{i \to j}$ from vertex i to j and, in the opposite direction, the message $\sigma_{j \to i}$ from

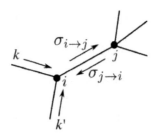

Fig. 6.5 Messages through edge ij and the nighbouring edges. Messages $\sigma_{k\to i}$ and $\sigma_{k'\to i}$ determine the value of message $\sigma_{i\to j}$, Eq. (6.40). Messages $\sigma_{k\to i}$, $\sigma_{k'\to i}$, and $\sigma_{j\to i}$ determine the value of σ_i, Eq. (6.41).

vertex j to i (Figure 6.5).[6] The messages can take the values 0 and 1. For example, in the message $\sigma_{i\to j}$, vertex i tells vertex j the following:

- if $\sigma_{i\to j} = 0$, then the word is 'I know nothing about you, but you are certainly not connected to an infinite cluster through me'.
- if $\sigma_{i\to j} = 1$, then the word is 'I know nothing about you, but you are certainly connected to an infinite cluster through me'.

Let us introduce the number $\sigma_i = 0, 1$ for each vertex i: $\sigma_i = 0$ if vertex i does not belong to the giant connected component and $\sigma_i = 1$ if vertex i belongs to the giant connected component. Assume that no nearest neighbours of vertices i and j coincide with each other, as in Figure 6.5. Then we immediately arrive at the equation for the messages and at the expression for σ_i in terms of messages:

$$\sigma_{i\to j} = v_i \left[1 - \prod_{k\in\partial i\backslash j} (1 - \sigma_{k\to i}) \right], \tag{6.40}$$

$$\sigma_i = v_i \left[1 - \prod_{k\in\partial i} (1 - \sigma_{k\to i}) \right]. \tag{6.41}$$

Here $\partial i\backslash j$ denotes the set of the nearest neighbours of vertex i excluding vertex j. Due to our strong assumption (that no nearest neighbours of vertices i and j coincide with each other), these equations, in general, are approximate. Furthermore, they turn out to be approximate even if the graph has longer, non-infinite cycles. If the damage is uniformly random, with a fraction p of retained vertices, then one can average these equations over

---

[6]Note that in these notations, $i \to j \equiv j \leftarrow i$ and $\sigma_{i\to j} \equiv \sigma_{j\leftarrow i}$.

all configurations $\{v_i\}$, $i = 1, 2, \ldots, N$, of the damage with the probability $P(\{v_i\}) = \prod_i p^{v_i}(1 - p)^{1-v_i}$, which gives

$$\langle\sigma\rangle_{i\to j} = p\left[1 - \prod_{k\in\partial i\backslash j}(1 - \langle\sigma\rangle_{k\to i})\right], \qquad (6.42)$$

$$\langle\sigma\rangle_i = p\left[1 - \prod_{k\in\partial i}(1 - \langle\sigma\rangle_{k\to i})\right]. \qquad (6.43)$$

Here the average values $\langle\sigma\rangle_{i\to j}$ and $\langle\sigma\rangle_i$ are real numbers, $0 \leq \langle\sigma\rangle_{i\to j}, \langle\sigma\rangle_i \leq 1$. For a given graph, start with an arbitrary configuration of $2E$ average messages and iterate Eq. (6.42) to convergence. Substitute the $2E$ resulting average messages into Eq. (6.43), from which get the relative size of a giant connected component $S_{\text{site}} = \sum_{i=1}^{N}\langle\sigma\rangle_i/N$. In essence, the message-passing algorithm iteratively updates the messages along the effectively directed edges to convergence.

For the bond percolation problem, the message-passing equations are organized similarly. Let $e_{ij} = 0, 1$ when edge (ij) is unoccupied or occupied, respectively. Then

$$\sigma_{i\to j} = \left[1 - \prod_{k\in\partial i\backslash j}(1 - e_{ki}\sigma_{k\to i})\right], \qquad (6.44)$$

$$\sigma_i = \left[1 - \prod_{k\in\partial i}(1 - e_{ki}\sigma_{k\to i})\right] \qquad (6.45)$$

and, for the average values,

$$\langle\sigma\rangle_{i\to j} = \left[1 - \prod_{k\in\partial i\backslash j}(1 - p\langle\sigma\rangle_{k\to i})\right], \qquad (6.46)$$

$$\langle\sigma\rangle_i = \left[1 - \prod_{k\in\partial i}(1 - p\langle\sigma\rangle_{k\to i})\right], \qquad (6.47)$$

leading to Eqs. (6.29) and (6.30) when the network is uncorrelated.

If the network is random and uncorrelated, then averaging $\langle\sigma\rangle_{i\to j}$ over the network ensemble gives exactly the probability Z introduced in the preceding sections, and averaging $\langle\sigma\rangle_i$ gives S_{site}. Averaging Eqs. (6.42) and (6.43) leads to exact Eqs. (6.27) and (6.28), respectively, from Section 6.2.

Thus the message-passing algorithm is indeed exact for infinite locally tree-like networks.

Let us look at, for example, the bond percolation problem from a slightly different angle, following Karrer, Newman, and Zdeborová (2014). We again consider a (non-random) graph, from which we delete a fraction $1 - p$ of uniformly randomly chosen edges. For each vertex i, let $\pi_i(s)$ be the probability that vertex i is within a finite component of s vertices averaged over all configurations of deleted vertices. For each edge (ij), let $\pi_{j \leftarrow i}(s)$ and $\pi_{i \leftarrow j}(s)$ be two messages, where $\pi_{j \leftarrow i}(s)$ is the probability that, following the edge in the direction $j \to i$, we can reach s vertices (excluding vertex j), and $\pi_{i \leftarrow j}(s)$ is the probability that, following the edge in the direction $i \to j$, we can reach s vertices (excluding vertex i). Again assuming that no nearest neighbours of vertices i and j coincide with each other, one can write the equations for, for example, $\pi_{i \leftarrow j}(s)$:

$$\pi_{i \leftarrow j}(0) = 1 - p,$$

$$\pi_{i \leftarrow j}(s \geq 1) = p \sum_{s_k:\, k \in \partial j \setminus i} \left[\prod_{k \in \partial j \setminus i} \pi_{j \leftarrow k}(s_k) \right] \delta\left(s - 1, \sum_{k \in \partial j \setminus i} s_k \right). \quad (6.48)$$

Similarly, the expression for $\pi_i(s)$ in terms of $\pi_{i \leftarrow j}(s)$ has the form

$$\pi_i(s) = \sum_{s_k:\, k \in \partial i} \left[\prod_{k \in \partial i} \pi_{i \leftarrow k}(s_k) \right] \delta\left(s - 1, \sum_{k \in \partial i} s_k \right). \quad (6.49)$$

With the generating functions

$$G_i(z) = \sum_{s=1}^{\infty} \pi_i(s) z^s, \quad (6.50)$$

$$H_{i \leftarrow j}(z) = \sum_{s=1}^{\infty} \pi_{i \leftarrow j}(s) z^s, \quad (6.51)$$

Eqs. (6.48) and (6.49) take the compact form

$$H_{i \leftarrow j}(z) = 1 - p + pz \prod_{k \in \partial j \setminus i} H_{j \leftarrow k}(z), \quad (6.52)$$

$$G_i(z) = \prod_{k \in \partial i} H_{i \leftarrow k}(z), \quad (6.53)$$

respectively. The set of $2E$ self-consistent equations, Eq. (6.52), is used as the update equations for the message passing algorithm.[7] For each value of z, iterating Eq. (6.52), starting with a random initial set, to convergence and substituting the resulting generating functions $H_{i \leftarrow j}(z)$ ($2E$ functions of z) into Eq. (6.53) enables one to find the full statistics $\pi_i(s)$ for all vertices. The value $z = 1$ is of particular interest, since $G_i(1)$ is the probability that vertex i belongs to a finite component, and so $S_{\text{bond}} = \sum_i [1 - G_i(1)]/N$. Furthermore, $H_{i \leftarrow j}(1)$ is the probability that, following the edge in the direction $i \rightarrow j$, we can reach only a finite number of vertices. For the value $z = 1$, Eqs. (6.52) and (6.53) take the form

$$H_{i \leftarrow j}(1) = 1 - p + p \prod_{k \in \partial j \backslash i} H_{j \leftarrow k}(1), \tag{6.54}$$

$$S_{\text{bond}} = \frac{1}{N} \sum_i \left[1 - \prod_{k \in \partial i} H_{i \leftarrow k}(1) \right]. \tag{6.55}$$

The reader can see that Eqs. (6.54) and (6.55) are equivalent to the equations for the bond percolation problem derived earlier, Eqs. (6.46) and (6.47), respectively, where $1 - H_{i \leftarrow j}(1) = \sigma_{i \leftarrow j} \equiv \sigma_{j \rightarrow i}$ and $1 - G_i(1) = \sigma_i$.

6.3.1   Percolation threshold

To find the percolation threshold p_c, one can inspect the linearized Eq. (6.54) (or Eq. (6.46)) for small deviations $u_{i \leftarrow j} = 1 - H_{i \leftarrow j}(1)$,

$$u_{i \leftarrow j} = p \sum_{k \in \partial j \backslash i} u_{j \leftarrow k}. \tag{6.56}$$

Introducing the $2E \times 2E$ *non-backtracking matrix* B, also called the Hashimoto matrix (Hashimoto, 1989) (see Appendix F) with the elements[8]

---

[7]Here we actually have a special case of message-passing algorithms, *belief propagation*, finding conditional probability distributions for edges and vertices.

[8]In other words,

$$B_{i \leftarrow j, k \leftarrow l} = \begin{cases} 1 \text{ if } j = k \text{ and } i \neq l, \\ 0 \text{ otherwise.} \end{cases}$$

The non-zero entry of the non-backtracking matrix indicates that there is an intermediate vertex between two different vertices. In this matrix, each edge in an undirected network is treated as a pair of directed edges $i \leftarrow j$ and $j \leftarrow i$. A *non-backtracking walk*, by definition, has no returns by one step back. The number of the non-backtracking walks of length ℓ from the edge $k \leftarrow l$ to the edge $i \leftarrow j$ is

$$B_{i \leftarrow j, k \leftarrow l} = \delta_{j,k}(1 - \delta_{i,l}), \tag{6.57}$$

enables one to represent Eq. (6.56) in the matrix form:

$$\mathbf{u} = p B \mathbf{u}, \tag{6.58}$$

where $\mathbf{u}$ is a $2E$-vector with entries $u_{i \leftarrow j}$. Then the message-passing algorithm provides the percolation threshold value

$$p_c = \frac{1}{\lambda_1}, \tag{6.59}$$

where λ_1 is the largest eigenvalue of the non-backtracking matrix.[9] We can go further and express the slope C of the dependence $S \cong C(p - p_c)$ at the critical point in terms of the components of the principal eigenvector $\mathbf{u}_1$ of the non-backtracking matrix (Timár, da Costa, Dorogovtsev, and Mendes, 2017a).

6.3.2 Non-backtracking expansion

A percolation phase transition can occur only in infinite networks. On the other hand, the non-backtracking matrices of finite graphs typically have $\lambda_1 > 1$ leading to an impossible value $p_c < 1$ if we believe Eq. (6.58), which demonstrates the approximate nature of this equation. The point is that the message-passing algorithm, neglecting finite cycles, actually unfolds an arbitrary finite (non-tree) graph into an infinite deterministic tree for which it provides the exact value of the percolation threshold.[10] Figure 6.6 explains

$$n_{\text{non-backtracking walks}}(i \leftarrow j, k \leftarrow l; \ell) = (B^{\ell-1})_{i \leftarrow j, k \leftarrow l}.$$

Note that the length $\ell \geq 2$ includes the source and target edges. In particular, for a tree graph, $\text{Tr} B^n = \sum_i \lambda_i^n = 0$ for any power $n \geq 1$, since closed non-backtracking walks are impossible in it. Hence all eigenvalues of the non-backtracking matrix are zero in a tree.

B is a non-symmetric non-negative matrix. According to the Perron–Frobenius theorem, its principal eigenvalue λ_1 is real and positive, and all elements of the corresponding eigenvector have the same sign, if B is irreducible, which is usually the case. For example, for a random q-regular graph, the non-backtracking matrix has $q - 1$ non-zero entries in each row and column. Consequently, for this network,

$$\lambda_1 = q - 1.$$

[9]Since the largest eigenvalue of the non-backtracking matrix of a tree graph equals zero, $\lambda_1 = 0$, the message passing approximation, Eq. (6.59) gives divergent p_c. This hints at the absence of the percolation transition in trees, which is indeed the case.

[10]This construction is also known in the computer science community as a computation tree (Weiss and Freeman, 2001). We call it the non-backtracking expansion of a finite graph.

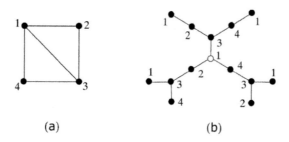

(a)   (b)

Fig. 6.6 Non-backtracking expansion of the finite graph (a) into the infinite deterministic tree (b). Here the expansion starts from vertex 0 although it could start from any vertex or edge. Only first three layers of the resulting tree are shown. Adapted from Timár, da Costa, Dorogovtsev, and Mendes (2017a).

this unfolding—the non-backtracking expansion of a graph. Clearly, the non-backtracking expansion of a tree coincides with it. The branching b of an infinite tree is defined as the limiting

$$b \equiv \lim_{\ell \to \infty} \frac{E(\ell+1)}{E(\ell)}, \qquad (6.60)$$

where $E(\ell)$ is the number of edges between vertices at distances ℓ and $\ell+1$ from an arbitrary vertex.[11] Then the branching of the non-backtracking expansion of a graph is equal to the largest eigenvalue λ_1 of the non-backtracking matrix (see Appendix F).

Figure 6.7 shows the second way to construct an infinite counterpart of a finite graph, proposed by Faqeeh, Melnik, and Gleeson (2015), for which this message-passing algorithm gives the exact result. Consider m clones of a labelled finite graph G of N vertices and E edges, cut each edge (ij) in all clones into two stubs i and j, and rejoin pairs of stubs from all clones uniformly at random in such a way that stub i in one clone can only join stub j in any clone. This generates a microcanonical ensemble, a random graph in spirit of the configuration model, whose members have equal statistical weights. Finally, tend m to infinity, which guarantees the local tree-likeness and the exactness of the message-passing algorithm.

6.3.3 Using message passing

Message-passing, in particular, belief-propagation, algorithms are widely used for a rapidly expanding range of problems: treatment of various probabilistic graphical models, spreading processes including disease spreading

---

[11]We consider only those trees for which this limit is independent of a chosen centre.

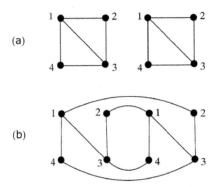

Fig. 6.7 Construction of a locally tree-like graph by uniform, label-conserving reconnection of edges between m copies of a graph ($m \to \infty$), following Faqeeh, Melnik, and Gleeson (2015). (a) Two copies of a sample graph ($m = 2$ for demonstration purposes). (b) One of configurations entering into the resulting statistical ensemble with the same statistical weight as all others.

(Altarelli, Braunstein, Dall'Asta, Lage-Castellanos, and Zecchina, 2014a; Altarelli, Braunstein, Dall'Asta, Wakeling, and Zecchina, 2014b), cooperative systems—Bethe–Peierls approximation, cavity method, and others. One may ask when these techniques are computationally more efficient than direct simulations and when they are less efficient. For percolation problems, the update equations of the message-passing algorithms contain p, and so these algorithms enable one to escape averaging over different configurations of removed vertices or edges, in contrast to simulations, saving time. This is a significant advantage. Let us, however, estimate the number of operations required to solve the message-passing equations by iterations (computational complexity). Roughly, this number is of the order of $N \langle q \rangle \times \langle \ell \rangle (N) \sim N \ln N$ (see, e.g., Timár, da Costa, Dorogovtsev, and Mendes, 2017a for details). On the other hand, to determine in simulations whether a vertex belongs to the giant (actually, the largest) connected component or not, we must explore the neighbourhood of this vertex. For that, we must visit a number of vertices exceeding the typical size of a finite connected component by a factor determined by the precision with which we want to find the size of the giant component. The typical cluster size depends on the distance from the critical point but not on the system size. Consequently, at a given distance from the critical point, for a given desired accuracy, the size of the giant component can be obtained in simulations in a constant time independent of N. Thus, at least for large network sizes, direct simulations are more

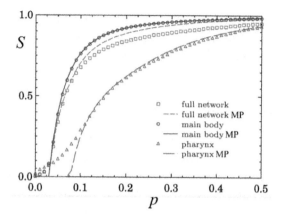

Fig. 6.8 Simulation results and message-passing (MP) solutions for the size of the giant connected component for the neural network of the hermaphrodite *C. elegans*, combining chemical and electrical synapses, in the bond percolation problem. Self-loops are removed from the network, and multiple edges are reduced to single ones. The network is treated as undirected. Message passing provides a rather bad approximation for the network as a whole, but a surprisingly good one for the modules separately. Adapted from Timár, da Costa, Dorogovtsev, and Mendes (2017*a*).

computationally effective than message-passing algorithms.

Message-passing algorithms may fail in two situations: in networks with numerous short cycles and in highly modular networks. The first example is a percolation problem for a finite dimensional lattice, in particular, the square lattice. Clearly, message passage provides the same result for this regular lattice as for the infinite random 4-regular graph, which significantly differs from the correct one. The percolation threshold for the square lattice is markedly higher than for the random 4-regular graph.

The second example, namely the neural network of the roundworm *C. elegans*, demonstrates the pitfalls and the power of the message-passing approximation. This small neural network of 448 vertices contains two obvious communities: of the pharynx, 51 vertices, and of the main body, 397 vertices, of the roundworm, separated by only two edges. Figure 6.8 compares the bond percolation simulation results and the corresponding message-passing solutions for the whole network and also for the two subnetworks. The message passing approximation is not particularly good for the whole network (see the points sufficiently far from the 'percolation threshold'), but it is notably accurate for the two subnetworks, given their small sizes and high clustering coefficients ($C = 0.47$ for the pharynx and $C = 0.26$ for the main body). This demonstrates that modularity can be an essentially more important factor in determining the accuracy of this approximation than short cycles.

6.4 Beyond tree-likeness

Section 6.3 explained that the application of the message passing algorithm to a finite graph can be treated as replacing of the original loopy graph by its infinite non-backtracking expansion (Figure 6.6b). This figure allows us to grasp the major drawback of this approximation falsely giving the infinite diameter for any finite graph if it is only not a tree. Due to neglecting finite cycles, the non-backtracking expansion contains numerous (actually infinite number) replicas of the vertices of the original loopy graph. Hence the update equations of message-passing algorithms, in particular, Eq. (6.48), overestimate sizes of finite sets of vertices to which edges lead. Karrer, Newman, and Zdeborová (2014) showed that this overestimation leads to the following inequality valid for any network:

$$p_c^{(\text{exact})} \geq \frac{1}{\lambda_1}, \qquad (6.61)$$

where $p_c^{(\text{exact})}$ is the exact value of the percolation threshold for a network, and λ_1 is the largest eigenvalue of the non-backtracking matrix.[12] In other words, the message-passing approximation provides the lower bound for the percolation threshold.[13]

The problem of finite cycles is actually one of the major and difficult issues in complex networks. A number of approaches exist, and here we discuss only two of them. The first (Yoon, Goltsev, Dorogovtsev, and Mendes, 2011) exploits a generalization of the random network with triangles proposed by Newman (2009) (see also Karrer and Newman (2010b), discussed in Section 4.9). This is a maximally random graph with a given sequence of motifs, constructed in spirit of the configuration model. Consider a set V_1 of vertices and a given set $\mathcal{M}$ of motifs (finite induced subgraphs) from which we want to construct our network.[14] If some of these motifs are not symmetric, like the graph G in Figure 6.6a, whose vertices 1 or 3 cannot

---

[12]See Radicchi (2015) for the empirical evidence of this inequality.

[13]The exception is a tree graph for which $1/\lambda_1 = \infty$.

[14]Recall the definition of an induced subgraph (Figure 2.5).

Fig. 6.9 Motifs to which vertex i belongs in a random graph with a given sequence of motifs.

be permutated with vertices 2 or 4 without changing the structure of connections, then keep in $\mathcal{M}$ the necessary number of copies of such motifs, one copy for each non-equivalent vertex. For example, keep two copies of motif G. Furthermore, some motifs can have special vertices marked as not belonging to V_1. Let each vertex from the set V_1 belong to its given multiset of motifs taken from $\mathcal{M}$, so we have the sequence of multisets of motifs playing the role similar to the sequence of degrees in the standard configuration model. Figure 6.9 shows an example of such multiset—a member of this sequence. The uniform ensemble of all possible configurations of these motifs interconnecting the vertices of set V_1, taken with equal statistical weights, is just our random network.[15] Note that the full set V of vertices in this network may occur to be wider than V_1, like in Figure 6.9.

Similarly to the configuration model, when this random network is infinite and sparse, it has a locally tree-like structure, apart from the cycles within motifs. So it has infinite cycles and the finite cycles of the motifs. Due to this convenient structure, the percolation problems and the Ising model on the top of this network were solved exactly in the framework of the message-passing algorithm.[16]

Another, instructive way of accounting for the effect of cycles is to consider an arbitrary graph and repair the update equations of the message passing algorithm by removing at least some of the replica vertices from the infinite non-backtracking expansion tree (Cantwell and Newman, 2019). This approximation accounts only for short cycles, say of length 3 and 4,

---

[15]This model is also applicable to hypergraphs. Indeed, the motifs can be treated as hyper-edges of various types (colours) to distinguish different motifs with the same number of vertices from each other.

[16]A special case of such networks, having only disjoint 3-cycles (triangles), was solved by Serrano and Boguñá (2006b). In these networks, the local clustering satisfies the condition $C(q) < 1/(q-1)$ (Section 4.7).

considering close neighbourhoods of edges and vertices, which is already sufficient for the precise description of percolation in numerous networks. Recall that the update equations of the standard message-passing algorithm, Eq. (6.52), were for the generating function of the probability $\pi_{i \leftarrow j}(s)$ that vertex j belongs to a finite component of s vertices once edge $(i \leftarrow j)$ is removed. To cope with broader neighbourhoods of vertices, this probability has to be generalized. Namely, let $\pi_{i \leftarrow j}(s)$ be the probability that vertex j belongs to a finite component of s vertices once the edges of a given neighbourhood of vertex i are removed. Then the update equation, Eq. (6.52), should be generalized to the following form, here shown for $z = 1$:

$$H_{i \leftarrow j}(1) = G_{i \leftarrow j}(\{H_{j \leftarrow k}(1)\}, p), \quad \text{where each } k \in \mathcal{N}(j) \backslash \mathcal{N}(i), \quad (6.62)$$

that is, the full set of vertices k in the list $\{H_{j \leftarrow k}(1)\}$ consists of all vertices from the neighbourhood of vertex j excluding the neighbourhood of vertex i. By $\mathcal{N}(i)$, we denote the full set of vertices in a given neighbourhood of vertex i, including the vertex itself. In the bond percolation problem, where each edge is occupied with a probability p, the function $G_{i \leftarrow j}(\cdot, p)$ is the average of the paths between vertices within $\mathcal{N}(j) \backslash \mathcal{N}(i)$, accounting for all occupancy configurations of the edges within this neighbourhood. For a more serious discussion of this function, see the original paper by Cantwell and Newman. The difficulty is that typically, for larger neighbourhoods than the closest one, the $G_{i \leftarrow j}$ function can be found only numerically, by approximating the average by Monte Carlo sampling of configurations. Once the full set of these $2E$ functions is obtained, the iteration of the update equations, Eq. (6.62), is straightforward.

After the readers learn about these refined approaches to networks with cycles, they should wonder what the effect of short cycles, say, triangles is, on the percolation properties of networks, which such techniques can reveal. Colomer-de-Simón and Boguñá (2014) explored bond percolation in networks, constrained in the following way. They were maximally random scale-free networks with a fixed average local clustering for a wide range of degrees. For networks of this sort with the degree distribution exponent $\gamma = 3.1$, Figure 6.10 shows the results of simulations: the dependence of the relative size S of a giant connected component on the concentration p of occupied edges for different values of the average local clustering. Notice that for high values of clustering, the curves $S(p)$ show a trace of what looks like a second phase transition located markedly above the real percolation threshold. With increasing clustering, the real percolation threshold p_c decreases, while the second threshold p_{c2} rises. The reason for observing these

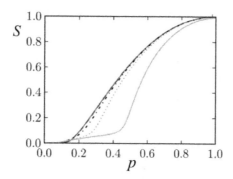

Fig. 6.10 Relative size S of a giant connected component vs. the fraction p of retained edges in the bond percolation problem for a maximally random scale-free network with a given average local clustering. The network size is 50,000 vertices, the degree distribution exponent $\gamma = 3.1$. The curves from top to bottom are for the average local clustering $\langle C \rangle = 0.001, 0.005, 0.03, 0.1$, and 0.25. Adapted from Colomer-de-Simón and Boguñá (2014).

two thresholds can be figured out if we recall the condensation of triangles in the Strauss model discussed in Section 4.11. Similarly to the Strauss model, triangles in the highly clustered networks of Colomer-de-Simón and Boguñá tend to form a core with a structure so markedly different from the 'periphery' that the core and periphery 'percolate' at different points, p_c and p_{c2}, respectively.[17] Although real percolation occurs at p_c, the second threshold turns out to be surprisingly sharp, which indicates that this 'core' and the periphery are very weakly interconnected. See somewhat more theoretical considerations of the influence of triangles on percolation in Gleeson, Melnik, and Hackett (2010).

Thus, researchers manage to handle infinite cycles in networks and can treat short cycles, but the treatment of intermediate cycles, say of length 50, is still a challenge. In fact, it is not clear when such cycles could significantly influence the percolation properties. Taking into account long (though not infinite) cycles is necessary for finding the precise statistics of shortest path lengths in finite networks. For finite uncorrelated networks this problem was solved by Dorogovtsev, Mendes, and Samukhin (2003b).

---

[17]This 'm-core' can be uncovered by a specific pruning algorithm progressively deleting edges with low edge multiplicity m_{ij} (number of triangles to which this edge belongs) (Section 2.5).

6.5 Bow-tie structure of directed networks

In general, directed networks have a richer set of giant connected components than undirected networks. A giant connected component obtained ignoring the directedness of edges is a *giant weakly connected component*. There is a *giant strongly connected component*, which consists of the vertices mutually reachable by directed paths. The vertices of *giant out-component* are reachable from the strongly connected component by directed paths. A *giant in-component* contains all the vertices from which the strongly connected component is reachable. By this definition, the giant strongly connected component is the intersection of the giant in- and out-components, although in the literature the giant strongly connected component is often not included in the giant in- and out-components. Apparently, the presence of a giant strongly connected component is vitally important for the function of directed networks. Broder, Kumar, Maghoul, Raghavan, Rajagopalan, Stata, Tomkins, and Wiener (2000) coined the term 'bow-tie diagram' for this set of giant components, see Figure 6.11a.[18] One can see that, in addition to the giant in-,out, and strongly connected components, the weakly connected component contains tendrils explained in this figure. The full set of tubes and tendrils turned out to be a more complicated than in the original bow-tie diagram (Figure 6.11b), as Timár, Goltsev, Dorogovtsev, and Mendes (2017b) found. Broder, *et al.* measured the sizes of all these components using the map of a sufficiently large part of the WWW (about 200 million pages). They found that the giant strongly connected component contained about 30% of the pages in the weakly connected component, while all tendrils contain about 25%.[19]

Let us denote the relative sizes of giant weak, in-, out-, and strongly connected components by S_W, S_{IN}, S_{OUT}, and S_{SCC}, respectively, and compute these numbers for uncorrelated networks with a given in-, out-degree distribution $P(q_i, q_o)$. Clearly, $\langle q_i \rangle = \langle q_i \rangle$. For the sake of concreteness, here we consider the site percolation problem and use the technique similar to that in Sections 6.1 and 6.2. We follow Dorogovtsev, Mendes, and Samukhin (2001b) and introduce (i) the probability X that following a uniformly randomly chosen edge we reach a finite number of vertices, and (ii) the proba-

---

[18]Similarly, one can define in-, out-, and strongly connected component for each individual vertex, and they can be giant or finite.

[19]The average length of the shortest direct path between two web pages in these measurements was about 16. Remarkably, the maximum separation of nodes observed in this network was very large, namely, about 1000 clicks!

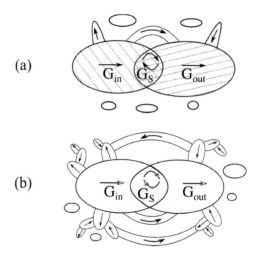

Fig. 6.11 (a) Organization of giant connected components in directed networks as Broder, Kumar, Maghoul, Raghavan, Rajagopalan, Stata, Tomkins, and Wiener (2000) presented it. (b) The full hierarchy of the tubes and tendrils. Adapted from Timár, Goltsev, Dorogovtsev, and Mendes (2017*b*).

bility Y that choosing an edge at random and moving against its direction we reach a finite number of vertices. Then, in a way similar to Sections 6.1 and 6.2, we derive the following equations for X and Y:[20]

$$X = 1 - p + p \sum_{q_i, q_o} \frac{q_o P(q_i, q_o)}{\langle q_o \rangle} X^{q_i},$$

$$Y = 1 - p + p \sum_{q_i, q_o} \frac{q_i P(q_i, q_o)}{\langle q_i \rangle} Y^{q_o}. \tag{6.63}$$

Here we consider networks only with directed edges. Boguñá and Serrano (2005) explored the more cumbersome, though qualitatively similar, case of uncorrelated networks having both directed and undirected edges.[21] Linearizing these equations we find that giant in-, out-, and strongly connected components exist in the region

---

[20]For the message-passing update algorithm for directed networks, see Timár, Goltsev, Dorogovtsev, and Mendes (2017*b*) and Goltsev, Timár, and Mendes (2017). The update equations of this algorithm also can be used to derive Eq. (6.63).

[21]Note that many real directed networks, like the WWW, have numerous reciprocal edges.

$$p > p_c = \frac{\langle q_o \rangle}{\langle q_i q_o \rangle} > p_c^{(W)} = \frac{\langle q_i + q_o \rangle}{\langle (q_i + q_o)^2 \rangle} \tag{6.64}$$

(Newman, Strogatz, and Watts, 2001), where $p_c^{(W)}$ in the birth point for the giant weakly connected compoonent. If the in- and out-degrees of individual vertices are independent, that is, $P(q_i, q_o) = P_i(q_i)P_o(q_o)$, then $p_c = 1/\langle q_i \rangle = 1/\langle q_o \rangle$, and in this situation there is no hyper-resilience for giant in-,out-, and strongly connected components in contrast to a giant weakly connected component. On the other hand, if, for example, for each individual vertex, its in- and out-degrees coincide, $q_i = q_o$, then $p_c = 2p_c^{(W)}$, and all the giant components are hyper-resilient against uniformly random damage. In a similar way to Sections 6.1 and 6.2, we can derive the expressions for the relative sizes of giant in-, out-, and strongly connected components:

$$S_{\text{IN}} = p \sum_{q_i, q_o} P(q_i, q_o)(1 - X^{q_i}),$$

$$S_{\text{OUT}} = p \sum_{q_i, q_o} P(q_i, q_o)(1 - Y^{q_o}),$$

$$S_{\text{SCC}} = p \sum_{q_i, q_o} P(q_i, q_o)(1 - X^{q_i})(1 - Y^{q_o}). \tag{6.65}$$

If the joint in-, out-degree distribution decays sufficiently rapidly, then, substituting the solution of Eq. (6.63) into Eq. (6.65), we get the following critical singularities:

$$S_{\text{IN}}, S_{\text{OUT}} \propto (p - p_c),$$

$$S_{\text{SCC}} \propto (p - p_c)^2. \tag{6.66}$$

In particular, when $P(q_i, q_o) = P_i(q_i)P_o(q_o)$, the relative size S_{SCC} coincides with the product of S_{IN} and S_{OUT} with the factor $1/p$,[22]

$$S_{\text{SCC}}(p) = \frac{1}{p} S_{\text{IN}}(p) S_{\text{OUT}}(p). \tag{6.67}$$

The part of a giant weakly connected component, exterior for the union of giant in-, out, and strongly connected components, consists of tendrils and tubes, which can be naturally divided into layers as Figure 6.11b explains.

---

[22]Relation between the sizes of these three giant components strongly varies in various real directed networks; see Vitali, Glattfelder, and Battiston (2011).

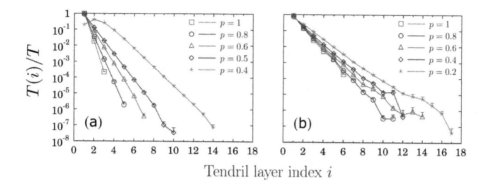

Fig. 6.12 Ratio of the size (number of vertices) $T(i)$ of the tendril layer i to the total size $T = \sum_i T(i)$ of tendrils versus i for different amount of damage in (a) a directed Erdős–Rényi graph ($N = 10^6$, $\langle q_i + q_o \rangle = 5$) and (b) a WWW domain ($N = 875{,}713$, $\langle q_i + q_o \rangle = 11.659$), the bond percolation problem. The lowest value of p in each panel is close to the percolation threshold in the corresponding network. The data was averaged over 100 realizations. Adapted from Timár, Goltsev, Dorogovtsev, and Mendes (2017b).

Timár, Goltsev, Dorogovtsev, and Mendes (2017b) obtained the statistics of tendrils in different layers, without distinguishing tendrils and tubes. Figure 6.12 demonstrates the decay with increasing i of the relative number of vertices in tendrils of layer i in two directed networks: a directed Erdős–Rényi graph, whose joint in-, out-degree distribution is the product of two Poisson distributions, for in- and out-degrees, and for a relatively small domain of the WWW. This decay is exponential in both the networks, and the tendrils and tubes of the first layer, directly connected to giant in-, out, and strongly connected components, contain the great bulk of vertices.

6.6   Finite connected components

Let us consider statistics of finite connected components focusing on uncorrelated undirected networks with a given degree distribution. In addition to the probability $\mathcal{P}(s)$ that a uniformly randomly chosen vertex belongs to a finite component of size s,

$$\sum_{s=1}^{\infty} \mathcal{P}(s) = 1 - S, \qquad (6.68)$$

we introduce the probability $\mathcal{P}_1(s)$ that an end vertex of a uniformly randomly chosen edge occurs in a finite component of size s after this edge is removed. Let the generating functions of the probability distributions $\mathcal{P}(s)$ and $\mathcal{P}_1(s)$ be $H(z)$ and $H_1(z)$. In particular, $H(1) = 1 - S$ and $H_1(1) = X$, which is the probability, introduced in Section 6.1, that an end vertex of a randomly chosen edge occurs in a finite component after this edge is removed. In a way, similar to Section 6.1, one can derive the following equation for $H_1(z)$ and the expression for $H(z)$ in terms of $H_1(z)$:

$$H_1(z) = zG_1(H_1(z)), \tag{6.69}$$

$$H(z) = zG(H_1(z)) \tag{6.70}$$

(Newman, Strogatz, and Watts, 2001). Notice the factor z on the right-hand sides of these equations. This factor comes from the fact that the vertex itself belongs to a connected component (see the formula for the generating function of $f(s-1)$, Eq. (C.10)). These equations also follow from the equations of the message passing algorithm, Eqs. (6.52) and (6.53). For $z = 1$, Eqs. (6.69) and (6.70) are reduced to Eqs. (6.11) and (6.12) from Section 6.1. Then the average size of a finite connected component to which a uniformly randomly chosen vertex belongs takes the form:

$$\langle s \rangle_{\text{finite comp}} = \frac{\langle s \rangle_{\mathcal{P}}}{1 - S} = \frac{H'(1)}{1 - S} = \frac{1 - S + G'(H_1(1))H_1'(1)}{1 - S}, \tag{6.71}$$

where $\langle s \rangle_{\mathcal{P}}$ is the first moment of the distribution $\mathcal{P}(s)$. In the absence of a giant connected component, $S = 0$, this gives

$$\langle s \rangle_{\text{finite comp}} = \langle s \rangle_{\mathcal{P}} = 1 + \frac{G'(1)}{1 - G_1'(1)}$$

$$= 1 + G'(1)H_1'(1) = 1 + \frac{\langle q \rangle^2}{\langle q(2 - q) \rangle}, \tag{6.72}$$

where the equality $H_1'(1) = 1/[1 - G_1'(1)]$ follows from Eq. (6.69) and we used Eq. (C.9). Here we assume that $\langle q^2 \rangle < \infty$. The denominator equals zero at the point of the birth of a giant connected component, Eq. (C.22). In a similar way one can show that in the phase with a giant connected component, near the critical point, $\langle s \rangle_{\text{finite comp}} \cong \langle s \rangle_{\mathcal{P}} \cong \langle q \rangle^2 / \langle q(q - 2) \rangle$. Recall that in percolation problems, the average size of a finite component to which a uniformly randomly chosen vertex belongs plays the role of susceptibility (Section 3.5), and the deviation from the critical point, $\langle q(q - 1) \rangle - \langle q \rangle$, plays the role of the control parameter. Hence we observe the Curie–Weiss law for the susceptibility, that is, the critical exponent γ_s for susceptibility equals 1, as is common for continuous phase transitions above the upper critical dimension. The critical amplitudes below and above the transition are equal, which is common for percolation problems.[23]

The asymptotics of the distribution $\mathcal{P}(s)$ can be obtained by inspection of the singularity of the generating function $H(z)$, closest to $z = 1$, as in Eqs. (C.26)–(C.28). Equation (6.70) shows that this asymptotics is actually determined by the singularity of $H_1(z)$, because the function $G(z)$ is analytic in the circle $|z| \leq 1$.[24] The singularity of the function $H_1(z)$ corresponds to the zero of the derivative of its inverse function $z = H_1^{-1}(u)$. This derivative can be found from Eq. (6.69), rewritten in the form:

$$z = \frac{u}{G_1(u)}, \tag{6.73}$$

from which the following condition for the singularity follows:

$$G_1(u) - uG'(u) = 0. \tag{6.74}$$

Since at the critical point, $G(1) = G_1(1) = G_1'(1) = 1$, the point of singularity here is corresponds to $u = 1$, and so the point of singularity $z = H_1^{-1}(1) = 1$. For the sake of brevity, we consider only the critical point. In the neighbourhood of the singularity, $1 - z \ll 1$, $u = 1 + \Delta$ with $\Delta \ll 1$, we get the equalities

$$1 - (1 - z) = H_1^{-1}(1 + \Delta) = \frac{1 + \Delta}{1 + 1 \cdot \Delta + G_1''(1)\Delta^2/2 + \ldots}$$

$$= 1 - \frac{1}{2}G_1''(1)\Delta^2 + \ldots. \tag{6.75}$$

This leads to the singularity

$$H_0(z) \sim H_1(z) \overset{1-z \ll 1}{\sim} (1 - z)^{1/2}, \tag{6.76}$$

if $G_1''(1) = \langle q(q-1)(q-2)\rangle/\langle q\rangle \neq 0$; see Eq. (C.9). This square root singularity of the generating function $H_0(z)$ corresponds to the asymptotics

---

[23]For comparison, for ferromagnetic systems, the critical amplitude of the Curie–Weiss law in the normal phase equals twice the critical amplitude in the phase with the long-rang order. For the susceptibility for directed networks, see Goltsev, Timár, and Mendes (2017). Interestingly, its critical amplitude in the phase without a giant strongly connected component equals triple the critical amplitude in the phase with this component.

[24]Indeed, it is clear that the function $G(z) = \sum_{q=0}^{\infty} P(q)z^q$ is absolutely convergent in the circle $|z| \leq 1$, because $P(q) \geq 0$ and $\sum_{q=0}^{\infty} P(q) = 1 < \infty$.

$$P(s) \sim s^{-3/2} \qquad (6.77)$$

of the distribution $P(s)$ at the critical point, check Appendix C, which, in its turn, corresponds to the asymptotics $n(s) \sim s^{-5/2}$ of the size distribution of finite components. If we move a little from the critical point into the phase with a giant connected component, then these distributions are corrected by an exponentially decaying factor (compare to the classical random graphs, Eq. (3.42)). On the other hand, if we move from the critical point into the phase without a giant connected component, then such an exponential cut-off occurs only for networks with degree distributions decaying faster than a power law. For scale-free networks, in the entire subcritical region, the asymptotics of $P(s)$ is power law.

The critical asymptotics, Eq. (6.76), holds for all uncorrelated networks with $\langle q^3 \rangle < \infty$, except quite exotic ones having $\langle q(q-1)(q-2)\rangle = 0$.[25] This asymptotics of the critical distribution of cluster sizes is common to many problems for complex networks. When $\langle q^3 \rangle$ diverges but $\langle q^2 \rangle$ is finite, which corresponds to $3 < \gamma < 4$ for networks with power-law degree distributions $P(q) \sim q^{-\gamma}$, we have the leading terms $1 - \text{const } \Delta^{\gamma-2} + \dots$ on the right-hand side of Eq. (6.75), which leads to $H_0(z) \sim (1-z)^{1/(\gamma-2)}$, and hence to

$$P(s) \sim s^{-[1+1/(\gamma-2)]}, \qquad (6.78)$$

and hence $n(s) \sim s^{-[2+1/(\gamma-2)]}$ (Cohen, Havlin, and Ben-Avraham, 2003b).

In Section 3.5 we already indicated how the sizes of connected components in the classical random graphs scale with the size of a large network, N. What's about uncorrelated networks with a power-law degree distribution $P(q) \sim q^{-\gamma}$? Let us rank the sizes of the connected components in descending order, $s^{(\alpha)}$, $\alpha = 1, 2, \dots$. Then, if $3 < \gamma \le 4$,

- for $p < p_c - CN^{-(\gamma-3)/(\gamma-1)}$, the size of the largest component scales with N as $s^{(1)}(N) \sim N^{1/(\gamma-1)}$ (Janson, 2008).
- within the scaling window $|p - p_c| \lesssim N^{-(\gamma-3)/(\gamma-1)}$, for any finite α, $s^{(1 \le \alpha < \infty)}(N) \sim N^{(\gamma-2)/(\gamma-1)}$ (Kalisky and Cohen, 2006).

---

[25]Beautifully, from Eqs. (6.69) and (6.70) one can even derive an explicit expression for $P(s)$ in terms of the generating function of any sufficiently rapidly decaying degree distribution,

$$P(s=1) = P(q=0), \quad P(s>1) = \frac{\langle q \rangle}{(s-1)!} \left\{ \frac{d^{s-2}}{dz^{s-2}}[G_1(z)]^s \right\}_{z=0}.$$

Newman (2007, 2010).

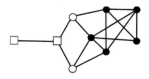

Fig. 6.13 3-core of a small sample graph (filled vertices). The pruning algorithm first removes two square vertices having degrees less than 3, and secondly it removes two open vertices appear with degrees less than 3 after the first round of removal.

- for $p < p_c + CN^{-(\gamma-3)/(\gamma-1)}$, the largest connected component is the giant one, $s^{(1)}(N) \sim N$, and the smaller components scale with N as $s^{(1<\alpha<\infty)}(N) \sim \ln N$.

Here C is a constant. On the other hand, if $\gamma \geq 4$, the connected components scale with N like in the classical random graphs; see Section 3.5 with one exception. Namely, in the subcritical region, still $s^{(1)}(N) \sim N^{1/(\gamma-1)}$. The width of the scaling window and $s^{(\alpha)}(N)$ within it follow from standard relations of percolation theory, Eqs. (C.13), (C.15), and (C.19) after we substitute $\beta = 1/(\gamma-3)$ into them. One can see that the size of the largest cluster in the subcritical region is determined by the most connected vertex having degree of the order of $N^{1/(\gamma-1)}$.

6.7 k-cores

While each two vertices in a connected component are linked by at least a single path, which is the minimal possible connectivity, easy to break, vertices in the k-cores of networks are tied stronger, with a few interconnecting paths between two vertices, providing some reserve of connectivity. The k-core of a graph is its largest subgraph with all vertices of degree at least k (Seidman, 1983; Bollobás, 1984) (see Figure 6.13). In general, the k-core subgraph, if it is present in a graph, can consist of a number of components, one of which can occur to be infinite in an infinite graph. Typically the focus is just on such a giant component, which is, for the sake of brevity, usually also called the 'k-core'. In physics, the k-cores are encountered in a range of jamming phenomena (Schwarz, Liu, and Chayes, 2006; Biroli, 2007).

The k-core of a graph can be easily found by the *pruning algorithm* explained by Figure 6.13. At each step, remove all vertices with degrees smaller than k until no such vertices remain. The remaining subgraph, if it is not null, is just the k-core. Clearly, the 1-core of a graph is the graph with the removed isolated vertices. It is also clear that the $(k \geq 1)$-core of a tree is null.

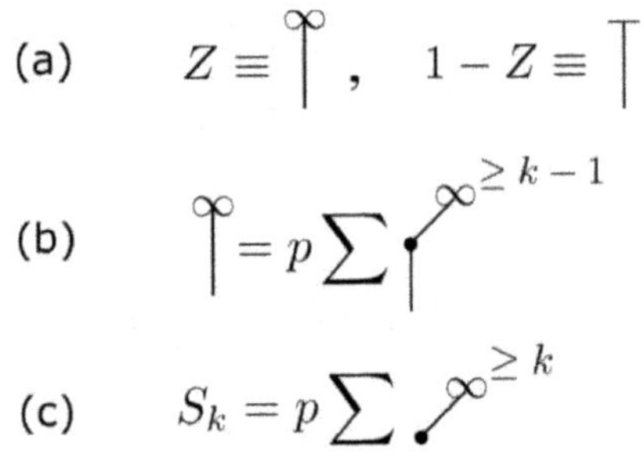

(a) $\qquad Z \equiv \uparrow, \qquad 1 - Z \equiv \uparrow$

(b) $\qquad \uparrow = p \sum \nolimits^{\geq k-1} $

(c) $\qquad S_k = p \sum \nolimits^{\geq k} $

Fig. 6.14 Graphical representation of equations for the order parameter. (a) The probability Z that a uniformly randomly chosen edge is a root of an infinite $(k-1)$-ary subtree, if we follow the edge in this direction, where Z plays the role of the order parameter; the complementary probability $1 - Z$ that an edge is not a root of an infinite $(k-1)$-ary subtree, in this direction. (b) and (c) Representations of Eqs. (6.79) and (6.80), respectively. Compare with Figure 6.2. For the sake of compactness, we do not show an arbitrary number of non-Z branches.

The full set of k-cores form a hierarchy of progressively more connected subgraphs with each next k-core embedded into the $(k-1)$-core from this hierarchy, up to the highest k, for which such a subgraph exists. The subset of vertices of the k-core that do not belong to the $(k+1)$-core form the k-shell. In this way, a network, can be decomposed into the set of k-shells, like an onion or a Russian nesting doll (Matryoshka) with the highest k-shell coinciding with the highest k-core. This decomposition (Alvarez-Hamelin, Dall'Asta, Barrat, and Vespignani, 2006; Carmi, Havlin, Kirkpatrick, Shavitt, and Shir, 2007) enables one to classify vertices according to the quality of their connectivity.

Let us focus on k-cores in infinite uncorrelated networks with an arbitrary degree distribution $P(q)$. Due to the locally tree-like structure, if such a network has a $(k{\geq}1)$-core, it should be single and giant. The theory of this k-core can be developed in the same framework as in Sections 6.1 and 6.2 (Dorogovtsev, Goltsev, and Mendes, 2006). The key notion in this theory is the so-called *infinite $(k-1)$-ary subtree* (Fernholz and Ramachandran, 2003). The m-ary tree is a tree, where all edges (except leaves) have branching at least m in each of the two directions. The exception for the leaves is not essential for us, since we are interested only in the infinite subtrees of this sort.[26] We introduce the probability Z that an end vertex of a uniformly

[26] The term 'infinite subtree' is not perfectly rigorous, since on the infinity this 'subtree' actually contains (infinite) cycles, which does not spoil our calculations.

randomly chosen edge is a root of an infinite $(k-1)$-ary subtree, if we follow the edge in this direction.[27] Strictly speaking, we should write Z_k, but we omit the subscript k for the sake of compactness. Figure 6.14 explains the structure of the self-consistent equation for Z, and of the expression for S_k, which is the probability that a vertex belongs to the k-core. In particular, Figure 6.14c shows that a vertex in such a network belongs to the k-core if at least k of its nearest neighbours are the roots of infinite $(k-1)$-ary subtrees. The probability that a vertex of degree q has exactly $n \geq k$ nearest neighbours in the k-core equals $\binom{q}{n} Z^n (1-Z)^{q-n}$. Let vertices be occupied with a probability p. Then the analytical representation of the equations in Figure 6.14b, c takes the form:

$$Z = p \sum_{n=k-1}^{\infty} \sum_{m=n}^{\infty} \frac{(m+1)P(m+1)}{\langle q \rangle} \binom{m}{n} Z^n (1-Z)^{m-n}, \qquad (6.79)$$

$$S_k(n) = p \sum_{q=n}^{\infty} P(q) \binom{q}{n} Z^n (1-Z)^{q-n}, \quad S_k = \sum_{n=k}^{\infty} S_k(n), \qquad (6.80)$$

where $S_k(n)$ is the probability that a vertex chosen uniformly at random has exactly $n \geq k$ nearest neighbours in the k-core, and S_k is the relative size of the k-core. Then the degree distribution $P_k(n)$ of the k-core subgraph equals

$$P_k(n) = \frac{S_k(n)}{S_k}, \qquad (6.81)$$

$\sum_{n \geq k} P_k(n) = 1$,[28] and hence, comparing Eqs. (6.79) and (6.80), we see that the average degree of a vertex in the k-core subgraph, $\langle q \rangle_k = \sum_{n \geq k} n P_k(n)$, is related to the average degree $\langle q \rangle$ of a vertex in the original network in the following elegant way:

$$S_k \langle q \rangle_k = Z^2 \langle q \rangle. \qquad (6.82)$$

We can also arrive at this equality by considering the numbers of edges of three types: edges between vertices within the k-core, between two vertices

---

[27] For the sake of brevity, we often say: 'an edge is a root of a tree'.

[28] The degree distribution of a k-core narrows with increasing k. This effect is especially salient for the highest k-core in a network with a fat-tailed degree distribution, where the highest threshold degree $k_{highest}$ is sufficiently large, and so all vertices in this k-core have degrees close to this value.

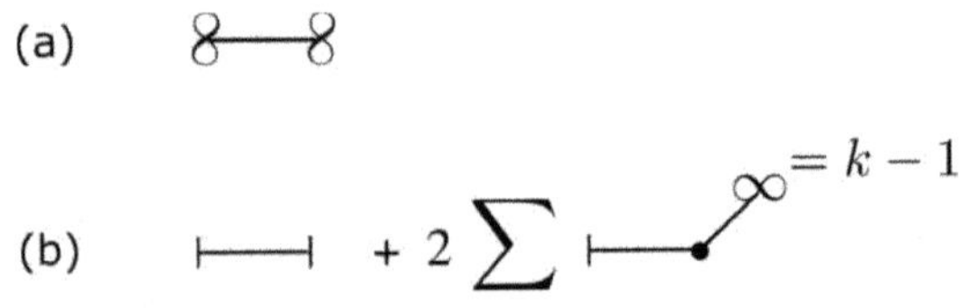

Fig. 6.15 Probability that an edge is within the k-core (a) or it connects vertices outside of the k-core (b). Notice that the vertex in the second contribution to (b) does not belong to the k-core, because it has not sufficient number of nearest neighbours—roots of infinite $(k-1)$-ary subtrees.

outside of the k-core, and between one vertex in the k-core and the second outside, the numbers E_k, E_0, and E_{0k}, respectively, where

$$E_k + E_0 + E_{0k} = E. \tag{6.83}$$

Recall that E is the total number of edges in a network. Figure 6.15 shows the contributions to the first two of these numbers:[29]

$$\frac{E_k}{E} = Z^2, \tag{6.84}$$

$$\frac{E_0}{E} = (1 - Z)^2 + 2(1 - Z) \sum_{q=k}^{\infty} \frac{qP(q)}{\langle q \rangle} \binom{q-1}{k-1} Z^{k-1}(1 - Z)^{q-k}. \tag{6.85}$$

Equation (6.84) confirms Eq. (6.82), while Eq. (6.85) leads to the following relation:

$$\frac{E_0}{E} = (1 - Z)^2 + 2\frac{1 - Z}{Z} \frac{kS_k(k)}{\langle q \rangle} = (1 - Z)^2 + 2Z(1 - Z) \frac{kS_k(k)}{\langle q \rangle_k S_k}, \tag{6.86}$$

allowing us to measure the fraction of vertices having exactly k nearest neighbours in the k-core by counting the number of edges. These weakest vertices of the k-core form the set, which is, together with edges of the k-core connecting these vertices, called the *corona* (Schwarz, Liu, and Chayes, 2006). The corona appears to be the key notion for understanding the k-cores.

Let us put aside the corona for a while and first discuss the direct consequences of Eqs. (6.79) and (6.80). Equation (6.79) can be rewritten in another form:

[29] In Eqs. (6.84)–(6.86), we assume that all vertices are occupied, $p = 1$. For $p < 1$, one should use Eqs. (6.84)–(6.86) with substituted characteristics of the damaged network, namely, the number of edges, the average degree, etc.

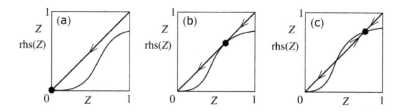

Fig. 6.16 Graphical solution of Eq. (6.79), or, equivalently, Eq. (6.87) for Z when $k \geq 3$. (a) The normal phase, that is, the k-core is absent. The solid curve shows the right-hand side of the equation. (b) The critical point, $p = p_c$. A new solution $0 < Z^* < 1$ emerges, for which $Z^* = \text{rhs}(Z^*, p_c)$ and $1 = \text{rhs}'(Z^*, p_c)$. At this point, the second derivative of the right-hand side of the equation is negative, $\text{rhs}''(Z^*, p_c) < 0$. (c) The phase with the k-core. The stable solution corresponding to the largest of the three fixed points should be chosen. Looking at arrows, one can grasp which of the fixed points (solutions) is stable. Compare with Figure 6.3 for the giant connected component problem.

$$Z = p - p \sum_{n=0}^{k-2} \sum_{m=n}^{\infty} \frac{(m+1)P(m+1)}{\langle q \rangle} \binom{m}{n} Z^n (1 - Z)^{m-n}. \qquad (6.87)$$

The reader can easily check that, for the 2-core, this equation coincides with Eqs. (6.13) and (6.28) for the giant connected component problem. Hence the 2-core of an uncorrelated network emerges at the same point as its giant connected component, and the transition is continuous. Near this critical point,

$$S_2 \cong p \frac{\langle q(q-1) \rangle}{2} Z^2 \cong \frac{1}{p} \frac{\langle q(q-1) \rangle}{2 \langle q \rangle^2} S^2, \qquad (6.88)$$

where we took into account that the relative size of the giant connected component is proportional to Z near the critical point, $S \cong p \langle q \rangle Z$. In particular, if the network is scale-free, with the degree distribution exponent $\gamma \geq 4$, then this leads to the square critical behaviour of the 2-core size:

$$S_2(p) \propto (p - p_c)^2. \qquad (6.89)$$

The 2-core is simply the giant connected component with all tree branches removed, and Eqs. (6.88) and (6.89) demonstrate that near the critical point these tree branches form the great bulk of the giant connected component.

6.7.1   Hybrid transition

For $k \geq 3$, Eq. (6.79) or, equivalently, Eq. (6.87) has a solution, markedly distinct from the giant connected component problem. Figure 6.16 graphically represents the solution of this equation, $Z = \text{rhs}(Z, p)$, for the order parameter Z in the case of $k \geq 3$. This figure demonstrates that the non-zero solution $0 < Z^* < 1$ emerges discontinuously at the critical point p_c satisfying two conditions

$$Z^* = \text{rhs}(Z^*, p_c), \tag{6.90}$$

$$1 = \text{rhs}'(Z^*, p_c), \tag{6.91}$$

and at this point,[30]

$$\text{rhs}''(Z^*, p_c) < 0. \tag{6.92}$$

In these equations, we denote $\text{rhs}'(Z, p) \equiv \partial \text{rhs}(Z, p)/\partial Z$ and $\text{rhs}''(Z, p) \equiv \partial^2 \text{rhs}(Z, p)/\partial Z^2$. One can easily check that these three conditions lead to the square-root singularity $Z(p \geq p_c) \cong Z^* + C\sqrt{p - p_c}$ at the critical point, where C is a constant proportional to $|\text{rhs}''(Z^*, p_c)|$. Substituting this solution into Eq. (6.80), we obtain the relative size of the k-core

$$S_k(p) \cong S_k^* + C_k\sqrt{p - p_c}, \tag{6.93}$$

where S_k^* is the height of the jump from zero and C_k is a constant. Note that this square-root singularity is common for arbitrary degree distributions. Thus we have a combination of a discontinuity and the square-root singularity at the same point.[31] The phase transitions of this sort are called *hybrid phase transitions*.[32] The relative size of the corona, $S_k(k)$ as a function of p shows a similar combination of a discontinuity and the square-root singularity at p_c.

It is useful to relate the hybrid transition to an ordinary first-order phase transition in statistical physics. Figure 6.17 shows how magnetization and zero-field susceptibility vary with temperature in a ferromagnetic system in the case of the first-order phase transition. A system with the first-order phase transition can be in two states: stable and metastable. At the temperature T^*, the metastable state with a non-zero magnetization disappears, which is called the *stability limit*, and at higher temperatures, only zero

---

[30] Note that, instead of p, we can use another control parameter, for example, the average degree of a vertex, $\langle q \rangle$.

[31] Notably, Chalupa, Leath, and Reich (1979), who found this transition in the k-core problem, originally called this problem the 'bootstrap percolation', which is actually not the case. In graph theory, the emergence of the k-core was described by Luczak (1991) and Pittel, Spencer, and Wormald (1996).

[32] A singularity in hybrid transitions is not necessary square root, but, in fact, the square root is typical.

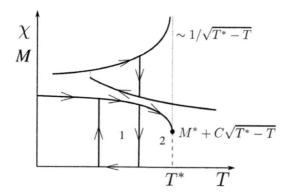

Fig. 6.17 Magnetization M and susceptibility χ at zero magnetic field versus temperature T for a ferromagnetic material with the first-order phase transition. The stability limit is at the temperature T^*. Arrows mark the heating and cooling scenarios. The numbers 1 and 2 indicate slow and fast heating. The curve for susceptibility is shown for fast heating.

magnetization is possible. A hysteresis phenomenon takes place in a restricted region below T^*, here the system can occur in one of two states, stable or metastable, depending on how quick the heating or cooling. Near T^*, the state with zero magnetization is stable. If we increase temperature sufficiently rapidly, the system has no time to escape the metastable state with non-zero magnetization, and we approach the stability limit sufficiently closely to observe the singularity identical to the one for the k-core.[33] This fast heating plays the role of a specific constraint keeping the system in the metastable state up to the stability limit. Furthermore, equations for the order parameter (magnetization) in the mean-field theory of a first-order phase transition appear to be similar to the equation for Z for the k-core, Eq. (6.79), or, equivalently, Eq. (6.87), and Figure 6.16, which explains the identical singularities. Thus we can naturally treat the hybrid transition as the stability limit of a first-order phase transition, where the definition of the k core as the *maximum* subgraph with vertices of degree at least k provides the constraint retaining the non-zero solution (the fixed point $0 < Z^* < 1$ instead of $Z = 0$) up to this point.

Figure 6.18 shows the dependence of the sizes of the k-core and the corona on the occupation probability p for the Erdős–Rényi graphs. For networks with non-Poisson degree distributions, this dependence looks similar. One

---

[33]The overheated water is a well-known example of this phenomenon.

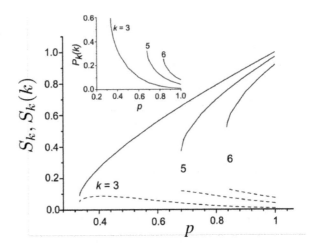

Fig. 6.18 Dependence of the sizes of the k-core ($k = 3$, 5, and 6) and the corona, S_k (solid lines) and $S_k(k)$ (dashed lines), respectively, on the occupation probability p in the Erdős–Rényi graphs with the mean degree $\langle q \rangle = 10$. Both S_k and $S_k(k)$ have a square-root singularity at the k-core percolation thresholds, but in the curves $S_{k=5,6}(k)$ this singular addition is hardly noticeable. Notice the non-monotonous dependence $S_k(k)$. The inset shows the fraction $P_k(k)$ of the corona vertices in the k-core versus p. Adapted from Goltsev, Dorogovtsev, and Mendes (2006).

can see that when we progressively remove vertices at random, first the higher cores disappear, one by one, and the 2-core is the last to vanish. Figure 6.19 shows, for the sake of comparison, the relative sizes S_k of all k-cores in Erdős–Rényi random graphs with a few values of average vertex degree and in scale-free networks with different exponents γ.[34]

---

[34]For the sake of further reference, in an uncorrelated network with a power-law degree distribution $P(q_0 \leq q \leq q_{\text{cut}}) \propto q^{-\gamma}$, $2 < \gamma < 3$, the relative size of a k-core decays with k as

$$S_k \sim (k/q_0)^{-(\gamma-1)/(3-\gamma)}$$

for k much smaller than the threshold value k_{highest} for the highest k-core,

$$k_{\text{highest}} \sim q_{\text{cut}}(q_{\text{cut}}/q_0)^{-(\gamma-2)}.$$

The relative size of the highest k-core is

$$S_{k_{\text{highest}}} \sim (q_{\text{cut}}/q_0)^{-(\gamma-1)}.$$

Notice that only if q_{cut} does not grow with the network size N, the highest k-core contains a finite fraction of vertices in the infinite network.

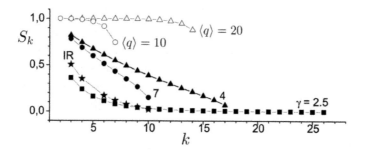

Fig. 6.19 Relative sizes of all k-cores, S_k, in the Erdős–Rényi graphs with $\langle q \rangle = 10$ and 20, in scale-free networks with the degree distribution exponents $\gamma = 2.5, 4$, and 7, and in an uncorrelated network with the degree distribution of the router-level Internet map (IR). The minimum degree in the scale-free networks equals 1. In the case of $\gamma = 2.5$, the maximum degree in the network is set to 2,000, and $\langle q \rangle = 2$; for $\gamma = 4$ and 7, the average vertex degrees $\langle q \rangle$ equal 30 and 50, respectively. Adapted from Dorogovtsev, Goltsev, and Mendes (2006).

6.7.2   Corona

Importantly, when we delete a vertex in a k-core of a locally tree-like network, together with this vertex, all the corona clusters, having at least one nearest neighbour, disappear from the k-core. (The originally deleted vertex can be a member of a corona cluster.) The remaining vertices stay in the k-core.[35] Indeed, if a single member of a corona cluster loses one edge, then it disappears from k-core, then its nearest neighbours in the corona cluster lose their edges and disappear from the k-core, and so on, until this avalanche sweeps away the corona cluster from the k-core completely. Let us denote the number of vertices in these corona clusters together with the original deleted vertex by s_{crn}, omitting the index k for the sake of compactness. Then decreasing the occupation probability p by Δp, we, on average, decrease the size of the k-core by the following number of vertices:

$$NS_k(p) - NS_k(p - \Delta p) = NS_k(p)\Delta p \langle s_{\text{crn}} \rangle(p), \tag{6.94}$$

leading to the relation:

$$\frac{d \ln S_k(p)}{dp} = \langle s_{\text{crn}} \rangle(p). \tag{6.95}$$

The average $\langle s_{\text{crn}} \rangle(p)$, which is the average size of such an avalanche induced by removing a uniformly randomly chosen vertex from the k-core, is rather

---

[35]Note that this is not true in networks with finite cycles.

close to the average size of the corona cluster to which a uniformly randomly chosen vertex belongs. Hence, similarly to the giant connected component problem, this quantity can be treated as a susceptibility for k-cores. Note however that in locally tree-like networks, this susceptibility is defined only in the phase with a k-core, in contrast to the giant connected component problem. Taking into account Eq. (6.93), we readily get the following singularity

$$\langle s_{\mathrm{crn}} \rangle (p) \propto \frac{1}{\sqrt{p - p_c}}, \tag{6.96}$$

markedly distinct from the Curie–Weiss law $1/|p-p_c|$ for the average size of a finite component to which a uniformly randomly vertex belongs, Eq. (6.72). This singularity, $1/\sqrt{p - p_c}$, Eq. (6.96), coincides with the singularity of susceptibility at the stability limit of a first-order phase transition (Figure 6.18).

Similarly to the ordinary percolation problem, the critical distribution of the sizes of corona clusters to which a randomly chosen vertex belongs has the power-law asymptotics

$$\mathcal{P}(s_{\mathrm{crn}}) \propto s_{\mathrm{crn}}^{-3/2}, \tag{6.97}$$

though for an arbitrary degree distribution in this case (Goltsev, Dorogovtsev, and Mendes, 2006). Hence the size distribution of corona clusters decays as $s_{\mathrm{crn}}^{-5/2}$ at the critical point. Like in ordinary percolation, this guarantees that the largest corona cluster scales with N as $s_{\mathrm{crn}}^{(\max)}(p_c) \sim N^{2/3}$ at this point. Thus the scenario looks as follows: when we progressively decrease p from large values, the corona clusters grow in size, merge, become percolate exactly at p_c, and at the same point disappear together with the k-core. At this point, as one can derive from Eq. (6.79) or Eq. (6.87), the following relation holds:

$$1 = \frac{k(k-1)P_k(k)}{\langle q \rangle_k}, \tag{6.98}$$

where $P_k(k)$ is the fraction of k-core's vertices in the corona, and $\langle q \rangle_k$ is the average degree of a vertex in the k-core subgraph. Compare this equality with the Molloy–Reed criterion for the giant connected component, Eqs. (4.14) and (6.20).[36]

---

[36] Equations (6.96)–(6.98) can be obtained in another way. We assume that the k-core in an uncorrelated network is uncorrelated. The degree distribution $P_k(q)$ of this subgraph is known, Eq. (6.81), which enables us to investigate the statistics of components consisting of vertices with degrees k by using Eqs. (6.69) and (6.70). In this way we get an equation

Lee, Choi, Stippinger, Kertész, and Kahng (2016*a*) and Lee, Jo, and Kahng (2016*b*) explored a distinct susceptibility introduced as the limit of the fluctuations of the k-core size, $\lim_{N\to\infty} [N(\langle S_k^2 \rangle - \langle S_k \rangle^2)]$. In ordinary percolation, both these susceptibilities, namely the average size of a finite component to which a randomly chosen vertex belongs and the limit of the fluctuations of the k-core size, demonstrate the Curie–Weiss law (Section 3.5). In contrast to that, in the case of the ($k{\geq}3$)-cores, the observed singularity

$$\lim_{N\to\infty} [N(\langle S_k^2 \rangle - \langle S_k \rangle^2)] \propto \frac{1}{p - p_c}, \qquad (6.99)$$

where $p > p_c$, markedly differs from the singularity $\langle s_{\mathrm{crn}} \rangle(p) \propto 1/\sqrt{p - p_c}$. In general, this combination of the Curie–Weiss law and the inverse square root singularity for two such susceptibilities should occur in all the hybrid transitions where a giant component emerges according to Eq. (6.93).

6.7.3 Heterogeneous k-core

In the heterogeneous k-core problem (Cellai, Lawlor, Dawson, and Gleeson, 2011, 2013*a*; Baxter, Dorogovtsev, Goltsev, and Mendes, 2011), a threshold degree value k varies from vertex to vertex. Let k_i, $i = 1, 2, ..., N$ be the full set of threshold degrees for all vertices of a graph. The *heterogeneous k-core* of a graph is its largest subgraph with each its vertex i of degree at least k_i. In general, the heterogeneous k-core consists of a giant component (*giant heterogeneous k-core*) and numerous finite ones. The heterogeneous k-core is the result of the progressive 'deactivation' of initially active vertices with

for the generating function $H_{1k}(x)$ of the probability that an end of a randomly chosen edge in the k-core belongs to a finite corona cluster of a given size,

$$H_{1k}(x) = 1 - \frac{kP_k(k)}{\langle q \rangle_k} + x\frac{kP_k(k)}{\langle q \rangle_k}[H_{1k}(x)]^{k-1}.$$

The additional term $1 - kP_k(k)/\langle q \rangle_k$, compared to Eq. (6.69), is the probability that the end of an edge does not belong to the corona. We estimate the generating function $H_k(x)$ for the size distribution of a corona cluster attached to a vertex in the k-core,

$$H_k(x) \approx \sum_q P_k(q)[H_{1k}(x)]^q,$$

and finally obtain

$$\langle s_{\mathrm{crn}} \rangle(p) = \frac{dH_k(x)}{dx}\bigg|_{x=1} = \frac{kP_k(k)}{1 - k(k-1)P_k(k)/\langle q \rangle_k} \propto \frac{1}{\sqrt{p - p_c}}.$$

185

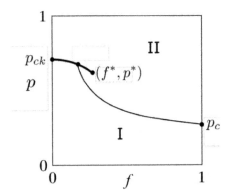

Fig. 6.20 Phase diagram for heterogeneous k-core in the $f-p$ plane. The threshold equals 1 with probability f and $k \geq 3$ with the complementary probability $1 - f$, p is the occupation probability for vertices. The giant heterogeneous k-core is present in region II, emerging continuously at the threshold marked by the thin black curve. The hybrid, discontinuous transition occurs at the points marked by the heavy black line. At the end point of this line, (f^*, p^*), this transition disappears. Adapted from Baxter, Dorogovtsev, Goltsev, and Mendes (2011).

degrees lower then their individual threshold values, that is, the heterogeneous k-core pruning process. In particular, the threshold degrees k_i can be drawn from some probability distribution. For locally tree-like networks, this problem was solved in a way similar to the homogeneous k-core one. For example, let a threshold degree for a heterogeneous k-core be 1 with a probability f and $k \geq 3$ with the complementary probability $1 - f$, which is a simple representative case, and p be the occupation probability for vertices. The probabilities f and p can be used as the control parameters. Figure 6.20 shows the qualitative form of the phase diagram for a heterogeneous k-core in the $f-p$ plane for uncorrelated networks with sufficiently rapidly decreasing degree distributions. The points of the continuous transition associated with the emergence of a giant heterogeneous k-core form the thin line on this diagram. Notice the thick line formed by points of the discontinuous (hybrid) transition. The discontinuity is combined with a square root singularity similar to Eq. (6.93) for all points of the thick line except the end point, (f^*, p^*), for which a discontinuity disappears and the singularity is a cube root one, namely,

$$S_k(f, p^*) - S_k(f^*, p^*) \propto (f - f^*)^{1/3},$$
$$S_k(f^*, p) - S_k(f^*, p^*) \propto (p - p^*)^{1/3}. \qquad (6.100)$$

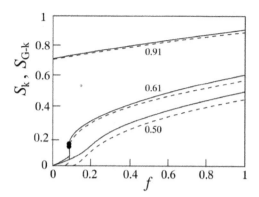

Fig. 6.21 Solid curves: relative sizes S_k of the heterogeneous k-core as a function of f for an Erdős–Rényi graph with average vertex degree 5 with the same $k = 3$, at three different values of the vertex occupation probability $p = 0.5$, 0.61, and 0.91. Dashed curves: relative sizes S_{G-k} of the giant component in this core vs. f at the same values of p. The dots on the curves with 0.61 indicates square-root singularities. Adapted from Baxter, Dorogovtsev, Goltsev, and Mendes (2011).

Here S_k is the relative size of a heterogeneous k-core. The same singularity at this point takes place for the relative size of a giant heterogeneous k-core, S_{G-k}. Figure 6.21 shows the dependence of S_k and S_{G-k} on f for a few values of the vertex occupation probability p for an Erdős–Rényi graph. Compare Figures 6.20 and 6.21 to get the approximate positions of these values of p on the phase diagram.

6.8   Dynamics of pruning

The pruning process resulting in the k-core is a simple paradigm for various cascading failures phenomena, where elements 'weaker' than a given threshold are progressively removed from a system, and so it deserves a thorough consideration. Let at each time step, $t = 1, 2, \ldots$, all vertices with degrees $q < k$ be removed from a network. The criterion for the removal of vertices is local, namely, their degrees, which greatly simplifies the problem. During the pruning, the network progressively shrinks to null or to the k-core, and the question is how the structure of the yet unpruned network evolves with time. It is convenient to fix the number of vertices, N, so, instead of pruning vertices with their edges, we prune only their edges, progressively increasing the number of isolated vertices. Clearly, this does not change the process. Let the network be infinite, locally tree-like and, furthermore, uncorrelated, where the last assumption is made here only for the sake of compactness. In

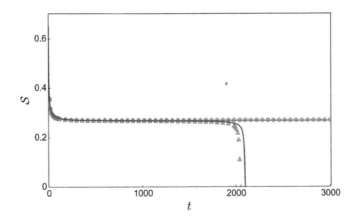

Fig. 6.22 Evolution of the relative size $S(t)$ of the network remaining after t steps of $k{=}3$ pruning of the Erdős–Rényi random graph with the initial average vertex degree $\langle q \rangle$ close to the critical one, $\langle q \rangle_c = 3.35092\ldots$. The upper and lower curves are for $\langle q \rangle = \langle q \rangle_c + 0.00001$ and $\langle q \rangle_c - 0.00001$, respectively. Solid lines are computed from Eqs. (6.101)–(6.104), and the points are the results of simulations of the Erdős–Rényi random graphs of 10^8 vertices. Adapted from Baxter, Dorogovtsev, Lee, Mendes, and Goltsev (2015).

addition, we assume that during the pruning process, the network remains uncorrelated. Let us introduce $r(t)$, the probability that, following a randomly chosen edge in the network at time t, we find that its end vertex has degree less than k:

$$r(t) = \sum_{q<k} \frac{qP(q,t)}{\langle q \rangle(t)}, \tag{6.101}$$

where $\langle q \rangle(t)$ is the average degree of the network at time t,

$$\langle q \rangle(t) = \sum_q qP(q,t). \tag{6.102}$$

The probability that a vertex of degree $q' \geq k$ at time t has q surviving edges at time $t+1$ is then $\binom{q'}{q}[1 - r(t)]^q [r(t)]^{q'-q}$, while a vertex of degree $q < k$ at time t has degree 0 at time $t+1$. Then the degree distribution evolves as follows:

188

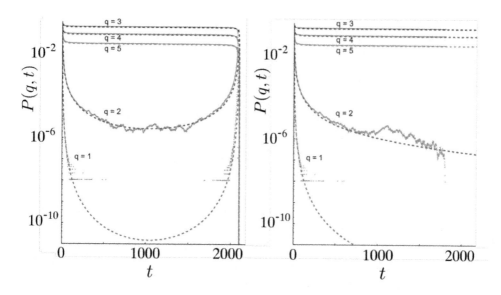

Fig. 6.23 Evolution of the network degree distribution $P(q,t)$ during the 3-core pruning process, for the Erdős–Rényi random graphs with the initial average vertex degrees $\langle q \rangle = \langle q \rangle_c - 0.00001$ and $\langle q \rangle_c + 0.00001$, left and right panels, respectively. The curves are theoretical, obtained from Eqs. (6.101)–(6.104), and also shown are traces from simulation runs (single realization for each of these two networks). Adapted from Baxter, Dorogovtsev, Lee, Mendes, and Goltsev (2015).

$$P(q > 0, t+1) = \sum_{q' \geq \max(q,k)} P(q',t) \binom{q'}{q} [1 - r(t)]^q [r(t)]^{q'-q}, \quad (6.103)$$

$$P(0, t+1) = \sum_{q'=0}^{k-1} P(q',t). \quad (6.104)$$

These equations together with Eqs. (6.101) and (6.102) completely describe the evolution of the structure of this network during the pruning process (Baxter, Dorogovtsev, Lee, Mendes, and Goltsev, 2015).[37] The difference $1 - P(0,t) = S(t)$ provides the relative size of the yet unpruned part of the network at time t.

The convenient structure of Eqs. (6.101)–(6.104) enables us to easily compute the full curve $S(t)$. Figure 6.22 demonstrates that below the critical point, the pruning process completes in a finite time, while above the critical point, $S(t)$ relaxes to the relative size of the k-core, S_k, only at infinity,

[37]See another approach to the pruning process in Iwata and Sasa (2009).

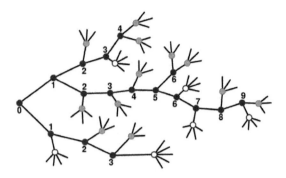

Fig. 6.24 A snapshot of the branching process of propagation of vertex pruning in a sample network during the plateau stage ($\langle q \rangle < \langle q \rangle_c$) of the 3-core pruning process. Vertex 0 is pruned, causing the corona vertices (vertices with degree 3) labeled 1 to lose edges. These two vertices are pruned in the next step, and so on, with further corona vertices removed in subsequent steps according to the numbered order. Grey and white circles represent the nearest-neighbouring vertices of degree 4 and greater than 4, respectively, that survive because their degrees exceed 3. The grey vertices after this pruning become of degree 3 and augment other corona clusters, which may then be pruned at a later time. Adapted from Baxter, Dorogovtsev, Lee, Mendes, and Goltsev (2015).

$S(t \to \infty) = S_k$, when the network is infinite. Notice the long plateau of $S(t) \approx S_k$ in the subcritical region, which occurs when $\langle q \rangle$ is close to $\langle q \rangle_c$. Figure 6.23 shows the evolution of the degree distribution $P(q, t)$ during the pruning. As we could expect, $P(q \geq k, t)$ evolves similarly to $S(t)$. Notice that, when $S(t)$ is close to S_k, $P(k-1, t)$ far exceeds $P(1 \leq q < k-1, t)$.

To understand the spreading of damage through the network as the pruning process evolves, let us recall that in the k-core, it is sufficient to remove a single connection to a corona cluster to destroy it completely. In the pruning process, the evolving clusters of vertices of degree k—corona clusters—play a similar role, providing a changing substrate for the spreading of damage. In locally tree-like networks, this is a branching process, but, unlike the classical Galton–Watson branching processes,[38] here we have not a single branch but rather a 'bush' of interacting branches. Figure 6.24 shows one of these branches. Notice that this wave of pruning within a corona cluster produces vertices of degree k augmenting other corona clusters and expanding

---

[38]The *Galton–Watson branching processes* generates a rooted branching tree with a random branching number (random number of descendants) for each edge.

the substrate for other such branches.

To get the branching of this process, we introduce $\widetilde{P}(n, t)$, the probability that a vertex removed at time t has n neighbours that will be removed at time $t + 1$, which is actually the branching distribution. The average branching is expressed in terms of this probability,

$$\langle b \rangle (t) = \sum_{n=0}^{k-1} \widetilde{P}(n, t). \tag{6.105}$$

In turn, one can express $\widetilde{P}(n, t)$ in terms of $P(q, t)$ and obtain the full picture of the evolution of $\langle b \rangle (t)$ for any $\langle b \rangle$. We do not do it here, and instead focus on the key plateau region (Figure 6.22), which is wide when $\langle b \rangle$ is close to the critical value. We have noted that within this region, the vertices with degrees smaller than $k - 1$ can be neglected (Figure 6.23) so the spreading of damage can be traced by the evolution of the concentration of vertices of degree $k - 1$, namely $P(k - 1, t)$, which enables us to get the average branching as the ratio:

$$\langle b \rangle (t) \cong \frac{P(k - 1, t + 1)}{P(k - 1, t)}. \tag{6.106}$$

Then Figure 6.22 shows that below the critical point, the average branching increases with time from small values, reaches the value $\langle b \rangle = 1$ in the middle of the plato region, and then grows further until all the edges are pruned. The overrun value 1 indicates the emergence of a giant corona cluster in this region of times. One can go further and, assuming that the branches of the pruning process only weakly interact with each other in the region of plateau, derive the following relation between the average branching, fraction of vertices within the corona clusters, and the average vertex degree of the network at a given instant:

$$\langle b \rangle (t) = \frac{k(k - 1)P(k, t)}{\langle q \rangle (t)}. \tag{6.107}$$

In the limit $t \to \infty$, this gives

$$\langle b \rangle (t \to \infty, \langle q \rangle \geq \langle q \rangle_c) \equiv \langle b \rangle_k = \frac{k(k - 1)P(k, t)}{\langle q \rangle_k}, \tag{6.108}$$

where $\langle q \rangle_k$ is the average degree of a vertex in the k-core. In particular, when $\langle q \rangle = \langle q \rangle_c$, we arrive at Eq. (6.98), that is, $\langle b \rangle_k = 1$, indicating that in the limit $t \to \infty$, the system of corona clusters is on the verge of merging

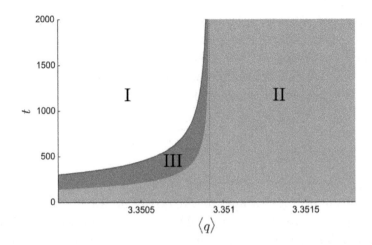

Fig. 6.25 Phase diagram for the k-core pruning process in the Erdős–Rényi random graph, plotted in the $\langle q \rangle$–t plane ($k = 3$). In region II at $\langle q \rangle > \langle q \rangle_c = 3.3509\ldots$, the pruning process reduces the network to the giant k-core as time approaches infinity. Only finite corona clusters are present in region II. A giant corona cluster is present in region III. The average branching is 1 on the border between regions II and III. The average branching is below 1 in region II and larger than 1 in region III. Only isolated vertices remain in region I. Adapted from Baxter, Dorogovtsev, Lee, Mendes, and Goltsev (2015).

into a giant corona. On the other hand, when $\langle q \rangle > \langle q \rangle_c$, $\langle b \rangle_k < 1$, and the corona clusters stay finite till the end of the pruning process. Figure 6.25 shows where these scenarios take place in the $\langle q \rangle$–t plane.

In the supercritical region, the average branching smaller than 1 leads to the exponentially decaying pruning process:

$$P(k - 1, t) \propto e^{-t/\tau}, \qquad (6.109)$$

where

$$\tau(\langle q \rangle) = \frac{1}{1 - \langle b \rangle_k} \propto \frac{1}{\sqrt{\langle q \rangle - \langle q \rangle_c}}. \qquad (6.110)$$

Note that, formally speaking, this exponential process will never complete in an infinite network. On the other hand, in the subcritical region, due to a giant corona, all the edges are pruned from the infinite network in a finite time T depending on $\langle q \rangle$. The middle of the plateau region, where the average branching equal 1, happens at $T/2$,

$$\langle b \rangle (t = T/2, \langle q \rangle < \langle q \rangle_c) = 1. \qquad (6.111)$$

192

The time T diverges as $\langle q \rangle \uparrow \langle q \rangle_c$,

$$T(\langle q \rangle) \propto \frac{1}{\sqrt{\langle q \rangle_c - \langle q \rangle}}, \tag{6.112}$$

and the ratio of the critical amplitudes of T and τ equals 2π,

$$\frac{T(\langle q \rangle_c - 0)}{\tau(\langle q \rangle_c + 0)} = 2\pi. \tag{6.113}$$

Finally, at the critical point, the relaxation is power law,[39]

---

[39]Let us consider the simplest dynamical system with a saddle point, demonstrating a similar set of behaviours. Imagine a particle moving in a one-dimensional potential $U(x)$ in a viscous medium:

$$\frac{\partial x}{\partial t} = -\frac{\partial U(x)}{\partial x},$$

where

$$U(x) = -\frac{1}{3}x^3 + cx.$$

The constant c is a control parameter, and the value $c = 0$ is the critical point. There are three scenarios for a particle falling from $-\infty$:

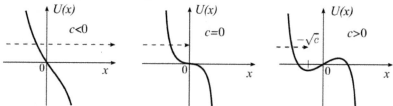

If $c > 0$, then the local minimum is at the point $x = -\sqrt{c}$, and the particle exponentially relaxes to this minimum,

$$x(t) - (-\sqrt{c}) \propto e^{-2\sqrt{c}t} \equiv e^{-t/\tau},$$

where

$$\tau = \frac{1}{2\sqrt{c}},$$

similarly to Eqs. (6.109) and (6.110). If $c < 0$, then the particle moves from $-\infty$ to $+\infty$ in a finite time,

$$T = \frac{\pi}{\sqrt{-c}},$$

similarly to Eq. (6.112). Hence

$$\frac{T(-c)}{\tau(c)} = 2\pi,$$

which coincides with Eq. (6.113). If $c = 0$, then the relaxation to the point $x = 0$ is power-law,

$$x(t) = 1/t,$$

compare to Eq. (6.114).

$$P(k-1,t) \propto t^{-2}. \tag{6.114}$$

This set of behaviours is also observed in more complicated pruning processes, where entire clusters of vertices are removed together, and the criterion for such a removal is non-local. This happens, in particular, in cascading failures of interdependent networks (Zhou, Bashan, Cohen, Berezin, Shnerb, and Havlin, 2014), which we discuss in Chapter 8.

6.9 s-cores in weighted networks

The k-core notion can be generalized to the weighted networks in a quite natural way. The s-core of a weighted graph is its largest subgraph with all vertices of strength at least s (Eidsaa and Almaas, 2013, 2016). (Recall that the strength s_i of vertex i in a weighted graph is the sum of the weights $w_{ij} \equiv A_{ij}$ of its adjacent edges, $s_i = \sum_j A_{ij}$, where we use the same notation A for the adjacency matrix of weighted graphs as for the unweighted ones and assume that the weights are non-negative real numbers.) Similarly to the k-cores, the s-cores are obtained by progressive pruning of all vertices of strength smaller than s from a weighted graph. Similarly to k-shells in the k-core decomposition, the s-cores enable one to rank vertices by ascribing the label s to each vertex belonging to s-core and not belonging to $(s+0)$-core, if the distribution of weights is continuous.[40] While the algorithm of the s-core pruning is quite straightforward and rapid, the analytical treatment of s-cores is challenging.

The s-cores of a weighted graph can be reproduced as relevant states of a ferromagnetic Ising model with heterogeneous local magnetic fields at zero temperature (Yoon, Goltsev, and Mendes, 2018). For the sake of simplicity, we assume that the distribution of weights is continuous. Let us place the spins $S_i = \pm 1$ on the vertices of a weighted graph, $i = 1, 2, \ldots, N$ and assume that the pairwise interactions between the spins are described by the Ising Hamiltonian with local magnetic fields $H_i = s_i - 2s$, where s_i is the strength of vertex i and s is the threshold number of a given s-core:

$$\mathcal{H} = -\frac{1}{2}\sum_{i,j} A_{ij}S_iS_j - \sum_i \left[\sum_j A_{ij} - 2s\right]S_i. \tag{6.115}$$

The factor $1/2$ here is due to the two symmetric entries $A_{ij} = A_{ji}$ contributing to the interaction between the spins S_i and S_j. In principle, we could

---

[40]For the s-decomposition of various real weighted networks, where this assumption may fail, see Eidsaa and Almaas (2013).

formally write $s-0$ instead of s, which is not essential in the case of the continuous distribution of weights. When $s = 0$, the positive local fields s_i align all spins into the ground state $S_i = +1$, $i = 1, 2, \ldots, N$. We start the process from this configuration of spins. The consecutive application of the additional negative field $-2s$ flips each spin i, $S_i = +1 \to -1$, that appears in the negative effective field

$$+\sum_j A_{ij}S_j + \sum_j A_{ij} - 2s = \sum_j A_{ij}(1 + S_j) - 2s < 0, \tag{6.116}$$

that is,

$$\sum_j A_{ij}\frac{1 + S_j}{2} = \sum_j A_{ij}\delta(S_j, 1) < s, \tag{6.117}$$

where $\delta(S_j, 1)$ is the Kronecker symbol. If we treat -1 spins as removed vertices in the s-core pruning process, Eq. (6.117) coincides with the pruning condition, where $\sum_j A_{ij}\delta(S_j, 1)$ corresponds to the vertex strength at a given moment of pruning. The final state of this progressive spin-flipping process is either (i) all spins are in state -1, that is, a s-core is absent, or (ii) a fraction of spins remain in state 1, which is the s-core, and the remaining spins are in state -1, indicating the effectively removed vertices.[41] The final state of this spin system is metastable when the s-core is present, and we are sufficiently close to the point of its birth. Otherwise this state is stable.

This idea is also applicable to the k-cores in the unweighted networks if we substitute $k-1+0$ for s in Eq. (6.115). Instead of $+0$, we could write $+\Delta$, where $0 < \Delta < 1/2$. The same approach can be applied to heterogeneous s- and k-cores, where a threshold for pruning varies for different vertices.

6.10 k-connected components

Section 2.8 touched upon the notion of a k-connected graph, whose any two vertices are linked by at least k independent paths. For a given graph, one can ask: what is its largest k(vertex or edge)-connected subgraph? Clearly,

---

[41]Instead of the Ising model, Eq. (6.115), equivalently, one can use the model of interacting 'spins' $X_i = 0, 1$ (inactive, active states of vertices) with the Hamiltonian:

$$\mathcal{H} = -\frac{1}{2}\sum_{i,j} A_{ij}X_iX_j + s\sum_i X_i.$$

Start from the initial configuration $X_i = 1$, $i = 1, 2, \ldots, N$. Then the final set of active spins, $X_i = 1$, form the s-core. In the k-core problem, substitute $k - 1 + 0$ for the field s in this Hamiltonian.

for $k = 1$, this is the largest connected component. For $k \geq 2$, if the graph is infinite and locally tree-like, this subgraph coincides with the k-core, already explored in the two preceding sections.[42]

It is reasonable to ask another, principally different, question: what is the best interconnected set of vertices in a network? Note that this time we speak not about a subgraph, but rather about a set of vertices. The question implies that paths interconnecting these vertices can run through arbitrary vertices in the network, including vertices not belonging to this set. To quantify such a set, the following notion was introduced. A *k(vertex or edge)-connected component* is the largest set of vertices in a network such that any two vertices in this set are interconnected by k (vertex or edge)-independent paths running through arbitrary vertices in the network. For $k = 2$ and 3, the terms *biconnected* and *triconnected components* are used.

For an infinite uncorrelated network, the relative size of the giant k-connected component, $\widetilde{S}_k$, can be straightforwardly found (Newman and Ghoshal, 2008). It is clear that the giant k-connected component is a subset of vertices of the giant connected component of this network, and hence the whole giant connected component is exploited for interconnecting vertices within this subset. In a locally tree-like network, a vertex belongs to the giant k-connected component if at least k of its edges lead to infinity. It is due to the local tree-likeness that these paths to infinity are independent. Then the formal expression for $\widetilde{S}_k$ in terms of the probability Z appears to be the same as for the size of the k-core in this network (Figure 6.14c and Eq. (6.80)). The difference from the k-core problem is that Z is now the probability that after we cut a uniformly randomly chosen edge, its end vertex is the root of an infinite tree, which is the same probability Z as for the giant connected component, the solution of Eq. (6.13) (Figure 6.2a). Then the birth point of the giant k-connected component coincides with that of the giant connected component, and the transition is continuous. Near the critical point, where Z is small, Eq. (6.80) results in the following relation between $\widetilde{S}_k$ and the relative size of the giant connected component S:

$$\widetilde{S}_k \cong \frac{1}{k!} \left\langle \prod_{j=0}^{k-1} (q-j) \right\rangle Z^k \cong \frac{1}{k! \langle q \rangle^k} \left\langle \prod_{j=0}^{k-1} (q-j) \right\rangle S^k, \qquad (6.118)$$

---

[42]Note that there is no difference between vertex- and edge-independent paths in locally tree-like networks.

where we used the relation $S \cong \langle q \rangle Z$, valid in the neighbourhood of the critical point. One can see that for $k = 1$, this gives S, as is natural, while $\tilde{S}_2$ coincides with the relative size of the 2-core, Eq. (6.88).

6.11   Giant component in correlated networks

Locally tree-like networks with degree–degree correlations between nearest-neighbouring vertices allow a treatment similar to that we applied to uncorrelated networks in Sections 6.1 and 6.2. For the sake of concreteness, here we focus on the site percolation problem. Let us introduce the probability Z_q that if one end of a randomly chosen edge in a network is attached to a vertex of degree q then, after we remove this edge, its second end belongs to an infinite cluster. For this probability, one can straightforwardly write an equation, similar to Eq. (6.28), and an expression for the relative size of a giant connected component, similar to Eq. (6.29),

$$Z_q = p \sum_{q'} P(q'|q)[1 - (1 - Z_{q'})^{q'-1}], \tag{6.119}$$

$$S = p \sum_{q} P(q)[1 - (1 - Z_q)^q]. \tag{6.120}$$

(Newman, 2002*a*; Vázquez and Moreno, 2003). Hence the percolation threshold p_c can be found from the equation

$$\mathbf{Z} = p_c B \, \mathbf{Z}, \tag{6.121}$$

where $\mathbf{Z}$ is the vector with the components Z_q, and B is the branching matrix, $B_{qq'} = (q' - 1)P(q'|q)$, introduced in Eq. (4.40). Then

$$p_c = \frac{1}{\lambda_1}, \tag{6.122}$$

where λ_1 is the largest eigenvalue of this matrix, equal to the global branching of the network. Due to the non-equalities for λ_1, Eq. (4.54), one can see that, for a given, sufficiently rapidly decaying degree distribution, the percolation threshold for networks with assortative correlations is lower than for uncorrelated networks, and that the percolation threshold for networks with disassortative correlations is the highest,

$$p_c^{(\mathrm{as})} < p_c^{(\mathrm{un})} = \frac{\langle q \rangle}{\langle q(q-1) \rangle} < p_c^{(\mathrm{dis})}. \tag{6.123}$$

Here the superscripts (as), (un), and (dis) are for assortative, uncorrelated, and disassortative networks, respectively.

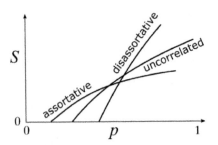

Fig. 6.26 The effect of degree–degree correlations on the emergence of a giant connected component. The size of a giant component vs. the retained fraction p of vertices after damaging the network.

Near the percolation threshold, if there is a gap between the largest and the second largest eigenvalues, the vector $\mathbf{Z}$ is proportional to the principal eigenvector $\mathbf{v}$ of the branching matrix, $\mathbf{Z} \cong c\mathbf{v}$, where c is a constant. In its turn, when the correlations are weak, the components of $\mathbf{v}$ are known, $v_q \approx g\,b(q) = g\sum_{q'}(q'-1)P(q'|q)$, Eq. (4.51), where g is a normalization constant. Substituting this vector into Eqs. (6.119) and (6.120) one can get the slope $[dS(p)/dp]_{p=p_c}$ at the critical point.[43] The result typically matches the combination of dependencies in Figure 6.26. Thus assortative correlations decrease p_c, making it harder to eliminate a giant connected component, but on the other hand, these correlations may also suppress the size of a giant connected component (Newman, 2002a).

The reader can find the comprehensive treatment of various pair degree–degree correlations and degree distributions in Goltsev, Dorogovtsev, and Mendes (2008), based on the spectral analysis of the branching matrix B. Notably, specific strong assortative correlations can move p_c to 0, inducing the hyper-resilience phenomenon, or, vice versa, strong disassortative correlations in some networks can move p_c away from 0, eliminating the hyper-resilience effect. Finally, strong correlations can even change the critical singularity $S(p) \propto (p-p_c)^\beta$.

6.12 k-clique percolation

A natural query made about a simple graph is how its cliques are interconnected. This question was addressed by Palla, Derényi, Farkas, and Vicsek

---

[43]One can also express the slope $[dS(p)/dp]_{p=p_c}$ in terms of the principal eigenvector of the non-backtracking matrix (Timár, da Costa, Dorogovtsev, and Mendes, 2017a), which accounts for a wide range of structural correlations in a network.

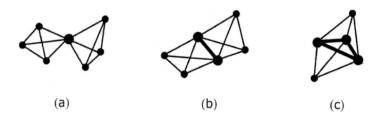

(a)　　　　　　　　　　(b)　　　　　　　　　　(c)

Fig. 6.27 (i) For $l = 1$, two 4-cliques are adjacent in configurations (a), (b), and (c); (ii) for $l = 2$, they are adjacent in configurations (b), and (c); (iii) for $l = 3$, only in configuration (c) are the 4-cliques adjacent.

(2005); Derényi, Palla, and Vicsek (2005); Palla, Derényi, and Vicsek (2007): all explored percolation in the set of k-cliques of a graph. In this scheme, two k-cliques in a simple graphs are considered to be adjacent if they share at least $1 \leq l \leq k - 1$ vertices (Figure 6.27). The set of all pairwise overlaps of k-cliques in a graph produce a network, where the k-cliques and their overlaps play the role of vertices and edges, respectively. The *k-clique percolation* problem concerns the birth of a giant component in this network of k-cliques.[44] The original works considered the particular case of the maximal overlap, $l = k - 1$, and so the term k-clique percolation was quite sufficient. Here we discuss a general l, satisfying $1 \leq l \leq k - 1$, which gives us the second number in the problem. For the sake of brevity, we assume that the original graph is the $G(N, p)$ model, where N is large, and restrict ourselves to heuristics.

What is the k-clique percolation threshold, $p_c(k, l, N)$? As is natural, we need numerous k-cliques for this percolation, so the original random graph cannot be sparse when $k > 2$. The average number of k-cliques in the $G(N, p)$ model equals

$$N_k = \binom{N}{k} p^{\binom{k}{2}} \cong \frac{N^k}{k!} p^{\binom{k}{2}}. \tag{6.124}$$

Then the condition $N_k \gg 1$ demands that

$$p_c(k, l, N) \gg (k!)^{2/[k(k-1)]} N^{-2/(k-1)}. \tag{6.125}$$

---

[44]This problem can be also formulated in terms of simplicial complexes if we recall that a simple graph is the skeleton graph of a simplicial complex; see Section 5.12. The k-cliques correspond to $(k-1)$-dimensional simplexes, and the minimal overlaps by l vertices, necessary for the adjacency of k-cliques correspond to the smallest $(k-1)$-dimensional faces between adjacent $(k-1)$-dimensional simplexes.

CONNECTED COMPONENTS

One can show that this k-clique network has a Poisson degree distribution (degree is the number of adjacent k-cliques of a k-clique) although it does not coincide exactly with the $G(N,p)$ model (Li, Deng, and Wang, 2015). For estimating $p_c(k,l)$, one can use the condition that the mean branching of this network should be equal to 1 as in the Molloy–Reed criterion. Clearly, the smallest possible overlaps should determine the percolation threshold, so we consider only overlaps by l vertices. To calculate the average branching in the network of k-cliques, we consider a neighbouring clique B of a randomly chosen clique A (the end of a randomly chosen edge in this specific network), and find the average excess number of cliques, to which k-clique B adjacent. The excess number of l-overlaps equals $\binom{k}{l} - 1$, where the subtracted 1 is due to the overlap with clique A. Each of these l-overlaps should be connected with one of $(k-l)$-cliques outside by $l(k-l)$ edges of the original graph. After we take into account Eq. (6.124), this gives the average branching, which should be equal to 1 at the percolation threshold:

$$\left[\binom{k}{l} - 1\right] \frac{N^{k-l}}{(k-l)!} p_c^{\binom{k-l}{2}+l(k-l)} \cong 1. \tag{6.126}$$

Hence the k-clique percolation threshold probability equals

$$p_c \cong \left[\frac{(k-l)!}{\binom{k}{l} - 1}\right]^{2/[(k-l)(k-1+l)]} N^{-2/(k-1+l)} \tag{6.127}$$

(Bollobás and Riordan, 2009; Fan and Chen, 2014; Li, Deng, and Wang, 2015), satisfying the inequality in Eq. (6.125). In particular: (i) when $k = 2$ and $l = 1$, Eq. (6.127) leads to the threshold for a classical random graph, $p_c = 1/N$; and (ii) when $k = l-1$, Eq. (6.127) reproduces the original result, $p_c = [(k-1)N]^{-1/(k-1)}$, of Palla, et al.

It is natural to consider two 'order parameters': (i) the relative number Ψ of k-cliques in the giant connected component of the k-clique adjacency network

$$\Psi \equiv \frac{N_k^*}{N_k}, \tag{6.128}$$

where N_k^* is the number of these cliques, while N_k is the total number of k-cliques; and (ii) the corresponding relative number Φ of vertices in the k-clique giant connected cluster in the original random graph of size N,

$$\Phi \equiv \frac{N^*}{N}, \tag{6.129}$$

where N^* denotes the total number of these vertices. In other words, Φ is the probability that a randomly chosen vertex belongs to k-clique giant connected cluster. The critical behaviour of the order parameter Ψ for the k-clique adjacency network should be similar to that of the giant component in classical random graphs, that is, the phase transition is continuous. Therefore, near the percolation threshold,

$$\Psi \cong c(k,l)\left[\frac{p}{p_c(k,l,N)} - 1\right], \tag{6.130}$$

where the coefficient $c(k,l)$ depends only on k and l. For estimating the critical singularity of the order parameter Φ, we take into account the average number of k-cliques sharing a uniformly randomly chosen vertex in the original graph,

$$\langle q \rangle_k = \frac{N_k k}{N} \cong \frac{N^{k-1}}{(k-1)!}p^{k(k-1)/2}. \tag{6.131}$$

At the critical point, $\langle q \rangle_k \sim N^{k-1-k(k-1)/(k-1+l)}$, diverging with N if $l > 1$, so the critical slope of $\Phi \sim \langle q \rangle_k \Psi$ diverges, and in the limit of $N \to \infty$, the order parameter Φ approaches the step function: [45]

$$\Phi \cong \theta\left(\frac{p}{p_c(k,l,N)} - 1\right). \tag{6.132}$$

Loosely speaking, at $p \sim p_c$, k-cliques are so numerous, $N_k \sim N^{kl/(k-1+l)} \gg N$ if $l > 1$, that, in the limit $N \to \infty$, any non-zero fraction of them, forming the giant k-clique connected component in the clique adjacency network, should almost surely cover all vertices in the original random graph. On the other hand, when $l = 1$ (overlaps by one vertex), $\langle q \rangle_k$ is size independent, $N_k \sim N$, $\Phi \sim \Psi$ near the critical point, and this transition in the original graph is continuous.

The size of the largest cluster of k-cliques at the critical point, $N_k^{(c)}$, should be of the order of $N_k^{2/3}$ k-cliques, similarly to the classical random graphs. Accounting for Eqs. (6.124) and (6.127) yields the number of cliques in this cluster in terms of N:

$$N_k^{(c)} \sim \begin{cases} N^{2kl/[3(k-1+l)]}, & l < 3(k-1)/(2k-3), \\ N, & l \geq 3(k-1)/(2k-3). \end{cases} \tag{6.133}$$

---

[45]More rigorously, if $k \geq 3$ and $l > 1$, then exactly at the critical point, $\Phi(p = p_c(k,l,N)) \xrightarrow{N\to\infty} a(k,l)$, where $a(k,l)$ is a number depending only on k and l, $0 < a(k,l) < 1$ (Li, Deng, and Wang, 2015).

The upper expression in this formula fails in the case of $l > 3(k-1)/(2k-3)$, when it grows faster than the number of vertices in the original random graph, N. This fast growth is impossible since the largest k-clique cluster at the critical point has a tree-like structure, which constrains the number of overlaps, and hence in this region, the number of cliques should scale as N. Inspecting the inequality $l \geq 3(k-1)/(2k-3)$, we see that it is fulfilled when $k \geq 3$ and $l > 1$, and so in all relevant situations, $N_k^{(c)} \sim N$. Then, due to the tree-like structure of the largest k-clique cluster at the critical point, the number of vertices in the original graph, belonging to these N cliques also scales as N. Hence in the limit $N \to \infty$, $\Phi(p = p_c) \to a(k, l)$, where $a(k, l)$ depends only on k and l, while $\Phi(p < p_c) \to 0$, and $\Phi(p > p_c) \to 1$.

6.13 Explosive percolation

The discontinuous birth of a k-core and other similar hybrid phase transitions appear to be a little bit dull. Indeed, as we saw, they show the universal combination of a discontinuity (jump of an order parameter) and, typically, the square root singularity at the critical point, see Eq. (6.93). Lots of efforts were made to find discontinuous transitions different from the standard hybrid transitions. An attempt by Achlioptas, D'Souza, and Spencer (2009) raised a big wave of studies in this area of research. The idea was to delay a percolation transition by hindering the merger of large connected components, hoping that this delay could make the transition discontinuous. Then, how can this be done? Let us recall how we described the statistics of connected components in the $G(N, p)$ random graphs by the aggregation process: at each step, select two components with probability proportional to their sizes and merge them together. The first thing that comes to mind is to select components for merging with probability proportional not to to the component size s, but, say, to s^α with $0 < \alpha < 1$.[46] It turns out, unfortunately, that while this preferential selection indeed hinders the merging of large components and delays the transition, it does not make this transition discontinuous. Achlioptas, et al. tried another option, namely, they applied the power of choice mechanism for the selection of components for merging, generating the so-called *Achlioptias processes*. Specifically, they use the following model. Initially the network consists of a large number N unconnected vertices, that is clusters (components). At each step sample two times (Figure 6.28):

---

[46]The size independent probability, $\alpha = 0$, would never produce a giant component.

Fig. 6.28 Evolution rules in the model by Achlioptas, D'Souza, and Spencer (2009) of explosive percolation. At each step, two sets of m vertices are chosen at random. Within each set, the vertex in the smallest cluster is selected, and these two vertices are interconnected.

(i) choose $m \geq 1$ vertices uniformly at random and compare the clusters to which these vertices belong; select the vertex within the smallest of these clusters;[47]

(ii) similarly choose uniformly at random the second set of m vertices and, again, as in (i), select the vertex belonging to the smallest of the m clusters;

(iii) add a link between the two selected vertices thus merging the two smallest clusters.

In particular, if $m = 1$, we arrive at ordinary percolation in the $G(N, p)$ model, in which at each step two randomly selected vertices are interconnected. On the other hand, when $m \geq 2$, this process is irreversible and so it does not produce an equilibrium ensemble at a fixed time, unlike $m = 1$. Note that this rule uses only local information in the sense that we do not need to know anything about components other than the selected ones. Achlioptas, D'Souza, and Spencer (2009) simulated this model for $N = 512,000$ and found so sharp a transition that they suggested that it is indeed discontinuous, and therefore named this phenomenon *explosive percolation*. This catchy term quickly became standard and now it is popping up with increasing frequency in numerous articles. Da Costa, Dorogovtsev, Goltsev, and Mendes (2010) soon found, however, that the transition in this model is actually continuous (second-order) although with so small a critical exponent β, $S \propto (t - t_c(m))^\beta$,[48] that it is virtually impossible to distinguish it from discontinuous for any network size accessible in numerical simulations (Figure 6.29). The theory of critical phenomena in this model allows to get exactly all critical exponents and scaling functions (da Costa, *et al.*,

---

[47]In the original work of Achlioptas, D'Souza, and Spencer (2009), m was set to 2.

[48]The time t is rescaled in such a way that it coincides with the average degree of a vertex, $t = \langle q \rangle$, as for the $G(n, p)$ random graph in Section 3.4.

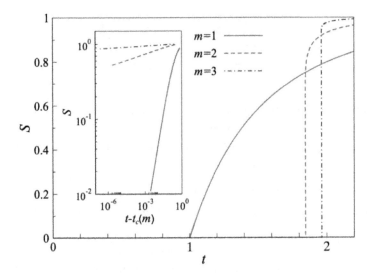

Fig. 6.29 Relative size of the percolation cluster S for $m = 1$, 2, and 3. Despite the visually abrupt behaviour of $S(t)$ at $m = 2$ and 3, the inset shows a power-law approach to the critical points. The slopes of the three curves in the inset (the values of the exponent β, $S \propto (t - t_c(m))^{\beta}$) are 1, $5.557106 \ldots \times 10^{-2}$, and $1.042872 \ldots \times 10^{-2}$. Adapted from da Costa, Dorogovtsev, Goltsev, and Mendes (2010). Note that the process is reversible for $m = 1$ and irreversible for $m \geq 2$.

2010, 2014, 2015), which means the complete description of the continuous phase transition and the solution of the problem for a physicist.[49]

The idea is to proceed in the spirit of the theory of the Erdős–Rényi random graph process described in Sections 3.4 and 3.5, but, in addition to the probability distribution $\mathcal{P}(s)$, to introduce the probability $\mathcal{Q}(s)$ that if we choose uniformly at random m vertices then the smallest of the clusters to which these vertices belong is of size s. The sum rule here is

$$\sum_s \mathcal{Q}(s) = 1 - S^m, \tag{6.134}$$

which reduces to $\sum_s \mathcal{P}(s) = 1 - S$ for $m = 1$, as is natural. The probability distribution $\mathcal{Q}(s)$ can be easily expressed in terms of the distribution $\mathcal{P}(s)$ that a randomly chosen vertex belongs to a cluster of size s. For that, let us introduce the cumulative distributions $\mathcal{P}_{\text{cum}}(s) \equiv \sum_{u=s}^{\infty} \mathcal{P}(u)$ and $\mathcal{Q}_{\text{cum}}(s) \equiv \sum_{u=s}^{\infty} \mathcal{Q}(u)$, so that

[49]The mathematically rigorous proof of the absence of the discontinuity in this transition can be found in Riordan and Warnke (2011).

$$\mathcal{P}(s) = \mathcal{P}_{\text{cum}}(s) - \mathcal{P}_{\text{cum}}(s+1), \tag{6.135}$$

$$\mathcal{Q}(s) = \mathcal{Q}_{\text{cum}}(s) - \mathcal{Q}_{\text{cum}}(s+1). \tag{6.136}$$

Then the probability that the smallest of these m clusters is not smaller than s equals

$$\mathcal{Q}_{\text{cum}}(s) + S^m = [\mathcal{P}_{\text{cum}}(s) + S]^m, \tag{6.137}$$

which should be substituted into Eq. (6.136) providing the expression of $\mathcal{Q}(s)$ in terms of $\mathcal{P}(s)$. We do not show this nonlinear formula here. It is enough to tell that $\mathcal{Q}(s)$, expressed in this way, is determined by the full set of $\mathcal{P}(s')$ with $s' \leq s$. The evolution equation for the aggregation process in this model is similar to that for the Erdős–Rényi random graph process, Eq. (3.23), with the difference that we should substitute $\mathcal{Q}(s,t)$ for $\mathcal{P}(s,t)$ in Eq. (3.23). This gives

$$\frac{\partial \mathcal{P}(s,t)}{\partial t} = \frac{1}{2}s \sum_{u+v=s} \mathcal{Q}(u,t)\mathcal{Q}(v,t) - s\mathcal{Q}(s,t). \tag{6.138}$$

Since $\mathcal{Q}(s)$, is determined by $\mathcal{P}(s')$ with only $s' \leq s$, Eq. (6.138) can be easily solved numerically allowing us to find $S = 1 - \sum_s \mathcal{P}(s)$ (Figure 6.29). Equation (6.138) for $\mathcal{P}(s)$ is bilinear only when $m = 1$, so the generating functions technique used in Section 3.4 ($m = 1$), is not applicable with $m \geq 2$. However, there exists another way to find exactly the critical behaviour, that is, the scaling forms of the distributions $\mathcal{P}(s,t)$ and $\mathcal{Q}(s,t)$, similar to Eq. (3.43). For that we substituted these forms into Eq. (6.138) and also into the expression of $\mathcal{Q}(s)$ in terms of $\mathcal{P}(s)$ and solved the resulting equations for the scaling functions and the critical exponents. The obtained values of the critical exponent β turned out to be much smaller than 1 when $m \geq 2$ and to decrease to 0 as $m \to \infty$, which results in a very 'sharp' second-order phase transition. This model is infinite-dimensional and the theory is of mean-field type, but knowing the critical exponents above the upper critical dimension, one can easily find the value of this dimension, d_{uc} as is shown in Appendix D. In this way, we got $d_{\text{uc}}(m)$ for arbitrary finite m, including $d_{\text{uc}}(1) = 6$, $d_{\text{uc}}(2) = 2.125\ldots$, and so on, and $d_{\text{uc}}(\infty) = 2$. This means that if we reformulate this model with $m \geq 2$ for a D-dimensional lattice, then for $D \geq 3$, our theory is asymptotically exact and even for a two-dimensional lattice, the critical exponents are close to those in our theory as d_{uc} is close to 2 for any $m \geq 2$. Other than small β for $m \geq 2$, the scaling behaviour in this model does not differ principally from what we described for $m = 1$ (classical random graphs) in Sections 3.4 and 3.5. It is still worthwhile to mention that in the normal phase $(t < t_c)$ the scaling functions of $\mathcal{P}(s,t)$ and $\mathcal{Q}(s,t)$ for $m \geq 2$ have a high peak in contrast to the dull exponential decay for $m = 1$, Eq. (3.46). This peak is produced by large finite components whose growth is suppressed by the power of choice rule.

We suggest that percolation models of this sort, based on a local choice (local optimization) cannot produce discontinuous phase transitions. On the other hand, if a model utilizes a global choice rule, that is, this rule uses some information about all components in the system—global information, then the discontinuous transition is possible. As a good example of such a system with a discontinuous transition, we briefly discuss the *restricted Erdős–Rényi network* (Panagiotou, Spöhel, Steger, and Thomas, 2011) whose theory was developed by Cho, Lee, Herrmann, and Kahng (2016) and Park, Yi, Choi, Lee, and Kahng (2019). Let g: $0 < g \leq 1$, be a parameter in this model. The process begins with $N \gg 1$ isolated vertices, and at each step two vertices are linked by an edge applying the following rule:

- At each step, rank all connected components in the network in ascending order, according to their sizes $s_1, s_2, \ldots, s_n$, so $\sum_i s_i = N$.[50] Divide these components into two sets: the smaller components $i = 1, 2, \ldots, k$ containing $N_A = \sum_{i=1}^{k} s_i$ vertices, where $N_A - s_k < gN \leq N_A$ (set A) and the larger components, containing the remaining vertices (set B).
- Choose uniformly at random a vertex in the set A of components and connect it to a uniformly randomly chosen vertex in the network (if this edge already exists, abandon this attempt).[51]

Since any two components in set B cannot merge together, the growth of large finite components is suppressed compared to the Erdős–Rényi graph process in Sections 3.4 and 3.5. It turns out that this suppression is essentially stronger than in the Achlioptas process,[52] and it is sufficient to produce a discontinuous transition at some moment $t_c(g)$. In the normal phase, the distribution $\mathcal{P}(s,t)$ has a power-low shape corrected by a bump formed by components of set B—the 'powder keg'—in the range of large s. This is similar to the Achlioptas process with the marked difference that this distribution cannot be described by using a scaling form like Eq. (3.43)

---

[50]If a few of the components have equal sizes, rank them in arbitrary order.

[51]Apparently, if we slightly modify the rule interconnecting only vertices within set A, the picture should not change much.

[52]Indeed, in the Achlioptas process, two large components still have some chance to merge together, while here this chance is zero.

and that the bump here contains a finite fraction of vertices. It was found that in a vanishingly narrow region around t_c the giant connected component with the relative size $S_c = 1 - g$ emerges taking all the vertices from set B (that is, from the bump in the distribution). The bump transforms into the giant component, and the distribution appears to be asymptotically power-law at this point,

$$\mathcal{P}_c(s) \cong As^{1-\tau(g)}. \tag{6.139}$$

Cho, Lee, Herrmann, and Kahng (2016) showed that the exponent τ increases from 2 to $2 + 1/2$ as the parameter g increases from 0 to 1. On the other hand, when the size of the largest cluster (giant component) exceeds $(1 - g)N$, set B disappears, and we arrive at the standard Erdős–Rényi process with the initial condition: $\mathcal{P}(s, t_c(g)) \equiv \mathcal{P}_c(s)$ and $S(t_c) = 1 - g$. So above t_c, the giant component should grow continuously. One can show that this initial condition leads to the singularity

$$S(t) - (1 - g) \cong B(t - t_c)^{\tau-2}. \tag{6.140}$$

Thus there is the combination of the discontinuity and the singularity, as in the standard hybrid transitions, but the exponent of the power-law singularity here differs from $1/2$.

6.14   Largest component in finite networks

Rigorously speaking, a giant connected component and other giant components of that sort are well defined only for infinite networks. For finite networks, their role is played by the largest components. In a finite random network, the largest component sizes differ in different realizations of the network, and for the description of their statistics, one should turn to the extreme value theory. According to the *large deviation principle* (den Hollander, 2000; Touchette, 2009) the distribution $\mathcal{P}(S, N)$ of the relative size S of the largest component (without specifying its kind) in a large random network asymptotically scales with N in the following way:

$$\mathcal{P}(S, N) = e^{-N\Phi(S)+o(N)}, \tag{6.141}$$

where the Landau symbol $o(N)$ indicates terms of order less than N and $\Phi(S)$ is the so-called *rate function*.[53] This general principle can be directly checked by measuring the statistics of largest components in simulated large

---

[53]Looking at this formula, readers who know statistical physics will recall the corresponding relation for fluctuations of the order parameter (Landau and Lifshitz, 2013).

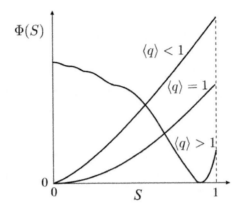

Fig. 6.30 Schematic view of the rate function $\Phi(S)$ according to findings of O'Connell (1997, 1998) for the largest connected component of an Erdő–Rényi graph. Notice that the rate function is convex only for $\langle q \rangle \leq 1$. Similar rate functions are generic for all types of largest components that demonstrate a continuous transition in infinite networks.

random networks; see Schawe and Hartmann (2019) for such a check for large Erdő–Rényi graphs, enabling one to obtain the rate function. This principle, Eq. (6.141), is valid for any value of a control parameter. In this section we focus on those specific components that demonstrate a continuous transition in infinite networks, including connected components, 2-cores, k-connected components, and so on. We discuss the range of the control parameter values corresponding to the phase with a giant component in an infinite network.

The rate function $\Phi(S)$ was found explicitly for the largest connected component in the $G(N, p)$ model (O'Connell, 1997, 1998). This exactly solvable case allows one to grasp the essence of the problem. Figure 6.30 schematically shows the rate function below, at, and above the critical point of this random graph. At the critical point, $\Phi''(0) = 0$. Above the critical point, in the neighbourhood of the minimum of the rate function, we can approximate it as $\Phi(S) = \text{const} + \frac{1}{2}\Phi''(\langle S \rangle)(S - \langle S \rangle)^2$, which asymptotically leads to the normal law for the distribution $\mathcal{P}(S)$. Computing the first and the second moments of this Gaussian distribution, we find the limit

$$N(\langle S^2 \rangle - \langle S \rangle^2) \overset{N \to \infty}{\cong} \frac{1}{\Phi''(\langle S \rangle)}. \qquad (6.142)$$

The left-hand side can be interpreted as susceptibility (see Section 3.5). Using the explicit shape of $\Phi(S)$ for an Erdő–Rényi graph, one can easily check that near the critical point this susceptibility has the asymptotics

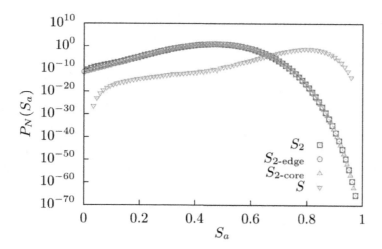

Fig. 6.31 The distributions of the relative sizes: of the largest connected component S, of the 2-core, $S_{2\text{-core}}$, of the biconnected (vertex and edge) components, S_2 and $S_{2\text{-edge}}$, respectively for the Erdő–Rényi graph of $N = 500$ vertices with $\langle q \rangle = 2$. S_a stays for any of these relative sizes. Notice over 70 orders of magnitude variation of the distributions. Adapted from Schawe and Hartmann (2019).

$1/(\langle q \rangle - 1)$ exactly coinciding with the critical asymptotics of the average size of a finite connected component to which a uniformly randomly chosen vertex belongs, Eq. (3.37).[54]

Hartmann (2011, 2017) and Schawe and Hartmann (2019) developed algorithms enabling them to obtain numerically the rate functions and distributions not only for largest connected components, but also for 2-cores and biconnected (vertex and edge) components in various random networks. Figure 6.31 shows some of their results for the Erdő–Rényi graph. Notice the closeness of the distributions for the largest components of the 2-cores and the biconnected components, which is natural given the local tree-likeness of a sparse Erdő–Rényi graph, as mentioned in Section 6.10.

For locally tree-like networks, the fluctuations of the largest component can be studied analytically by applying the message-passing techniques (Bianconi, 2018*b*, 2019). This techniques is also applicable to more difficult situations, where a phase transition in the corresponding infinite networks is discontinuous (hybrid) (Coghi, Radicchi, and Bianconi, 2018).

---

[54]Be careful doing this check: there is a typo in the formula for $\Phi(S)$ in the original work of O'Connell (1997). Instead, one can use the corresponding formulas from Hartmann (2011, 2017).

We touched upon only one of the problems involving fluctuations of the largest component in different realizations of a random network. One can ask other interesting questions even for infinite networks. For example, let us damage a network twice, deleting at random a given fraction of its vertices or edges, and inquire about correlations between the impacts of these two realizations of a random damage. What is a fraction of vertices occurring in a giant connected component in both cases? What is a fraction of vertices remaining out of a giant connected component in both samples? See the answers to these questions found by using message passing in Bianconi (2017).

6.15   Cores and controllability

A *matching (independent edge set)* of an undirected graph is a subset of its edges having no common vertices (Hartmann and Weigt, 2006). A *vertex cover* of an undirected graph is a subset of its vertices such that any edge of the graph is adjacent to at least one vertex from this subset. Finding a *maximum matching* (containing the largest possible number of edges) and a *minimum vertex cover* of a graph are among key combinatorial optimization problems. *Perfect matching* (a matching that matches all vertices) is related to statistics of dimers in physics, while the minimum vertex cover problem has a vivid representation as the task of hiring the minimum number of guards that succeed to watch all corridors in a museum (Weigt and Hartmann, 2000). For many important graphs, including the bipartite graphs, the maximum matching problem and the minimum vertex cover one are equivalent and can be solved rapidly, in a polynomial time. The solution can be found by the *Karp–Sipser greedy leaf removal algorithm* (Karp and Sipser, 1981). The leaves (vertices of degree 1) and their nearest-neighbouring vertices are progressively removed from an undirected graph in any order until it is possible. The edges of the removed dimers during this process form a maximum matching. It turns out that in sufficiently dense graphs, at some point of this leaf removal process, no leaves remain in a still not eliminated graph, the process becomes stacked, and extra efforts have to be applied to proceed further and complete the algorithm. In this situation, the algorithm slows down and it is an NP-complete problem.[55] In sparse networks, this problem is treatable in the framework of statistical physics by the cavity method—belief propagation (Zhou and

---

[55]These extra efforts, in essence, are limited to the so-called 'randomized edge removal'—a pair of randomly chosen neighbouring vertices is removed, which gives a

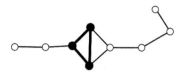

Fig. 6.32 The core (filled vertices and thick edges) of a small sample graph. Open vertices are removed by the greedy leaf removal algorithm.

Ou-Yang, 2003; Zdeborová and Mézard, 2006), and in the simplest case of an Erdős–Rényi random graph, the boundary between these two regimes occurs at the mean degree $\langle q \rangle = e = 2.718\ldots$. The same two regimes and the same boundary $\langle q \rangle = e$ in the case of an Erdős–Rényi random graph were found in the minimum vertex cover problem AKA searching for the maximum number of museum guards (Weigt and Hartmann, 2000).

The subgraph remaining after the completion of the greedy leaf removal algorithm is called the *core* of a graph (Figure 6.32). Bauer and Golinelli (2001*a*) developed the theory of cores in undirected uncorrelated networks, enabled them to compute the thresholds of the core percolation, the core sizes, and their characteristics.[56] Liu, Csóka, Zhou, and Pósfai (2012) extended this theory to the important case of directed uncorrelated networks. Let us outline the idea of these works in the more compact case of undirected uncorrelated networks.

Each vertex in an undirected graph strictly falls into one of three classes based on what will happen with it in the leaf removal process. One can show that this classification is unique for each vertex, although we shall not do it here since it is apparent for locally tree-like networks:

- α-removable vertices are those vertices that occur as leaves at some moment of the leaf removal process.
- β-removable vertices are the nearest neighbours of these leaves, and so they all will be removed together.
- None-removable vertices are just the vertices in the core, if it exists.

Let α be the probability that if we choose uniformly at random an edge and remove one of its end vertices, then the second end occurs as an α-

---

chance for the emergence of new leaves, enabling one to restart the greedy leaf removal process and continue it until it becomes stacked again, and so on.

[56]For possible generalizations of the core problem, see Zhao, Zhou, and Liu (2013) and Azimi-Tafreshi, Osat, and Dorogovtsev (2019). For the theory of cores in hypergraphs, see Coutinho, Wu, Zhou, and Liu (2020).

removable vertex. In other words, this is the probability that a random neighbour of a random vertex i is α-removable after vertex i is removed. Similarly, let β be the probability that if we choose uniformly at random an edge and remove one of its end vertices, then the second end occurs as an β-removable vertex. In other words, this is the probability that a random neighbour of a random vertex i is β-removable after vertex i is removed. Finally, $1 - \alpha - \beta$ is the probability that if we choose uniformly at random an edge and remove one of its end vertices, then the second end is neither α-removable nor β-removable. We use a convenient graphical representation of these probabilities shown in Figure 6.33a.

Figure 6.33b, c explains the form of the equations for α and β and the expression for the relative size of the core, S_{core}, under the assumption that a network is locally tree-like and, moreover, uncorrelated,

$$\alpha = \sum_q \frac{qP(q)}{\langle q \rangle} \beta^{q-1}, \tag{6.143}$$

$$\beta = \sum_q \frac{qP(q)}{\langle q \rangle} [1 - (1 - \alpha)^{q-1}], \tag{6.144}$$

$$1 - \alpha - \beta = \sum_q \frac{qP(q)}{\langle q \rangle} \{(1 - \alpha)^{q-1} - [(1 - \alpha) - (1 - \alpha - \beta)]^{q-1}\}, \tag{6.145}$$

$$S_{\text{core}} = \sum_q P(q) \sum_{s=2}^q \binom{s}{q} (1 - \alpha - \beta)^s \beta^{q-s}. \tag{6.146}$$

Equation (6.145) is actually superfluous, since it follows from Eqs. (6.143) and (6.144). In addition, clearly, the fraction of edges belonging to the core is expressed in terms of the probability $1 - \alpha - \beta$ as

$$\frac{E_{\text{core}}}{E} = (1 - \alpha - \beta)^2. \tag{6.147}$$

Using generating functions, we rewrite Eqs. (6.143) and (6.144) in the compact form

$$\alpha = G_1(\beta), \tag{6.148}$$

$$\beta = 1 - G_1(1 - \alpha), \tag{6.149}$$

which readily results in the equation for the probability α:

212

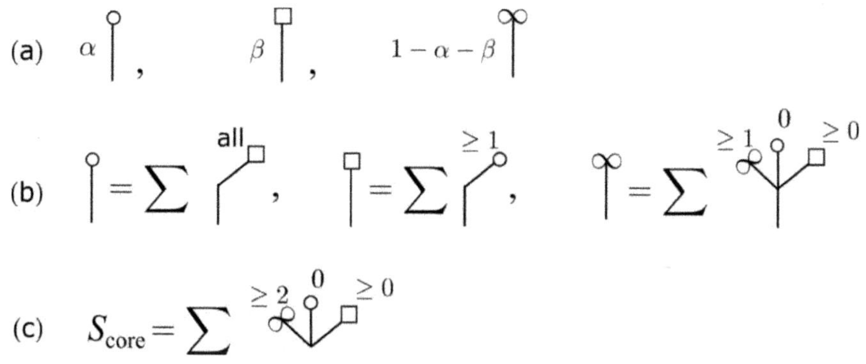

Fig. 6.33 Graphical representations of equations for the involved probabilities in the core problem. (a) Graphical notations for the probabilities α, β, and $1-\alpha-\beta$. (b) Equations for the probabilities α, β, and $1 - \alpha - \beta$, which are Eqs. (6.143), (6.144) and (6.145), respectively, in graphical form. (c) Expression for the relative size of the core in graphical form, Eq. (6.146).

$$\alpha = G_1(1 - G_1(1 - \alpha)). \tag{6.150}$$

Figure 6.34 explains solutions of this equation, why the smallest root must be chosen, and demonstrates that the phase transition is continuous. The critical point is obtained from Eq. (6.150) together with the second condition,

$$1 = \frac{d}{d\alpha}G_1(1 - G_1(1 - \alpha)) \tag{6.151}$$

(Figure 6.34b).

Analyzing these equations, one can quickly arrive at the following conclusions. Below the critical point, $1-\alpha-\beta = 0$ and $\alpha, \beta > 0$. Above the critical point, $1 - \alpha - \beta, \alpha, \beta > 0$. For the Erdős–Rényi random graph, the critical point is at $\langle q \rangle = e$. Near the critical point, S_{core}, $E_{\text{core}}/E \propto (\langle q \rangle - e)$. For all uncorrelated scale-free networks with a degree distribution $P(q) \sim q^{-\gamma}$, $\gamma > 2$, where $P(1) > 0$, the transition is similar,

$$S_{\text{core}}, \frac{E_{\text{core}}}{E} \propto \langle q \rangle - \langle q \rangle_c(\gamma), \tag{6.152}$$

where the critical value of the average degree diverges when $\gamma \downarrow 2$. This divergence, $\langle q \rangle_c(\gamma \downarrow 2) \to \infty$, that is, the disappearance of a core in these networks, is quite opposite to what happens with the point of the birth of a giant connected component in networks with fat-tailed degree distributions, where $\langle q \rangle_c(\gamma \leq 3) \to 0$, corresponding to the hyper-resilience of a giant connected component against random damaging.

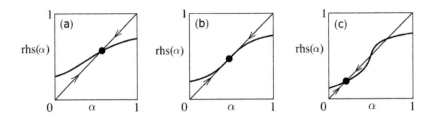

Fig. 6.34 Graphical solution of Eq. (6.150) for the probability α. (a) The phase without a core. (b) The critical point. (c) The phase with a core. Looking at the arrows, one can figure out which of the fixed points (solutions) is stable.

In directed graphs, the leaves are the vertices with a single edge, in- or out-, and the core is a subgraph, remaining after the progressing removal of all such leaves together with their nearest neighbours. For directed uncorrelated networks, the size of the core is computed in a similar way as above for uncorrelated undirected ones, but the phase transition for a core turns out to be markedly different (Liu, Csóka, Zhou, and Pósfai, 2012). Let the in- and out-degrees of the individual vertices of a network be independent, that is, the in-, out-degree distribution factorizes as $P(q_i, q_o) = P_i(q_i)P_o(q_o)$. Then the phase transition is continuous only if the distributions $P_i(q_i)$ and $P_o(q_o)$ coincide. Otherwise, the transition is hybrid, and the discontinuity increases with the difference between these two distributions.

The problem of cores in directed networks is particularly important, since it is directly related to the *structural controllability* of complex linear dynamical systems defined on directed network substrates (Liu, Slotine, and Barabási, 2011; Pósfai, Liu, Slotine, and Barabási, 2013). Let the N-component vector $\mathbf{x}(t)$ define the current state of a linear dynamical system of coupled N vertices, and the second ($M \leq N$)-component vector $\mathbf{u}(t)$ is for time-dependent inputs at some of these vertices. The evolution of the vector $\mathbf{x}(t)$ is described by the linear equation:

$$\frac{d\mathbf{x}(t)}{dt} = A\mathbf{x}(t) + B\mathbf{u}(t), \qquad (6.153)$$

where A is the $N \times N$ coupling matrix, which is in essence an adjacency matrix of a directed weighted network (coupling network), B is the $N \times M$ input matrix with real elements. Loosely speaking, the term 'controllability' means the possibility to guide the entire vector $\mathbf{x}(t)$ to a desired state by varying inputs—entries of $\mathbf{u}(t)$—of a limited set of 'driver vertices'. The adjective 'structural' indicates that the structure of the unweighted projection of a given coupling network guarantees that the linear dynamical system will be controllable if we properly choose the parameters of this system. The question is: what is the minimal number of driver nodes sufficient to control the entire system? Liu, *et al.* mapped this problem to finding the maximum matching for these unweighted projections of coupling networks, that is, to finding a largest set of edges without shared vertices by the greedy leaf removal algorithm, which we discussed earlier. The driver vertices in this mapping are vertices left unmatched. This scheme provides a way to the analytical treatment of the problem. The reader will find the details in the Supplementary Information to Liu, Slotine, and Barabási (2011) and in Zhao and Zhou (2019).

7

Epidemics and Spreading Phenomena

7.1 Bootstrap percolation

In this chapter we mainly focus on the results of activation processes in networks and on various combinations of activation and deactivation processes.

The *bootstrap percolation problem* is about the basic activation process on networks, in which vertices can be in active and inactive states. A vertex becomes active when the number of its active neighbours exceeds some threshold; and once active, a vertex never becomes inactive (Adler and Aharony, 1988; Adler, 1991). This is one of the spreading processes with discontinuous phase transitions (Bizhani, Paczuski, and Grassberger, 2012). Let us define bootstrap percolation on undirected graphs in more strict terms. In the initial state, a fraction f of vertices is active (seed vertices). These vertices are chosen uniformly at random. Each inactive vertex becomes active if it has at least k_b active nearest neighbours. Here we introduce the subscript 'b' to distinguish this threshold from a threshold in the k-core percolation problem. In heterogeneous bootstrap percolation, the thresholds are individual for different vertices, $k_{b,i}$, where $i = 1, 2, \ldots, N$. The main point of the interest is the final state of the process, in which the fraction $S_b \geq f$ of all vertices are active and, assuming that a network is infinite, the fraction $S_{Gb} \leq S_b$ of vertices with their connections form a giant active component.

It is reasonable to compare bootstrap percolation with another threshold process already discussed in Chapter 6, namely, with the heterogeneous k-core problem. Speaking in terms of active and inactive vertices, a k-core is the final result of the progressive deactivation of initially active vertices, where each vertex i becomes inactive if it has less than $k_{c,i}$ active nearest neighbours. These two processes are complementary if the individual thresholds $k_{b,i}$ and $k_{c,i}$ for each vertex i in a network are related with each other in the following way:

$$k_{b,i} = q_i + 1 - k_{c,i}, \tag{7.1}$$

where q_i is the degree of vertex i (Miller, 2016; Di Muro, Valdez, Stanley, Buldyrev, and Braunstein, 2019). One can easily obtain this relation. The point is that an active vertex at each moment of the bootstrap percolation process is inactive at the corresponding moment of the complementary k-core process.[1] So at this moment, the numbers of active nearest neighbours of vertex i, $a_{b,i}$ for bootstrap percolation and $a_{c,i}$ for the k-core process, should be related, $a_{c,i} = q_i - a_{b,i}$. The condition of the activation of inactive vertex i in the heterogeneous bootstrap percolation is $a_{b,i} \geq k_{b,i}$ while the condition of the deactivation of active vertex i in the heterogeneous k-core problem is $a_{c,i} = q_i - a_{b,i} \leq k_{c,i} - 1$. These two inequalities coincide when the equality in Eq. (7.1) is true.

In particular, initially active vertices in bootstrap percolation (seed vertices) can be treated as having zero threshold of activation, $k_{b,i} = 0$, which, according to Eq. (7.1), corresponds to the threshold $k_{c,i} = q_i + 1$ for deactivation of vertices in the heterogeneous k-core problem. On the other hand, the threshold $k_{c,i} = 1$ of vertices in the heterogeneous k-core problem, considered in Section 6.7, corresponds to the activation threshold $k_{b,i} = q_i$ in bootstrap percolation, that is, to activate such a vertex, all its nearest neighbours have to be active. Furthermore, in the final states of these two complementary processes with the thresholds satisfying Eq. (7.1), the relative number of active vertices in bootstrap percolation, S_b, and the relative size of the k-core, S_k, which is the relative number of vertices remaining active after the completion of the deactivation (pruning) process, are strictly related, $S_k = 1 - S_b$. There is no similar relation between the giant components of active vertices in these problems. Notice, however, that the less-interesting giant component of inactive vertices in the final state of the bootstrap percolation process coincides with the giant k-core in the complementary k-core problem. In Di Muro, Valdez, Stanley, Buldyrev, and Braunstein (2019), all these statements were supported by a theory for tree-like networks, based on equations similar to those for heterogeneous k-cores.[2] We do not show here the equations allowing us to obtain the final fraction of active vertices.

---

[1]Here we do not specify for bootstrap percolation how rapidly vertices become activated and so cannot describe quantitatively the temporal evolution of the activation process. In activation processes, at successive time steps, all vertices update their states in random asynchronous order in accordance with the threshold rule of a model or in parallel, synchronously. In bootstrap percolation, the final, steady state of the system is the same for both these types of updates.

[2]Note that in bootstrap percolation, this work takes into account a term for the giant component of active vertices missed in Baxter, Dorogovtsev, Goltsev, and Mendes (2011).

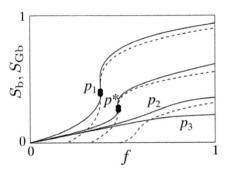

Fig. 7.1 Relative number S_b of active vertices in the final state of the bootstrap percolation process and relative size S_{Gb} of a giant component of active vertices vs. fraction f of seed active vertices in a network with rapidly decaying degree distribution (solid and dashed curves, respectively), a qualitative plot. The probabilities $1 \geq p_1 > p^* > p_2 > p_3$ at the curves are the occupation probabilities of vertices, controlling the structure of a network. In the range $1 < p < p^*$, a hybrid phase transition occurs with the square root singularity at the normal phase side of the transition, $f < f_c$. At the vertex occupation probability p^*, the discontinuity disappears, and the singularity is a cube root one. The dots on the curves indicate the points of these singularities. For the vertex occupation probability p_3, a giant component is absent at any f.

The reader can easily guess their structure from the equations for the Watts model in Section 7.2.

A hybrid transition in this problem is similar to that for heterogeneous k-cores with the difference that the square root singularity here is at the normal phase side of the transition, $f < f_c$, (Baxter, Dorogovtsev, Goltsev, and Mendes, 2010) (Figure 7.1). For this discontinuous transition, the subcritical clusters, consisting of vertices which are not seed vertices and which have exactly $k_b - 1$ active neighbours, play a special role. Activation of an extra neighbour of a vertex in such a cluster launches an avalanche activating the entire cluster. The average size of these avalanches, that is, an average size of a subcritical cluster to which a randomly chosen vertex belongs, diverges at the critical point f_c as

$$\langle s \rangle_{\text{sub}} \propto (f_c - f)^{-1/2}, \tag{7.2}$$

while the critical distribution of the sizes of avalanches decays as $\mathcal{P}(s) \sim s^{-3/2}$.

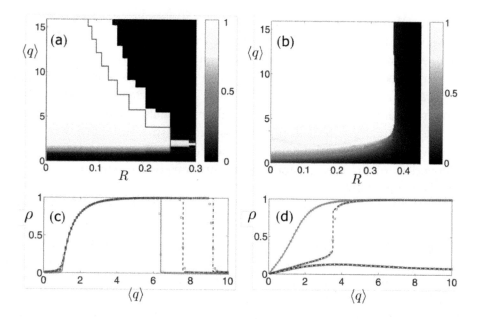

Fig. 7.2 Final density ρ of active vertices in the Watts model on an Erdős–Rényi graph with an average degree of a vertex $\langle q \rangle$ and (a,c) identical thresholds R for all vertices or (b,d) the Gaussian threshold distribution with mean R and standard deviation 0.2. (a) and (b) show ρ on the $R-z$ plane (value of ρ scales from 0, black, to 1, white) for the initially active fraction of vertices $\rho_0 = 10^{-2}$ and 0, respectively. The solid line in (a) shows the cascade boundary for the initially active fraction of vertices $\rho_0 \to 0$. (c) ρ vs. $\langle q \rangle$ for the uniform threshold $R = 0.18$ and $\rho_0 = 10^{-3}$, 5×10^{-3}, and 10^{-2} (solid, dashed, and dot-dashed lines, respectively). The arrow indicates the boundary of cascade for $\rho_0 \to 0$. (d) ρ vs. $\langle q \rangle$ for the Gaussian threshold distribution with mean $R = 0.2$, 0.362, and 0.38 (solid, dashed, and dot-dashed lines, respectively) and standard deviation 0.2, $\rho_0 \to 0$. Note that non-positive thresholds r present in a Gaussian distribution in (b) and (d) effectively increase the fraction of seed vertices from zero. Adapted from Gleeson and Cahalane (2007).

7.2   The Watts model

In a wide sense, the *Watts model* (sometimes the *Watts threshold model*) often refers to a class of models in which each vertex i in a network has an activation threshold r_i drawn from some distribution, and this vertex becomes active when a given function $\mathcal{F}(m_i, q_i)$ of the number of its active nearest neighbours, m_i, and the degree q_i equals at least r_i. Once active, a vertex never becomes inactive. Bootstrap percolation and several other percolation and epidemic problems sit inside of this class of models (Miller,

2016). In Watts's original work (2002), this model was formulated in the following, more narrow form. Initially, a fraction ρ_0 of vertices (seed vertices) is activated. This fraction can be even 0, which corresponds to a finite number of seed vertices. Each vertex has a random threshold r (real number) drawn from a distribution $f(r)$.[3] Each inactive vertex i becomes active if the fraction of its active nearest neighbours equals or exceeds the threshold, $m_i/q_i \geq r_i$. The activation process ends with a fraction ρ of vertices activated, $\rho_0 \leq \rho \leq 1$. The final steady state of the process in this model is independent of the type of update, synchronous or asynchronous (Gleeson and Cahalane, 2007).

Let us introduce the probability $F(r) \equiv \int_{-\infty}^{r} du f(u)$ that a vertex has a threshold equal to or less than r. Then if a vertex of degree q has m active nearest neighbours, it become active with the probability $F(m/q)$. We assume that a random network is uncorrelated and hence locally tree-like. To obtain the final fraction of active vertices ρ,[4] we introduce the probability Z that if we choose uniformly at random an edge, choose one of its end vertices, and delete the edge, then this vertex is active either because it is a seed vertex or because it has a sufficient number of active neighbours (without accounting for the second end of the deleted edge). The tree-likeness of the network readily leads to the following equation for Z (Watts, 2002; Gleeson and Cahalane, 2007):

$$Z = \rho_0 + (1 - \rho_0) \sum_{q=1}^{\infty} \frac{qP(q)}{\langle q \rangle} \sum_{m=0}^{q-1} F\left(\frac{m}{q}\right) \binom{q-1}{m} Z^m (1-Z)^{q-1-m}. \quad (7.3)$$

The probability ρ that a vertex is active also has two contributions. Either a vertex, say i, is active because it is one of seed vertices or this vertex was activated due to a sufficient number of already active nearest neighbours, that is of those neighbours that became activated by other vertices than vertex i. Then ρ is expressed in terms of Z in the following way:

$$\rho = \rho_0 + (1 - \rho_0) \sum_{q=1}^{\infty} P(q) \sum_{m=0}^{q} F\left(\frac{m}{q}\right) \binom{q}{m} Z^m (1-Z)^{q-m}. \quad (7.4)$$

[3]Vertices with $r \leq 0$ can be interpreted as initially active. Hence, without loss of generality, we can assume that $r > 0$.

[4]Note that we are not interested here in the giant component of active vertices, which can also be found.

In this model, the evolution scenarios resulting in a large fraction of active vertices while starting with a small fraction ρ_0 of seed vertices can be naturally interpreted as *global cascades*. The analysis of Eq. (7.3) for Z shows that for some of distributions $f(r)$ of threshold values, the global cascades may occur even for $\rho_0 \to 0$, that is, for a finite number of seed vertices (Figure 7.2). Without loss of generality, one can restrict the range of thresholds to $r > 0$. In addition, we assume that $F(0) = 0$. Then linearizing Eq. (7.3), one can get the following condition of a global cascade at $\rho_0 \to 0$:[5]

$$\sum_{q=1}^{\infty} \frac{q(q-1)P(q)}{\langle q \rangle} F(1/q) > 1 \qquad (7.5)$$

(Watts, 2002). Compare Eq. (7.5) with the Molloy–Reed criterion of the existence of a giant connected component in uncorrelated networks, $\sum_q q(q - 1)P(q)/\langle q \rangle > 1$, Eq. (6.25). The only difference is the factor $F(1/q)$ within the sum in Eq. (7.5), which is the probability that a vertex is 'vulnerable' in the sense that its threshold is sufficient for activation even by a single active nearest neighbour.

In the simplest case of a uniform threshold for all vertices, that is for $f(r) = \delta(r - R)$, where $\delta(x)$ is a delta-function, and hence $F(r) = \theta(r - R)$, where $\theta(x)$ is the theta step function, Figure 7.2a, (c) show a final fraction of active vertices in an Erdős–Rényi graph. The global cascade area in Figure 7.2a is restricted by a stepwise boundary obtained from Eq. (7.5) in the case of ρ_0, where the steps are due to the uniform threshold. Notice in Figure 7.2c that for $\rho_0 \to 0$, a finite final fraction of active vertices, ρ, emerges after a continuous transition, which is the lower border of the global cascade area. On the other hand, at the second special (higher) value of an average degree, ρ discontinuously drops from almost 1 to almost 0, which is the upper border of the global cascade area. For some threshold distributions $f(r)$, discontinuity in the dependence $\rho(\langle q \rangle)$ occurs as a jump up, or it may be even absent (see Figure 7.2d for a Gaussian threshold distribution).

The so-called *generalized epidemic process* (Dodds and Watts, 2004, 2005; Janssen, Müller, and Stenull, 2004) provides an alternative view of activation or complex contagion or epidemic phenomena discussed in these sections. Vertices in this process can be in one of the two states: active and inactive. Once active, a vertex never become inactive; once a vertex is active, it

---

[5]For the condition of a global cascade for a finite fraction of seed vertices, ρ_0, see Gleeson and Cahalane (2007).

attempts to activate once every one of its inactive nearest neighbours. The probability that a vertex becomes active depends of the number of these attempts of the active nearest neighbours. Let p_m, where $m = 0, 1, ..., q - 1$, be the conditional probability for a given vertex that if first m attempts of neighbours are all failed, then $(m+1)$-th attempt finally activates this vertex. Note that, in general, the probabilities p_m can be set differently for different vertices. In particular, one can introduce degree-dependent $p_m(q)$. The probability that a vertex is inactive after m attempts of activation ('attacks' from active neighbours) equals $\prod_{s=0}^{m-1}[1 - p_s]$, and, clearly, the probability that a vertex is active after m attempts of activation equals $1 - \prod_{s=0}^{m-1}[1 - p_s]$. The probability that the $(m+1)$-th 'attack' succeeds to activate a vertex equals $p_m \prod_{s=0}^{m-1}[1 - p_s]$.

Miller (2016) showed that the generalized epidemic process is equivalent to the Watts model treated in a wide sense. For a uniform probability $p_0 = p$ and $p_m = 0$ for $m \geq 1$ (that is, if the first attack fails, the later ones also fails), the generalized epidemic process reduces to site percolation. In this case, for any (non-zero) number of active neighbours of a vertex, the probability of its activation equals p. For a uniform probability $p_m = p$ for $m \geq 0$ the generalized epidemic process reduces to bond percolation, or, equivalently, to the SIR model, discussed in the following sections. If every attempt of activation of a vertex makes it more resilient (less susceptible), then p_m decreases with m. If every attempt of activation of a vertex makes it less resilient, then p_m increases with m.

It turns out that when p_m decreases with m or it is independent of m, or even when p_m increases with m sufficiently slowly, then the phase transition in this process (emergence of a giant cluster of active vertices) is continuous. See the theory in the supplementary material of Bizhani, Paczuski, and Grassberger (2012). On the other hand, for sufficiently rapidly increasing p_m, the transition appears to be discontinuous, of the first order. On a phase diagram of the system (the generalized epidemic process), the lines of these transitions, continuous and first order, meet in a special, tricritical point.

Although the Watts model and the generalized epidemic process include a range of activation and epidemic processes, in the following sections we treat all particular epidemic models directly, demonstrating useful basic approximations.

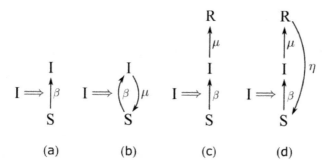

Fig. 7.3 Transitions between susceptible (S), infective (I), and removed (R) states of vertices in basic epidemic models. (a) The SI model. (b) The SIS model and contact process. (c) The SIR model. (d) The SIRS model. The transition $S \rightarrow I$ (transition rate β) is induced by infective nearest neighbours ($I \Rightarrow$). μ and η are the transition rates of spontaneous transitions between indicated states.

7.3   Main epidemic models

We consider 'compartmental' epidemic models with only two or three states of a vertex: *susceptible*, *infective*, and *removed* (recovered). This number of states is sufficient to produce a rich set of behaviours even without introducing activation thresholds. The standard terms of epidemiology—'susceptible', 'infective', and 'removed'—will be clear from the formulations of specific models. Succinctly, the 'susceptible', and 'infective' states of individual vertices usually correspond to the inactive and active states in the preceding sections, and a 'removed' vertex neither influences the states of other vertices in a network nor its state can be influenced by them.

Figure 7.3 explains the organization of the four basic epidemic models—the SI, SIS, SIR, and SIRS models—and of the contact process.[6] In each of the models, the transition from the susceptible (S) to the infective (I) states of a vertex is induced by infective nearest neighbours (transition rate β, also called effective spreading rate or transmission rate), see details in the next sections, where we also explain the difference between the SIS model and the contact process. The rest transitions between states are spontaneous with transition rates μ and η. Often, one of the spontaneous transition rates, for example, μ, is set to 1 by rescaling time and then the transition rate β

---

[6]More realistic epidemic models include one more state of vertices—*exposed*, representing the latent and, partly, incubation periods of diseases. In these models, before a susceptible vertex becomes infective it passes through the exposed state. This addition produces the SEI, SEIS, SEIR, and SEIRS epidemic models.

rescales to $\lambda \equiv \beta/\mu$. In the following sections, we denote the fractions of susceptible, infective, and removed (or recovered) vertices in a network at time t by $S(t)$, $I(t) \equiv \rho(t)$, and $R(t) \equiv r(t)$, respectively.

7.4 The SIS model and contact process

We first consider two epidemic processes that can be trapped into a so-called *absorbing state*, in which all the vertices are susceptible, cannot be infected, and hence the evolution finishes. The presence of the absorbing state strongly distinguishes these models from the processes which we discussed up to now. In the *SIS model*, a susceptible vertex becomes infective via contact with each of its infective nearest neighbours with rate λ, which is the control parameter, and an infective vertex spontaneously recovers with unit rate (Marro and Dickman, 2005; Henkel, Hinrichsen, and Lübeck, 2008). In the *contact process*, an infective vertex infects one of its susceptible nearest neighbours with rate λ, and an infective vertex spontaneously recovers with unit rate (Harris, 1974).[7] The difference between the SIS model and the contact process is only in the number of the acts of infection by an infective vertex. On regular lattices, these two processes are equivalent.

7.4.1 SIS model on a fully connected graph

The SIS model and the contact process are exactly solvable only on a fully connected graph, where they are equivalent. The solution demonstrates the essence of the processes with an absorbing state, and so we dwell on this case. Let the spontaneous transition rate from the infective vertex state to the susceptible one be 1. We predicate that if in the fully connected graph of N vertices, n vertices are infective, then a susceptible vertex becomes infective with rate $\lambda n/(N-1)$. Here the factor $1/(N-1)$ is introduced to ensure that the epidemic threshold be at finite value of λ (subtraction of 1 from N is not essential). Figure 7.4 shows the evolution of the *prevalence*, that is the fraction of infective vertices $\rho(t) = n(t)/N$, in this process for finite and infinite fully connected graphs (Deroulers and Monasson, 2004). In a finite graph, the process reaches the final, absorbing state with all vertices infective in a finite time (Figure 7.4a). For $\lambda > 1$, where $\lambda_c = 1$ is the critical spreading rate (epidemic threshold) above which the epidemic

---

[7]The contact process is more known in high energy physics as the Gribov process (Gribov, 1968), and it is equivalent to directed percolation (Stauffer and Aharony, 1991). In directed percolation, all bonds of $(D+1)$-dimensional lattice are directed along one axis, which plays the role of time in the contact process on a D-dimensional lattice.

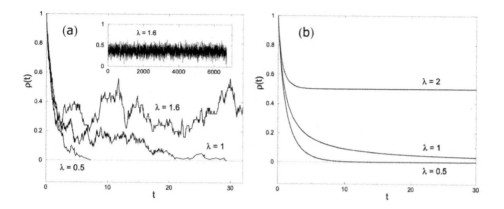

Fig. 7.4 Temporal evolution of a prevalence $\rho(t) = n(t)/N$ in the SIS model and, equivalently, the contact process on a fully connected graph of N vertices for different values of spreading rate λ. Initially, all the vertices are infective, $\rho(t = 0) = 1$. (a) $N = 100$. Evolution of a fraction of infective vertices for a single run of the process for each of three values of λ, namely, 1.6, 1, and 0.5. The inset shows the complete evolution of the prevalence for $\lambda = 1.6$. In a finite graph, the process reaches the absorbing state for any λ. (b) $N \to \infty$. For $\lambda > \lambda_c = 1$, $\rho(t)$ exponentially relaxes to $\rho^* = (\lambda - 1)/\lambda$. For the critical rate $\lambda_c = 1$ (epidemic threshold), the prevalence decays to zero as $\rho(t) \sim 1/t$. For $\lambda < \lambda_c = 1$, the prevalence exponentially decays to zero. Adapted from Deroulers and Monasson (2004).

is endemic in the infinite fully connected graphs, before falling into the absorbing state, the system spend some time, fluctuating, in a metastable state, see the inset in Figure 7.4a. Note that Figure 7.4a shows a single evolution run for each value of λ. In the range $\lambda > 1$, $N < \infty$, averaging over many runs should result in an exponentially decaying asymptotics of the prevalence, $\rho(t) \sim e^{-t/\tau(\lambda,N)}$, where τ is the lifetime of a metastable state. In the infinite size limit (Figure 7.4b) the prevalence exponentially relaxes to $\rho^* = (\lambda - 1)/\lambda$ when $\lambda > \lambda_c$, and the infection is endemic. Below the critical point, $\lambda < \lambda_c$, the prevalence exponentially decays to zero, and at the critical point, the prevalence decays to zero as $\rho(t) \sim 1/t$. Thus in finite networks, the evolution ends in the absorbing state for any λ, while in infinite networks, this happens only below the epidemic threshold. That is, in the models with the absorbing state on infinite networks below an epidemic threshold all vertices finally become infective. This is in contrast to, for example, percolation, where finite clusters are present below the percolation threshold. In Figure 7.4 the evolution starts with all vertices

225

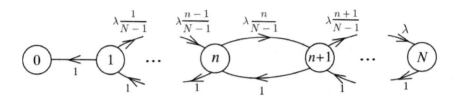

Fig. 7.5 Markov process described by Eq. (7.6). Transitions between states with different numbers of infective vertices are shown by arrows with transition rates indicated.

infective. Note however that in infinite networks even a tiny initial fraction of infective vertices results in the same final states as in Figure 7.4b both below (as is natural) and above the epidemic threshold.

These sets of behaviours can be obtained from the evolution equation for the probability $P_n(t)$ $(0 \le n \le N)$ that n vertices are infective at time t in this model on the fully connected graph of N vertices (Dickman and Vidigal, 2002; Deroulers and Monasson, 2004), namely

$$\frac{d}{dt}P_n(t) = \frac{\lambda(n-1)}{N-1}(N-n+1)P_{n-1}(t)$$

$$- \left[n + \frac{\lambda n}{N-1}(N-n) \right] P_n(t) + (n+1)P_{n+1}(t) \qquad (7.6)$$

with the boundary conditions $P_{-1}(t) = P_{N+1}(t) = 0$. This equation describes the Markov process depicted in Figure 7.5. For large N, the resulting probability $P_n(t)$ is, asymptotically,

$$P_n(t) \cong e^{N\Phi(n/N,t)}, \qquad (7.7)$$

where $\Phi(n/N, t \to \infty) \to \Phi^*(\rho)$, $\rho = n/N$, compare with Eq. (6.141). Here, from Eq. (7.6),

$$\Phi^*(\rho) = -\frac{1}{\lambda} + (1-\rho)[1 - \ln \lambda - \ln(1-\rho)], \qquad (7.8)$$

which shows the metastable state for $\lambda > \lambda_c = 1$ with the prevalence

$$\rho^*(\lambda) = \frac{\lambda - 1}{\lambda}. \qquad (7.9)$$

The lifetime τ of the metastable state is obtained from the equality:

$$\tau(\lambda, N) = \frac{1}{P_{n=1}(t \to \infty)}. \qquad (7.10)$$

For $\lambda > 1$, this leads to the following expression:

$$\tau(\lambda, N) = \frac{1}{P_{n=1}(t \to \infty)} \sim e^{-N\Phi^*(0)} = e^{N(\ln \lambda - 1 + 1/\lambda)}, \qquad (7.11)$$

which shows that the lifetime increases as $\tau \sim \lambda^N$ for large λ and N. Equation (7.6) also allows one to get the time asymptotics for $\rho(t)$ in the infinite graph, which we mentioned above.

7.4.2   Heterogeneous mean-field approximation

For more complex networks, the SIS model and the contact process are treatable approximately. The simplest approximation is provided by the so-called *heterogeneous mean-field theory*, where all the vertices of degree q at time t are assumed to have the same probability to be infective, $\rho_q(t)$. Then the prevalence equals

$$\rho(t) = \sum_q P(q)\rho_q(t). \qquad (7.12)$$

This approximation is applicable to infinite networks with degree–degree correlations for the neighbouring vertices (including a particular case of uncorrelated networks), described by a joint degree–degree distribution $P(q, q') = P(q|q')q'P(q')/\langle q \rangle$. The word 'applicable' here doesn't mean that the approximation is precise for these networks. For the SIS model, where an infective vertex becomes susceptible with the rate 1, and a susceptible vertex with n infective nearest neighbours becomes infective with the rate λn, the evolution equation for $\rho_q(t)$, has the following form (Pastor-Satorras and Vespignani, 2001 b; Pastor-Satorras and Vespignani, 2001 a):[8]

$$\frac{d}{dt}\rho_q(t) = -\rho_q(t) + \lambda q[1 - \rho_q(t)] \sum_{q'} P(q'|q)\rho_{q'}(t). \qquad (7.13)$$

One can understand the form of this equation intuitively. The first term on the right-hand side accounts for spontaneous recovery of infective vertices with the unit rate. The second term accounts for infecting susceptible vertices of degree q (which explains the factor $1 - \rho_q(t)$) with rate λ multiplied by the average number of infective nearest neighbours of such vertices, namely, $q \sum_{q'} P(q'|q)\rho_{q'}(t)$.[9] In the particular case of a fully connected graph, Eq. (7.13) and its analogues for the other epidemic models,

---

[8]It is also often called the 'dynamical mean-field reaction rate equation'.

[9]For a strict derivation of this equation see Catanzaro, Boguñá, and Pastor-Satorras (2005 a) or Bianconi (2018 a).

reduce to the so-called *mixed (or well-mixed) population model* with a uniform probability $\rho(t)$ to be infective for all vertices.

Assuming stationarity, reasonable in the limit $t \to \infty$, we readily get an equation for ρ_q,

$$\rho_q = \frac{\lambda q \sum_{q'} P(q'|q)\rho_{q'}}{1 + \lambda q \sum_{q'} P(q'|q)\rho_{q'}}. \tag{7.14}$$

To find the epidemic threshold λ_c below which $\rho_q = 0$, one should linearize Eq. (7.14) for small ρ_q, which provides the following matrix equation:

$$\lambda_c^{-1}\rho_q = \sum_{q'} qP(q'|q)\rho_{q'}. \tag{7.15}$$

Consequently, λ_c coincides with the inverse of the largest eigenvalue of the matrix $qP(q'|q)$. For uncorrelated networks, $P(q|q') = qP(q)/\langle q \rangle$, and Eq. (7.15) reduces to

$$\lambda_c^{-1}\rho_q = q \sum_{q'} \frac{q'P(q')}{\langle q \rangle}\rho_{q'}, \tag{7.16}$$

which readily leads to the expression for the epidemic threshold,

$$\lambda_c = \frac{\langle q \rangle}{\langle q^2 \rangle}, \tag{7.17}$$

finite when $\langle q^2 \rangle < \infty$, that is, when a power-law degree distribution $P(q) \sim q^{-\gamma}$ has exponent $\gamma > 3$, and to the absence of the epidemic threshold, $\lambda_c = 0$, if $\langle q^2 \rangle$ diverges. The prevalence $\rho(\lambda)$ can be easily obtained from Eq. (7.14). The phase transition associated with the emergence of the endemic infection is continuous. In uncorrelated networks with $\langle q^3 \rangle < \infty$, that is, $\gamma > 4$, near the epidemic threshold, $\rho(\lambda) \propto (\lambda - \lambda_c)$. Even without calculations, the reader can easily guess the full set of critical asymptotics of the prevalence $\rho(\lambda)$ for more slowly decaying degree distributions recalling the corresponding critical asymptotics for the size of the giant connected component from Section 6.2. In particular, for $\gamma = 3$, this approximation gives the singularity $\rho(\lambda) \propto e^{-\text{const}/\lambda}$. For $3 < \gamma < 4$, the critical singularity is $\rho(\lambda) \propto (\lambda - \lambda_c)^{1/(\gamma-3)}$. A more thorough analysis going beyond this rough approximation showed that the epidemic threshold of the SIS model and the contact process actually equals zero in infinite scale-free networks even for $3 < \gamma < \infty$ (Chatterjee and Durrett, 2009; Castellano and Pastor-Satorras, 2010). This occurs in the uncorrelated networks having hubs with degrees diverging with N.

Evolution equations of the heterogeneous mean-field approximation for the contact process, where an infective vertex generates an infective offspring (infects one of its susceptible nearest neighbours) with rate λ, are similar to the equations for the SIS model (Castellano and Pastor-Satorras, 2006):

$$\frac{d}{dt}\rho_q(t) = -\rho_q(t) + \lambda q[1 - \rho_q(t)] \sum_{q'} \frac{1}{q'} P(q'|q)\rho_{q'}(t). \qquad (7.18)$$

The only difference from Eq. (7.13) here is an extra factor $1/q'$ in the sum on the right-hand side of the equation. The resulting critical value λ_c coincides with the inverse of the largest eigenvalue of the matrix $qP(q'|q)/q'$. For uncorrelated networks, Eq. (7.18) takes the form

$$\frac{d}{dt}\rho_q(t) = -\rho_q(t) + \lambda \frac{1}{\langle q \rangle} q[1 - \rho_q(t)]\rho(t), \qquad (7.19)$$

where the prevalence $\rho(t) = \sum_q P(q)\rho_q(t)$. Therefore at stationarity,

$$\rho_q = \frac{\lambda q\rho/\langle q \rangle}{1 + \lambda q\rho/\langle q \rangle}, \qquad (7.20)$$

and hence the equation for the prevalence ρ is as follows:

$$\rho = \sum_q P(q)\frac{\lambda q\rho/\langle q \rangle}{1 + \lambda q\rho/\langle q \rangle}. \qquad (7.21)$$

This leads to the critical point $\lambda_c = 1$ of the contact process for any degree distribution with a finite first moment, which corresponds to $\gamma > 2$ for scale-free networks. For $2 < \gamma < 3$, Eq. (7.21) provides the critical singularity $\rho(\lambda) \propto (\lambda - 1)^{1/(\gamma-2)}$, and for $\gamma > 3$, $\rho(\lambda) \propto (\lambda - 1)$. Still, thoroughly accounting for hubs with degrees diverging with network size results in $\lambda_c = 0$ even for $3 < \gamma < \infty$ in infinite scale-free networks (Chatterjee and Durrett, 2009).

7.4.3 Quenched mean-field approximation

The evolution equations of the second widely used version of a mean-field theory for the SIS model, the so-called *quenched mean-field approximation* (Hethcote and Yorke, 1984; Van Mieghem, Omic, and Kooij, 2008), contain only the probabilities to be infective of individual vertices neglecting all correlations between the states (susceptible, infective) of different vertices. This approximation is applicable to arbitrary finite and infinite graphs (individual graphs, not statistical ensembles) described by an adjacency matrix

A_{ij}. Here again the applicability of this approximation doesn't mean that it is precise. Let the susceptible and infective states of vertex i be described by the 'spin' variable $\sigma_i = 0, 1$, respectively. The probability that vertex i is infective at time t equals $\rho_i(t) = \langle \sigma_i(t) \rangle$, and the prevalence for a network of N vertices equals the sum

$$\rho(t) = \frac{1}{N} \sum_i \rho_i(t). \tag{7.22}$$

Neglecting correlations between the states of different vertices $i, j, \dots$ means the factorization of correlation functions, $\langle \sigma_i(t) \sigma_j(t) \dots \rangle = \rho_i(t) \rho_j(t) \dots$, which leads to the following evolution equations of the quenched mean-field theory for the SIS model:

$$\frac{d}{dt} \rho_i(t) = -\rho_i(t) + \lambda[1 - \rho_i(t)] \sum_{j=1}^{N} A_{ij} \rho_j(t). \tag{7.23}$$

The reader can intuitively interpret the form of this equation in a way similar to Eq. (7.13) for the heterogeneous mean-field approximation.[10,11] If we assume that a network is annealed, with the edge weights corresponding to a joint degree–degree distribution $P(q, q')$ for the neighbouring vertices, that is,

$$A_{ij} = \frac{\langle q \rangle P(q_i, q_j)}{N P(q_i) P(q_j)} = \frac{q_j P(q_i | q_j)}{N P(q_i)} = \frac{q_i P(q_j | q_i)}{N P(q_j)}, \tag{7.24}$$

then Eq. (7.23) leads to the heterogeneous mean-field approximation (see Eq. (7.13)) for the SIS model on networks with such correlations. In particular, if the weights are

---

[10]For a strict derivation of this equation we again refer the reader to Catanzaro, Boguñá, and Pastor-Satorras (2005a) or Bianconi (2018a). Furthermore, we can take one step further accounting for the pairwise correlations between the nearest neighbours, which results in the next, pair approximation (Mata and Ferreira, 2013; Gleeson, 2013).

[11]There is a closely related discrete-time ansatz for the SIS process (Wang, Chakrabarti, Wang, and Faloutsos, 2003; Gómez, Arenas, Borge-Holthoefer, Meloni, and Moreno, 2010), which is often called the 'discrete-time Markov chain approach to the SIS spreading'. Let us assume that (i) at each time step, an infective vertex may infect each of its susceptible nearest neighbours with probability $\tilde{\beta}$, and (ii) at each time step each infective vertex becomes susceptible with probability $\tilde{\mu}$. Then the approximate recursions for the probabilities $\rho_i(t)$ can be written in the following form:

$$1 - \rho_i(t + 1) = [1 - (1 - \tilde{\mu}) \rho_i(t)] \prod_j [1 - \tilde{\beta} A_{ji} \rho_j(t)].$$

In the continuous time limit, for sufficiently small probabilities $\rho_j(t)$, this discrete equation reduces to Eq. (7.23).

$$A_{ij} = \frac{q_i q_j}{N \langle q \rangle}, \tag{7.25}$$

that is the annealed network mimics an uncorrelated one, then Eq. (7.23) leads to the heterogeneous mean-field approximation for the SIS model on uncorrelated networks.

At stationarity, Eq. (7.23) leads to the equation for the probabilities ρ_i:

$$\rho_i = \frac{\lambda \sum_{j=1}^{N} A_{ij} \rho_j}{1 + \lambda \sum_{j=1}^{N} A_{ij} \rho_j}. \tag{7.26}$$

Linearizing this equation, we conclude that the epidemic threshold λ_c in the quenched mean-field approximation equals the inverse of the largest eigenvalue λ_1 of the adjacency matrix of the graph.[12] According to the Perron–Frobenius theorem, this eigenvalue of the real symmetric matrix is positive, and all components v_{i1} of the principal eigenvector $\mathbf{v}_1$ have the same sign. By $\mathbf{v}_\alpha$ and λ_α, where $\alpha = 1, 2, \dots, N$, we denote the eigenvectors and the corresponding eigenvalues (in the descending order) of the adjacency matrix A_{ij}. It is convenient to normalize the eigenvectors, $|\mathbf{v}_\alpha|^2 = 1$, which form a complete orthogonal basis. This allows us to represent the probabilities ρ_i as a linear superposition,

$$\rho_i = \sum_\alpha c_\alpha v_{i\alpha}, \tag{7.27}$$

where the coefficients c_α are the projections of the vector $\vec{\rho}$ with the components ρ_i on the eigenvectors $\mathbf{v}_\alpha$. Substituting Eq. (7.27) into Eq. (7.26) leads to the equation for these coefficients:

$$c_\alpha = \lambda \sum_\beta \lambda_\beta c_\beta \sum_{i=1}^{N} \frac{v_{i\alpha} v_{i\beta}}{1 + \lambda \sum_\eta \lambda_\eta c_\eta v_i^\eta}. \tag{7.28}$$

In the close neighbourhood of the critical point, it is sufficient to take into account the principal eigenvector, that is, $\rho_i \cong c_1 v_{i1}$. Keeping only components of the principal eigenvector in Eq. (7.28) we readily obtain the prevalence near the epidemic threshold $\lambda_c = \lambda_1^{-1}$,

---

[12]The largest eigenvalue of the adjacency matrix of a random network is estimated as $\lambda_1 \sim \sqrt{q_{max}}$, where q_{max} is the maximum degree of a vertex in the network (see Eq. (9.13)). Consequently, in the quenched mean-field approximation, the epidemic threshold $\lambda_c \sim 1/\sqrt{q_{max}(N)}$ approaches zero when $q_{max}(N)$ diverges with N. The more thorough studies, mentioned earlier, support this conclusion (Chatterjee and Durrett, 2009; Castellano and Pastor-Satorras, 2010).

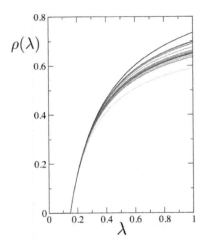

Fig. 7.6 Prevalence in the SIS model on the karate club network obtained by the quenched mean-field approximation. The lowest curve accounts for only the eigenvector of the adjacency matrix, corresponding to λ_1 in Eq. (7.28). Accounting for the eigenvectors corresponding to λ_1 and λ_2, we find the higher curve and so on. The most upper curve takes into account all eigenvectors of the adjacency matrix. They are all delocalized in this network. Adapted from Goltsev, Dorogovtsev, Oliveira, and Mendes (2012).

$$\rho(\lambda) \cong C_1 \lambda_1 (\lambda - \lambda_1^{-1}), \tag{7.29}$$

where the coefficient is expressed in terms of the components of the principal eigenvector,

$$C_1 = \frac{\sum_{i=1}^{N} v_{i1}}{N \sum_{i=1}^{N} v_{i1}^3}. \tag{7.30}$$

Taking into account the two largest eigenvalues in Eq. (7.28), λ_1 and λ_2, and the corresponding eigenvectors, one can obtain the next term of the expansion, $\rho(\lambda) \cong C_1 \lambda_1 (\lambda - \lambda_1^{-1}) + C_2 \lambda_1^2 (\lambda - \lambda_1^{-1})^2$, and so on. One can easily compute, say, the first 20 coefficients of this series for reasonably large graphs. Figure 7.6 shows the curves $\rho(\lambda)$ obtained in this way for the SIS model on the karate club network, (Section 9.5).

Note that the mean-field approximations, neglecting fluctuations, ignore the possibility that a system can finally fall into the absorbing state from a metastable one. In particular, for a finite network, the heterogeneous mean-field approximation erroneously predicts the endemic infection for spreading rates exceeding a finite critical value λ_c. Figure 7.4a clearly demonstrates that this is not true.[13] This heterogeneous mean-field approximation value, λ_c, for finite networks, can at best be interpreted as an estimate for the point of the emergence of a metastable state, and the corresponding $\rho(\lambda)$ as an estimate for the fraction of infective vertices in this state.

The quenched mean-field approximation for the contact process provides the following evolution equations for the probabilities $\rho_i(t)$:

$$\frac{d}{dt}\rho_i(t) = -\rho_i(t) + \lambda[1 - \rho_i(t)]\sum_{j=1}^{N}\frac{A_{ij}}{q_j}\rho_j(t), \tag{7.31}$$

where the only difference from the corresponding equations for the SIS model, Eq. (7.23), is an extra factor $1/q_j$ in the sum on the right-hand side of Eq. (7.31). Therefore, the critical value λ_c of the contact process in the quenched mean-field approximation equals the inverse of the largest eigenvalue λ_1 of the matrix A_{ij}/q_j, which is 1.

7.5 The SIR, SI, and SIRS models

In the *SIR model* (Figure 7.3) susceptible vertices become infective by the same rule as in the SIS model, and infective vertices spontaneously become removed (recovered) with a recovery rate μ. By rescaling time, one can set this rate to 1 and introduce the spreading rate $\lambda = \beta/\mu$ as in the SIS model. We first consider the heterogeneous mean-field approximation for this model. Let $s_q(t)$, $\rho_q(t)$, and $r_q(t)$ be the fractions of susceptible, infective, and recovered vertices, respectively, among the vertices of degree q. Clearly,

$$s_q(t) + \rho_q(t) + r_q(t) = 1. \tag{7.32}$$

The prevalence in this model is the fraction of all vertices that are recovered,

$$r(t) = \sum_q P(q)r_q(t). \tag{7.33}$$

Analogously to the SIS model, one can write the evolution equations of the heterogeneous mean-field approximation for the SIR model,

$$\frac{d}{dt}\rho_q(t) = -\rho_q(t) + \lambda q[1 - \rho_q(t) - r_q(t)]\sum_{q'}\frac{q'-1}{q'}P(q'|q)\rho_{q'}(t), \tag{7.34}$$

$$\frac{d}{dt}r_q(t) = \rho_q(t) \tag{7.35}$$

---

[13]This flaw is not unique. For example, recall that the message passing algorithm, discussed in Section 6.3, predicts a finite percolation threshold p_c in finite graphs.

(Moreno, Pastor-Satorras, and Vespignani, 2002; Boguñá, Pastor-Satorras, and Vespignani, 2003). The notable difference from Eq. (7.13) of the SIS model is the factor $(q'-1)/q'$ in the sum on the right-hand side of Eq. (7.34). Loosely speaking, this ratio appears due to the fact that an infected vertex in the SIR model cannot infect back its infector, and hence one of the q' edges is effectively blocked.

The number of removed vertices in the SIR model grows monotonically, and in the final state, this number cannot be smaller than the number of infective vertices at any moment of the evolution. Therefore, for observing a phase transition below which the fraction of removed vertices equals zero, one should start from a small seed, that is, a tiny fraction of infective vertices. Linearization of Eq. (7.34) readily gives the epidemic threshold λ_c of this model equal to the largest eigenvalue of the matrix $qP(q'|q)(q'-1)/q'$. For uncorrelated networks, this matrix has the largest eigenvalue $(\langle q^2 \rangle - \langle q \rangle)/\langle q \rangle$, and then in this case

$$\lambda_c = \frac{\langle q \rangle}{\langle q^2 \rangle - \langle q \rangle}. \tag{7.36}$$

The components v_q of the corresponding eigenvector are proportional to q. Consequently in the nearest neighbourhood of the epidemic threshold, $r_q \propto q$.

One can easily analyse the stationary solution of Eq. (7.34), which results at the same set of critical singularities of the prevalence for the SIR model on uncorrelated networks, as in the bond percolation problem for these networks (Section 6.2). Moreover, the SIR spreading process can be mapped to the bond percolation problem, although this mapping is non-trivial, significantly depending on details of the SIR process (Ludwig, 1975). In this mapping, the cluster of removed vertices in the epidemic outbreak that started from a seed infective vertex in the SIR model corresponds to the connected component to which the seed vertex belongs in the bond percolation problem. The occupation probability p of edges in the bond percolation problem coincides with the *transmissibility* T, that is, the probability that an infective vertex manages to infect a neighbouring susceptible vertex before removal (Newman, 2002b). For a given distribution $\mathcal{P}(\tilde{t})$ of the lifetimes of infective vertices, the transitivity is estimated by the integral

$$T = \int_0^\infty d\tilde{t} \mathcal{P}(\tilde{t}) (1 - e^{-\beta \tilde{t}}), \tag{7.37}$$

where $1 - e^{-\beta \tilde{t}}$ is the probability that an infective vertex infects a neigh-bouring susceptible vertex during time $\tilde{t}$ given the transmission rate is β. Assuming that the removal of infective vertices is the Poisson process, the distribution of lifetimes $\tilde{t}$ of infective vertices has the following shape:

$$\mathcal{P}(\tilde{t}) = \mu e^{-\mu \tilde{t}}, \tag{7.38}$$

where $\mu = 1/\langle \tilde{t} \rangle$ is the removal (or recovery) rate for infective vertices in the SIR model.[14] For this distribution, the transmissibility takes the form

$$T = p = \frac{\beta}{\beta + \mu} = \frac{\lambda}{\lambda + 1}. \tag{7.39}$$

For uncorrelated networks, $p_c = \langle q \rangle / (\langle q^2 \rangle - \langle q \rangle)$, Eq. (6.34), which results in the epidemic threshold

$$\lambda_c = \frac{\langle q \rangle}{\langle q^2 \rangle - 2\langle q \rangle}. \tag{7.40}$$

Two formulae for the epidemic threshold, Eqs. (7.36) and (7.40), markedly differ from each other. Note that both these two results are approximate, since the derivation of Eq. (7.40) also implies strong assumptions, including Eq. (7.37) (see a discussion in Pastor-Satorras, Castellano, Van Mieghem, and Vespignani, 2015).[15]

The *SI model* is a simple particular case of the SIR model with only two states of vertices—susceptible and infective—so that once infected, a vertex stays infective forever. The evolution equations of the heterogeneous mean-field theory for a fraction $\rho_q(t)$ of vertices of degree q in the SIS model have the form:

$$\frac{d}{dt}\rho_q(t) = \lambda q[1 - \rho_q(t)] \sum_{q'} P(q'|q) \left\{ \rho_{q'}(0) + \frac{q'-1}{q'}[\rho_{q'}(t) - \rho_{q'}(0)] \right\} \tag{7.41}$$

(Barthélemy, Barrat, Pastor-Satorras, and Vespignani, 2004, 2005b), where the probabilities $\rho_q(0)$ that a vertex of degree q is infected at the initial

---

[14]The Poisson process is a stochastic process in continuous time, in which single events of zero duration, independent of each other, occur with a constant rate μ, so that the probability that an event happens within a time window Δt equals $\mu \Delta t$. Equation (7.38) provides the distribution of inter-event times in this process.

[15]For a more refined message-passing treatment of a generalized version of the SIR model of epidemic disease accounting for arbitrary distributions of transmission and recovery times, see Karrer and Newman (2010a) and Wilkinson, Ball, and Sharkey (2017). This treatment is exact in locally tree-like networks.

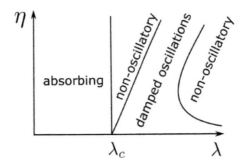

Fig. 7.7 Phase diagram of the SIRS model in the mean-field theory for a mixed population on the $\lambda-\eta$ plane, where η is the immunity decay rate. Transition rate from the infective state to the recovered one is set to 1.

moment enter the right-hand side of the equation. Notice that these seed infective vertices are better spreaders of a disease than vertices becoming infective during the outbreak, and they should be treated differently. The seed infective vertices can infect all their susceptible nearest neighbours, while the other infective vertices can infect their susceptible neighbours except those from whom they got the infection. This explains intuitively the terms within the sum on the right-hand site of Eq. (7.41). Even for a tiny fraction of seed infective vertices, the final state in this model is with all vertices infective for any (non-zero) transmission rate, and hence it has no epidemic threshold. We discuss the dynamics of the SI and SIR models in Section 7.6.

We have mentioned the SEIR and SEI models intending to account for latent and incubation periods by the additional 'exposed' state (Murray, 2007; Small and Tse, 2005). These slightly realistic models do not demonstrate qualitatively different features, compared to the SIR and SI models. On the other hand, the SIRS model, containing the spontaneous transition from the recovery state to the susceptible one with the immunity decay rate η (Figure 7.3) shows a richer set of behaviours including a damped oscillatory one (de Souza and Tomé, 2010). Figure 7.7 shows the phase diagram of this model in the mixed population regime, obtained by the mean-field theory. This phase diagram has the absorbing phase and the phase above the epidemic threshold with a final state having finite fractions of susceptible, infective, and removed vertices. The latter phase contains a region with a damped oscillatory behaviour between two regions with an exponential relaxation to the final state.

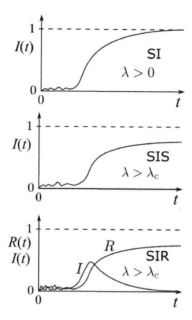

Fig. 7.8 Temporal evolution of epidemic outbreaks in the SI, SIS, and SIR models above an epidemic threshold according to Moreno, Pastor-Satorras, and Vespignani (2002). $I(t)$ and $R(t)$ are the fractions of infective and removed vertices at time t, respectively.

7.6   Epidemic outbreaks

Figure 7.8, sketching the temporal evolution of prevalence for three basic epidemic models above their epidemic threshold ($\lambda_c = 0$ for the SI model), provides a qualitative view of epidemic outbreaks. The evolution starts from a tiny fraction of infective vertices, uniformly randomly selected in a network, and its initial stage with a yet-small prevalence is noisy (Moreno, Pastor-Satorras, and Vespignani, 2002; Barthélemy, Barrat, Pastor-Satorras, and Vespignani, 2005b; Pastor-Satorras, Castellano, Van Mieghem, and Vespignani, 2015). During this period, if the initial number of infective vertices is finite, the SIS model has a chance to fall into the absorbing state. The next stage of the evolution of an epidemic is an exponential growth of prevalence. This stage determines the time scale of an epidemic outbreak. During the third period of the evolution, the models exponentially (with negative exponential) converge to their final endemic states.[16]

---

[16]Note that Figure 7.4b did not show the first two of these stages of the epidemic since the evolution started with all vertices infective.

Within the exponential expansion period of an epidemic, a prevalence grows as $\rho(t) \sim e^{t/\tau}$, where the characteristic time τ is related with a key number in epidemiology, a *reproduction number* $\mathcal{R}_0$,

$$\frac{d \ln I(t)}{dt} = \frac{1}{\tau} = \mathcal{R}_0 - 1. \tag{7.42}$$

The reproduction number $\mathcal{R}_0$ (or effective reproduction number, basic reproduction number, or basic reproduction ratio) is the average number of secondary cases produced by one infected vertex in a fully susceptible population (Van den Driessche, 2017). Note that in epidemiology the reproduction number is confidently used not only for characterization of the exponential stage of an epidemic but also for all its stages. The characteristic time τ of the exponential epidemic expansion can be easily obtained from the heterogeneous mean-field theory (Eqs. (7.41), (7.14), and (7.34) for the SI, SIS, and SIR models, respectively). For example, to obtain the characteristic time τ for the SI model, one should solve the linearized Eq. (7.41). Its solution $\rho_q(t)$, proportional to the components of the principal eigenvector of the matrix $qP(q'|q)(q'-1)/q'$ has the time-dependent factor $e^{t/\tau}$ with $\tau = 1/(\lambda\lambda_1)$, where λ_1 is the largest eigenvalue of this matrix. Therefore, for uncorrelated networks, where $\lambda_1 = (\langle q^2 \rangle - \langle q \rangle)/\langle q \rangle$, for the SI model we have

$$\tau = \frac{\langle q \rangle}{\langle q^2 \rangle - \langle q \rangle} \frac{1}{\lambda}, \quad \lambda_c = 0. \tag{7.43}$$

In a similar way, linearizing Eq. (7.14), one can get the characteristic time for the SIS model:

$$\tau = \frac{\lambda_c}{\lambda - \lambda_c}, \quad \lambda_c = \frac{\langle q \rangle}{\langle q^2 \rangle}, \tag{7.44}$$

and, linearizing Eq. (7.34), for the SIR model,

$$\tau = \frac{\lambda_c}{\lambda - \lambda_c}, \quad \lambda_c = \frac{\langle q \rangle}{\langle q^2 \rangle - \langle q \rangle}. \tag{7.45}$$

Notice that the only difference in these two identical-looking formulae for τ, Eqs. (7.44) and (7.45) for the SIS and the SIR models, is due to the different epidemic thresholds λ_c. Recall that we set the rates μ to 1 in these models by rescaling time. Thus for all these models, the time scale of the epidemic diverges at the epidemic threshold.

These results, obtained by applying the rough heterogeneous mean-field approximation, were supported by more refined techniques, in particular,

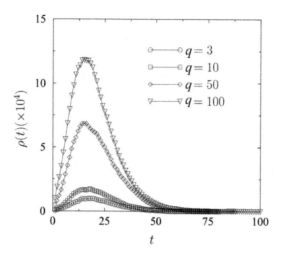

Fig. 7.9 Fraction of infective vertices vs. time for the SIR model in the Barabási–Albert network of $N = 10^6$ vertices with $\langle q \rangle = 6$. For each curve, the same number of initial infective vertices are uniformly randomly distributed among the vertices of a given degree q, where $q = 3$, 10, 50, and 100 for different curves. The transmission rate is $\lambda = 0.09$. Adapted from Moreno, Pastor-Satorras, and Vespignani (2002).

by message-passing algorithms enabling one to assess the full time dependence of an epidemic outbreak started from a given vertex in an individual graph (Lokhov, Mézard, Ohta, and Zdeborová, 2014; Min, 2018; Moore and Rogers, 2020; Min and Castellano, 2020). As is natural, at least in uncorrelated networks, the vertices with large degrees are the best spreaders ('superspreaders'), and the epidemic outbreaks launched by infective vertices of large degrees are most strong (as the simulations of Moreno, Pastor-Satorras, and Vespignani (2002) demonstrated; see Figure 7.9). This observation leads to natural strategies to prevent or, quite the reverse, to maximize epidemic spreading by removing (vaccinating) vertices with large degrees (targeted immunization) or, respectively, by infecting them (Pastor-Satorras and Vespignani, 2002). It is possible to select strongly connected vertices without knowing all degrees of vertices in a network. This immunization strategy, called acquaintance immunization, calls for the immunization of random acquaintances of uniformly randomly chosen individuals (Cohen, Havlin, and Ben-Avraham, 2003a). This strategy works well for uncorrelated networks. For many other networks it turns out to be not optimal,

and not cost-efficient (Holme and Litvak, 2017).[17] The mind-boggling number of publications—thousands—on, at first sight, all possible aspects of
the optimal immunization of networks with various architectures is actually
not quite startling for times 'when logic and proportion have fallen sloppy
dead'.[18] Still, the great majority of these numerous works is about the min-
imization of the total number of cases during an epidemic. It is becoming
clear nowadays, however, that it is more important to control the height of
an outbreak, even at the expense of its stretching.

Above the epidemic threshold of, say, the SIR model, the size of the
outbreak, which is the final fraction of removed vertices, R_f, is almost independent of size of the initial fraction of infective vertices, I_0, if this fraction
is sufficiently small. Below the epidemic threshold, where the prevalence decays exponentially, R_f is, clearly, proportional to I_0, $R_f(\lambda) = I_0 f(\lambda)$, where
$f(\lambda)$ is a function of λ diverging at the epidemic threshold. Let us discuss
the dependence of R_f on (small) I_0 at the epidemic threshold (Radicchi and
Bianconi, 2020; Krapivsky, 2021). Consider the mean-field theory equations
of the SIR model on an infinite well-mixed population:

$$\frac{d}{dt}S = -\lambda S I, \tag{7.46}$$

$$\frac{d}{dt}I = -I + \lambda S I, \tag{7.47}$$

where $S(t)$, $I(t)$, and $R(t)$ are the fractions of susceptible, infective, and
removed vertices, respectively, with the constraint $S + I + R = 1$. The
initial condition is fixed by a given $I(0) = I_0 = 1 - S_0$. Later we shall
assume that I_0 is small though finite, $I_0 \ll 1$, but first let it be arbitrary.
Dividing Eq. (7.47) by Eq. (7.46) and integrating over S one gets I as a
function of S,

$$I = 1 + \lambda^{-1} \ln(S/S_0) - S, \tag{7.48}$$

which, accounting for $I_f = 0$, leads to the transcendental equation for the
final fraction R_f of removed vertices,

$$R_f + S_0 e^{-\lambda R_f} = 1. \tag{7.49}$$

---

[17]Recall a simple way to select vertices of large degrees in an arbitrary undirected graph
by using only local information, mentioned in Section 4.2. Select all nearest neighbours of
a number of random vertices. If the number of vertices of degree q in the graph is $N(q)$,
then the frequency with which selected vertices have degree q equals $q N(q)/(N\langle q \rangle)$.

[18]Grace Slick, 'White Rabbit'.

Setting $S_0 \to 1$ in this equation, we see that the epidemic threshold in this model is $\lambda_c = 1$. For $I_0 = 1 - S_0 \ll 1$, below the epidemic threshold,[19] Eq. (7.49) gives the following size of the outbreak:

$$R_f(\lambda < 1) \cong \frac{I_0}{1 - \lambda}, \tag{7.50}$$

while for $\lambda = \lambda_c = 1$ and $I_0 \ll 1$, Eq. (7.49) provides the outbreak size

$$R_f(\lambda = 1) \cong \sqrt{2I_0}. \tag{7.51}$$

Substituting Eq. (7.48) into Eq. (7.46) results in the duration of the epidemic outbreak

$$T(\lambda = 1) \cong \frac{\ln(1/I_0)}{\sqrt{2I_0}}, \tag{7.52}$$

that is, in this approximation, $T(\lambda = 1)$ vanishes as $I_0 \to 0$. Above the epidemic threshold, the solution of Eq. (7.46) demonstrates an outbreak similar to Figure 7.9. Below the epidemic threshold, Eq. (7.46) indicates the exponential decay of the prevalence, $I(t) \cong I(0)e^{-(1-\lambda)t}$, while at the epidemic threshold, $I(t) \sim I_0$ and $R(t) \sim I_0 t$ at least for $t \ll T(\lambda = 1)$. This differs markedly from the evolution of the average number of vertices removed after infecting a single vertex, increasing as t^2 (Lauritsen, Zapperi, and Stanley, 1996).

Importantly, at the epidemic threshold, for a finite number $n_0 = NI_0$ of initial infective vertices, the number of infective vertices $n(t) = NI(t)$ strongly fluctuates during an outbreak (Figure 7.10). The statistics of outbreaks for a finite n_0 at the epidemic threshold of the SIR model was found exactly both for infinite and finite, $N < \infty$, well-mixed populations (Krapivsky, 2021).[20] The obtained moments of the distributions of size n and duration T of an outbreak agree with findings and arguments of Radicchi and Bianconi (2020). For large N, these distributions have a scaling form, with the first moments

$$\langle n \rangle = N^{2/3} f\left(\frac{n_0}{N^{1/3}}\right), \quad \text{where } f(x \ll 1) \cong 1.45 \ldots x, \ f(x \gg 1) \cong \sqrt{2x}, \tag{7.53}$$

$$\langle T \rangle = \frac{1}{3} N^{1/3} \ln N \, g\left(\frac{n_0}{N^{1/3}}\right), \quad \text{where } g(x \ll 1) \cong x, \ g(x \gg 1) \cong \sqrt{2/x}. \tag{7.54}$$

[19]More strictly, here λ must be outside the scaling window $\lambda_c - \lambda \sim \sqrt{I_0}$.

[20]The solution of this problem is particularly simple if a population is infinite. In this case at the epidemic threshold, the SIR process is the critical branching process (that is, with average branching equal 1), where at each step each infective vertex either infects another vertex (duplicates) or dies with equal probability.

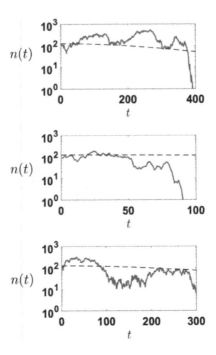

Fig. 7.10 Variation of the number of removed vertices $n(t)$ during three runs of the critical SIR process ($\lambda = \lambda_c = 1$) for a well-mixed population of size $N = 10^7$ started with 128 infective vertices. Adapted from Radicchi and Bianconi (2020).

Notice that while the average size of an outbreak monotonously increases with n_0, which is not surprising, the average duration is a non-monotonous function of n_0 with a maximum, and for sufficiently large n_0, $\langle T \rangle \propto 1/\sqrt{U n_0}$. Both these distributions are wide when $n_0 \ll N^{1/3}$, as the following formulae show,

$$\frac{\sqrt{\langle n^2 \rangle - \langle n \rangle^2}}{\langle n \rangle} = \tilde{f}\left(\frac{n_0}{N^{1/3}}\right), \text{ where } \tilde{f}(x \ll 1) \cong \frac{1.37 \ldots}{\sqrt{x}}, \quad \tilde{f}(x \gg 1) \cong \frac{2^{-1/4}}{x^{3/4}}, \quad (7.55)$$

$$\frac{\sqrt{\langle T^2 \rangle - \langle T \rangle^2}}{\langle T \rangle} = \frac{1}{\ln N} \tilde{g}\left(\frac{n_0}{N^{1/3}}\right), \text{ where } \tilde{g}(x \ll 1) \sim \frac{1}{\sqrt{x}}, \quad \tilde{g}(x \gg 1) \sim \text{const.} \quad (7.56)$$

In particular, for the critical outbreaks initialized by a single infective vertex,

$$\langle n \rangle \cong 1.45 \ldots N^{1/3}, \qquad \frac{\sqrt{\langle n^2 \rangle - \langle n \rangle^2}}{\langle n \rangle} \cong 1.37 \ldots N^{1/6}, \qquad (7.57)$$

$$\langle T \rangle \cong \frac{1}{3} \ln N, \qquad \frac{\sqrt{\langle T^2 \rangle - \langle T \rangle^2}}{\langle T \rangle} \sim \frac{N^{1/6}}{\ln N}. \qquad (7.58)$$

The size distribution $\mathcal{P}(n)$ for critical outbreaks has the asymptotics typical for critical branching processes. For $n_0 = 1$ and $N \to \infty$,

$$\mathcal{P}(n) \cong \frac{1}{\sqrt{4\pi}} n^{-3/2}. \tag{7.59}$$

For finite N, this distribution has a cutoff $n_{\text{cut}}(N) \sim N^{2/3}$ conforming Eq. (7.57). As is natural, the asymptotics $\mathcal{P}(n) \sim n^{-3/2}$ and this cut-off are the same as in the critical probability distribution that a vertex belongs to a cluster of size n in percolation above the upper critical dimension (see Appendix D).

7.7 Epidemics in a metapopulation

A *metapopulation* is a complex network of coupled populations. Of numerous metapopulation models of disease spreading and various processes (Colizza, Pastor-Satorras, and Vespignani, 2007), here we touch upon the one explored by Brockmann and Helbing (2013). Let a metapopulation consist of M well-mixed populations of large sizes N_n, where $n = 1, 2, \ldots, M$. Let $S_n = N_n^{(\text{susceptible})}/N_n$, $I_n = N_n^{(\text{infective})}/N_n$, and $R_n = N_n^{(\text{removed})}/N_n$ be, respectively, the fractions of susceptible, infective, and removed (recovered) individuals in population n, and $S_n + I_n + R_n = 1$. The model treats the epidemic process within each of the populations in the spirit of the SIR model. In addition, it assumes that there is a given stationary flux of individuals (irrespective of their states) between different populations. Let F_{mn} be the flux of individuals from population n to population m, and the full set of the flows $\{F_{mn}\}$ be known. This set of flows between populations forms a directed weighted network. It is convenient to introduce a matrix P with the entries $P_{mn} = F_{mn}/\sum_m F_{mn}$, so that $0 \leq P_{mn} \leq 1$, $P_{mm} = 0$. That is, the entry P_{mn} is the fraction of travellers that leave population n and arrive at population m. In fact, the model is defined by the following phenomenological evolution equations for the full set of $I_n(t)$ and $S_n(t)$, $n = 1, 2, \ldots M$:

$$\frac{d}{dt} I_n = \beta_n S_n I_n \sigma(I_n/\epsilon) - \mu_n I_n + \gamma \sum_m P_{mn}(I_m - I_n), \tag{7.60}$$

$$\frac{d}{dt} S_n = -\beta_n S_n I_n \sigma(I_n/\epsilon) + \gamma \sum_m P_{mn}(S_m - S_n). \tag{7.61}$$

The meaning of the rates β_n and μ_n is clear from the SIR model. Usually, for the sake of simplicity, these rates are assumed to be uniform,

$\beta_n = \beta$, $\mu_n = \mu$. The Laplace terms on the right-hand sides of these equations are due to the flows of individuals described by the matrix P. If $\beta = \mu = 0$, then Eqs. (7.60) and (7.61) are reduced to the diffusion equations, where the matrix P plays the role of the adjacency matrix of a weighted directed network. The so-called mobility parameter γ is the average mobility rate, $\gamma = \Phi / \sum_n N_n$, where $\Phi = \sum_{m,n} F_{mn}$ is the total flux between all populations. The additional factor $\sigma(I_n/\epsilon)$ is the sigmoid function $\sigma(x) = 1/(x^{-\eta} + 1)$ with a sufficiently large parameter η, accounting for the local invasion threshold ϵ. Here, for the sake of simplicity, this threshold is assumed to be uniform.

Based on the entries of the matrix P, Brockmann and Helbing (2013) introduced the effective distance from population n to a 'nearest neighbouring' population m, assuming that $P_{mn} > 0$,

$$d_{mn} \equiv 1 - \ln P_{mn} \geq 1. \tag{7.62}$$

A small P_{mn}, that is, a small fraction of the flux from n directed to m, implies that the effective distance from n to m is large, and, quite the reverse, if the entire flux from n goes to m, then the effective distance is the shortest possible, that is, 1. Note that, in general, $d_{mn} \neq d_{mn}$. This definition enables one to compute the effective directed length $\ell^{(\mathrm{eff})}[\mathcal{R}(k, n)]$ of a given path $\mathcal{R}(k, n)$ from population n to population k as the sum of effective distances of its legs. Then the *effective distance* from population n to population k is naturally introduced as the effective length of the optimal path from n to k, that is, of the path with the smallest effective length,

$$d_{kn}^{(\mathrm{eff})} \equiv \min_{\mathcal{R}(k,n)} \ell^{(\mathrm{eff})}[\mathcal{R}(k, n)]. \tag{7.63}$$

Note again that, in general, $d_{kn}^{(\mathrm{eff})} \neq d_{nk}^{(\mathrm{eff})}$.

Using this metric, the effective distance, and optimal paths, one can construct the *optimal path tree* $\mathcal{T}_n$ for any vertex (population) n in the network. The root of this tree is vertex n and the optimal paths to the remaining vertices form its branches. In a graphical representation, it is convenient to show radial distances from the root to the other vertices in this tree, proportional to the corresponding effective distances, like in Figure 7.11. In this picture, vertices equally distant from the root n sit in a circle with centre n. Assuming that population n is the origin of an epidemic started at time $t = 0$, one can study how, in this representation, the contagion, $I_k(t)$, spreads over this tree. Brockmann and Helbing (2013) numerically solved

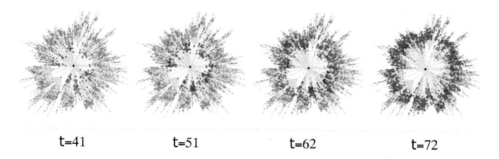

| t=41 | t=51 | t=62 | t=72 |

Fig. 7.11 Four instances of a contagion process (global pandemic) in a metapopulation constructed using the global mobility network connecting large populations. The optimal path tree representation of this network, with Hong Kong—the root—chosen as the origin of the pandemic, is constructed as is explained in the text. Radial distances to vertices from the root are proportional to the corresponding effective distances. Dark dots indicate populations with high fraction of infective individuals at a given time, measured in days for the used set of epidemic parameters. Adapted from Brockmann and Helbing (2013).

the evolution equations of the model, Eqs. (7.60) and (7.61), with the matrix P describing the global mobility network of passenger flows along direct connections between 4069 airports located in large populations. Figure 7.11 demonstrates a few instances of this epidemic started from a large hub. The marked sharpness of the wave front of the contagion process is largely due to the sigmoid terms in Eqs. (7.60) and (7.61).

It was observed that the wave front in this picture is spreading ballistically, that is, the average time t_a of the arrival of the front to a population is proportional to the effective distance to this population from the root, the origin of the epidemic,

$$t_a = \frac{1}{v^{(\text{eff})}} d^{(\text{eff})}, \tag{7.64}$$

where $v^{(\text{eff})}$ plays the role of the effective spreading speed. Clearly, the effective distance depends only on the matrix P, $d^{(\text{eff})} = d^{(\text{eff})}(P)$, while the time of arrival depends on P and also on the epidemic parameters of the model, $t_a = t_a(P, \mu, \beta, \gamma, \epsilon, \eta)$. Interestingly, the resulting effective spreading speed appears to depend only on the epidemic parameters, $v^{(\text{eff})} = v^{(\text{eff})}(\mu, \beta, \gamma, \epsilon, \eta)$. The ballistic propagation of the front of contagion in this model gives a chance to guess the origin of the epidemic by analysing its late stages.

7.8 Rumour spreading

Section 7.2 discussed the Watts threshold model applicable to social contagion phenomena. Here we consider more traditional models of rumour (or, one can say, information) spreading without thresholds for activation. The principal difference of rumour spreading as a social contagion phenomenon from disease spreading is that a rumour spreader can be influenced by an attempted recipient, unlike a virus spreader. Indeed, if you discover that somebody already knows the rumour or they show a little excitement about it, then you may stop gossiping further. This back influence on a rumour spreader is taken into account in the classical *Daley–Kendall model* (Daley and Kendall, 1964, 1965) and in its later variation, the *Maki–Thompson model* (Maki and Thompson, 1973). Each individual in a population in these models can be in one of three states: ignorant (S), spreader (I), stifler (R), similarly to the epidemic SIR model. As in the SIS and SIR model, an ignorant individual in both models becomes a spreader via contact with each of the neighbouring spreaders with a transmission rate β. The difference between the models is in the second rule. In the Maki–Thompson model, when a spreader contacts another spreader or a stifler, it becomes a stifler with rate α. In contrast to this, in the original Daley–Kendall model, when a spreader contacts another spreader or a stifler, each of involved spreaders become a stifler with rate α, effectively doubling the rate of emergence of stiflers when two spreaders meet.[21]

Let us consider the easier Maki–Thompson model on the infinite fully connected graph (well-mixed population) using the mean-field theory. We even add an extra channel—spontaneous transformation of a spreader to a stifler with rate μ, like in the SIR model. Without complicating our calculations, this addition allows a convenient comparison with the epidemic SIR model. In the spirit of the SIR model, we denote the fractions of ignorants, spreaders, and stiflers by $S(t)$, $I(t)$, and $R(t)$. Then the mean-field equations for the resulting model, can be written as

$$\frac{d}{dt}S = -\beta SI, \tag{7.65}$$

$$\frac{d}{dt}I = -\mu I + \beta SI - \alpha I(1 - S). \tag{7.66}$$

---

[21] A generalization of the Daley–Kendall and Maki–Thompson models was described by Nekovee, Moreno, Bianconi, and Marsili (2007).

In addition, we have the constraint $S + I + R = 1$ and the initial condition $R(0) = 0$, $S(0) \equiv S_0$, $I(0) \equiv I_0 = 1 - S_0$. For $t \to \infty$, the system approaches the stationary state in which $I(\infty) = 0$ and $R(\infty) \equiv R_f$, $S(\infty) = 1 - R_f$. Equations (7.65) and (7.66) are easily integrated in a way, similar to Eqs. (7.46) and (7.47) for the SIR model in Section 7.6, resulting in the equation for the final fraction R_f of ignorants:

$$R_f + S_0 \exp\left[\frac{\alpha(1 - S_0) - (\alpha + \beta)R_f}{\alpha + \mu}\right] = 1, \qquad (7.67)$$

compare with Eq. (7.49). If, initially, the fraction of spreaders I_0 tends to zero, that is, $S_0 \to 1$, then this equation takes the form:

$$R_f + \exp\left(-\frac{\alpha + \beta}{\alpha + \mu}R_f\right) = 1, \qquad (7.68)$$

showing that the critical value of the transmission rate, above which the final fraction of ignorants is non-zero, equals $\beta_c = \mu$, and that in the neighbourhood of this point, $R_f \cong 2(\beta - \mu)/\alpha$. Thus the standard Maki–Thompson model ($\mu = 0$) has no threshold for rumour spreading in contrast to the SIR model, but accounting for spontaneous transformation of a spreader to a stifler recovers the SIR model scenario. This result for the Maki–Thompson model was rigorously obtained by Sudbury (1985), who found the final 'proportion of people whom the rumour has not reached' $S_f = 1 - R_f = 0.2031\ldots$ for $\beta/\alpha = 1$. Instead of the term $-\alpha I(I + R) = -\alpha I(1 - S)$ on the right-hand side of Eq. (7.66), the Daley–Kendall model provides the term $-\alpha I(2I + R)$, which does not allow integration the mean-field equations explicitly. Nonetheless, the qualitative conclusions for the Daley–Kendall model on the fully connected graph are the same as for the Maki–Thompson one.[22] One can even introduce a more general rule: when a spreader contacts a stifler, it itself becomes a stifler with rate $\alpha > 0$; when a spreader contacts another spreader, one of them become a stifler with rate $\zeta > 0$. Then the term on the right-hand side of Eq. (7.66) becomes $-I(\zeta I + \alpha R)$, which doesn't change the scenario described.

For complex networks, these rumour-spreading models can be treated in the framework of a heterogeneous mean-field theory (Moreno, Nekovee, and

---

[22]See Daley and Gani (2001) and Gani (2000) for a strict approach to these stochastic models on a fully connected graph, exploiting the master equations for the probability $P_{n_S,n_I}(t)$ that there are n_S ignorants and n_I spreaders at time t in the system, where $n_S + n_I + n_R = N$. This theory is in the spirit of the one for the SIS model in Section 7.4, based on Eq. (7.6).

Pacheco, 2004a), similarly to the SIR model, Eqs. (7.34) and (7.35). Let $s_q(t)$, $\rho_q(t)$, and $r_q(t)$ be the fractions of ignorants, spreaders, and stiflers, respectively, among the vertices of degree q, where $s_q(t) + \rho_q(t) + r_q(t) = 1$. Let us place the generalized rumor-spreading model introduced, with the rates β, α, ζ, and μ on an infinite degree–degree correlated network with the conditional probability distribution $P(q'|q)$. The equations of the heterogeneous mean-field approximation describing this system read

$$\frac{d}{dt}\rho_q(t) = -\mu\rho_q(t) + \beta q[1 - \rho_q(t) - r_q(t)]\sum_{q'}\frac{q'-1}{q'}P(q'|q)\rho_{q'}(t)$$

$$-q\rho_q(t)\sum_{q'}P(q'|q)[\zeta\rho_{q'}(t) + \alpha r_{q'}(t)],$$

$$\frac{d}{dt}r_q(t) = \mu\rho_q(t) + q\rho_q(t)\sum_{q'}P(q'|q)[\zeta\rho_{q'}(t) + \alpha r_{q'}(t)]. \tag{7.69}$$

These equations can be explored numerically, and the verdict is that, if $\mu = 0$, then, similarly to the case of a fully connected graph, rumours spread over these networks for any non-zero transmission rate β, and the resulting final density of stiflers differs from zero.

Interestingly, for these rumour-spreading processes, the network heterogeneity and presence of hubs have a different effect on the final fraction of 'informed' vertices, R_f, than it is typical for the epidemic and percolation models having finite thresholds in networks with rapidly decaying degree distributions. While the presence of hubs in networks, typically decreases or even eliminates such epidemic and percolation thresholds and increases the prevalence and the size of a giant connected component, the hubs may even obstruct the rumour spreading in the Daley–Kendall and Maki–Thompson models. This makes homogeneous networks a better medium for the dissemination of information (Liu, Lai, and Ye, 2003; Moreno, Nekovee, and Vespignani, 2004b). The point is that after a hub heard a rumour and became a spreader, it quickly met another spreader or stifler, turning into a stifler. Thus hubs are bad gossipmongers in these models.

7.9 Opinion formation: The voter model

To quote Sood and Redner (2005), 'the voter model is perhaps the simplest and most completely solved example of cooperative behavior.' This model and its versions give the most basic description of opinion formation. Let

248

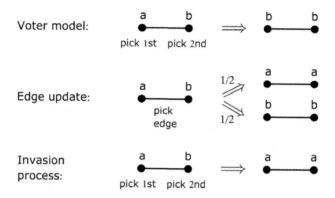

Fig. 7.12 Update rules of the voter model, the voter model with edge update, and the invasion process. The initial states a and b of the vertices can be anyone of 0, 1. Adapted from Sood, Antal, and Redner (2008).

each vertex i be in one of two states (opinions), $\sigma_i = 0, 1$.[23] The evolution starts with some random configuration of opinions. In the standard *voter model* (voter model with vertex update), at each time step (time increment $1/N$):

- choose a vertex uniformly at random and ascribe to this vertex the state of its randomly chosen neighbour.

Figure 7.12 explains this update rule. After N such updates, giving together time increment 1, each vertex is updated, on average, once.[24] In finite graphs, this process has two absorbing states: (i) all vertices are in the state $\sigma = 0$ and (ii) all vertices are in the state $\sigma = 1$. On the other hand, in the great majority of infinite graphs, consensus is never reached.[25]

The edge update version of the voter model implies the following rule. At each time step:

- choose an edge uniformly at random, choose one of its end vertices, and ascribe to this vertex the state of the second end vertex.

Finally, in the update rule of the *invasion process*, at each time step:

- choose a vertex uniformly at random and its random neighbour, and ascribe to this neighbour the state of the originally chosen vertex

[23] Equally, one can use spins: spin down or spin up, $S_i = 2\sigma_i - 1$.

[24] We consider only this stochastic update. The alternative, synchronous update can produce different results.

[25] Only the infinite one-dimensional voter model evolves to consensus.

(Figure 7.12). Loosely speaking, in the voter model, an individual (voter) adopts the neighbour's opinion, in the voter model with edge update, the neighbours together try to arrive at a joint opinion, and in the invasion process, an individual (invader) persistently propagates his opinion to the neighbours. Clearly, all three models on regular networks are equivalent.

There are three key questions for each of these models. (i) For an infinite network, what is the configuration of opinions in the limit $t \to \infty$? (ii) For a finite network, what is the exit probability $P_1^{(\text{exit})}$ that the process will end in the state, say, with all σ_i equal to 1? (iii) Finally, what is the average time T_N to reach an absorbing state (consensus) for a model on a network of size N? Let us discuss these three problems, following Sood and Redner (2005) and Sood, Antal, and Redner (2008).

To answer the first two questions, we first have to get the flip probability $P^{(\text{flip})}(i; \vec{\sigma})$ that the state of vertex i changes, within the time increment 1, from its state σ_i, provided the states of its nearest neighbours are $\{\sigma_k\}$, where $k \in \partial i$, are known. Here the state vector $\vec{\sigma}$ denotes the state of the system (all vertices–individuals). This probability can be written as

$$P^{(\text{flip})}(i; \vec{\sigma}) = \sum_k \frac{1}{N g^{(\text{VM,EU,IP})}} A_{ik}(\sigma_i - \sigma_k)^2. \tag{7.70}$$

Here, $g^{(\text{VM})} = q_i$, $g^{(\text{EU})} = \langle q \rangle$, and $g^{(\text{IP})} = q_k$ for the voter model with vertex update, with edge update, and the invasion process, respectively. The factor $1/N$ is present since we choose a given vertex (or an edge for the edge update model) with probability $1/N$ during the time interval 1. For, say, the voter model, we choose a neighbour of this vertex with probability $1/q_i$, which explains g in the denominator. Equally easily the reader can grasp the term g for the voter model with edge update and the invasion process.

Let us consider three quantities (densities) for these models:

$$d^{(\text{VM})} = \frac{\langle q\sigma \rangle}{\langle q \rangle} = \frac{1}{N\langle q \rangle} \sum_i q_i \sigma_i = \sum_q \frac{qP(q)}{\langle q \rangle} \sigma(q), \tag{7.71}$$

$$d^{(\text{EU})} = \langle \sigma \rangle = \frac{1}{N} \sum_i \sigma_i = \sum_q P(q)\sigma(q), \tag{7.72}$$

$$d^{(\text{IP})} = \frac{\langle \sigma/q \rangle}{\langle 1/q \rangle} = \frac{1}{N\langle 1/q \rangle} \sum_i \frac{1}{q_i} \sigma_i = \sum_q \frac{P(q)/q}{\langle 1/q \rangle} \sigma(q), \tag{7.73}$$

where $\sigma(q)$ is the fraction of vertices of degree q in state 1, and $P(q)$ is the degree distribution of a network.

Using the flip probability, one can easily find how the average opinion $\langle \sigma_i \rangle$ of vertex i changes after one full update of a system (time increment $N(1/N) = 1$), if the state of the system just before the update was $\vec{\sigma}$,

$$\Delta \langle \sigma_i \rangle = \left\langle (1 - 2\sigma_i) P^{(\text{flip})}(i; \vec{\sigma}) \right\rangle . \tag{7.74}$$

The term $1 - 2\sigma_i$ gives 1 for a flip up, that is, from the state $\sigma_i = 0$, and -1 for a flip down, that is, from the state $\sigma_i = 1$, which explains the form of this relation. Then the changes of the densities $d^{(\text{VM,EU,IP})}$ for the three models after one full update of a system can be written as

$$\Delta d^{(\text{VM})} = \frac{1}{N} \sum_{i,k} \frac{1}{q_i} q_i (\sigma_k - \sigma_i) = 0, \tag{7.75}$$

$$\Delta d^{(\text{EU})} = \frac{1}{N} \sum_{i,k} \frac{1}{\langle q \rangle} (\sigma_k - \sigma_i) = 0, \tag{7.76}$$

$$\Delta d^{(\text{IP})} = \frac{1}{N} \sum_{i,k} \frac{1}{q_i q_k} (\sigma_k - \sigma_i) = 0, \tag{7.77}$$

taking into account the equality $(1-2\sigma_i)(\sigma_i-\sigma_k)^2 = \sigma_k-\sigma_i$ valid for $\sigma = 0, 1$ and the $i \leftrightarrow k$ antisymmetry of the summands. Thus the densities $d^{(\text{VM})}$, $d^{(\text{EU})}$, and $d^{(\text{IP})}$, Eqs. (7.71)–(7.73), are conserved along the evolution of the corresponding models (Suchecki, Eguíluz, and San Miguel, 2004, 2005). Their values are determined by the initial conditions, that is, by an initial configuration of opinions.

For a finite network, where the evolution finishes in one of the absorbing states, this conservation implies the equality $d^{(\text{VM,EU,IP})} = P_1^{(\text{exit})} \times 1 + (1 - P_1^{(\text{exit})}) \times 0$, which implies the following result for the exit probability:

$$P_1^{(\text{exit})} = \begin{cases} \dfrac{\langle q\,\sigma \rangle}{\langle q \rangle} & \text{voter model,} \\[2mm] \langle \sigma \rangle & \text{edge update,} \\[2mm] \dfrac{\langle \sigma/q \rangle}{\langle 1/q \rangle} & \text{invasion process.} \end{cases} \tag{7.78}$$

The average time to reach consensus T_N can be obtained for an uncorrelated network of N vertices by using the heterogeneous mean-field approximation. For the sake of brevity, we consider only the voter model and denote $d^{(\text{VM})} \equiv d$. The states of a system for this network is described by a vector $\hat{\sigma}$ whose components $\sigma(q)$ are the average fractions of vertices of degree q

that have opinion 1. We intend to obtain the average time $T_N(\hat{\sigma})$ to reach consensus starting from a given state vector $\hat{\sigma}$. The equation for the average time, incorporating transition probabilities between different states, can be written as

$$T_N(\hat{\sigma}) = \sum_q P\left(\hat{\sigma} \to \hat{\sigma} + \frac{1_q}{NP(q)}\right)\left[T_N\left(\hat{\sigma} + \frac{1_q}{NP(q)}\right) + \frac{1}{N}\right]$$

$$+ \sum_q P\left(\hat{\sigma} \to \hat{\sigma} - \frac{1_q}{NP(q)}\right)\left[T_N\left(\hat{\sigma} - \frac{1_q}{NP(q)}\right) + \frac{1}{N}\right]$$

$$+ \left[1 - \sum_q P\left(\hat{\sigma} \to \hat{\sigma} + \frac{1_q}{NP(q)}\right) - \sum_q P\left(\hat{\sigma} \to \hat{\sigma} - \frac{1_q}{NP(q)}\right)\right]\left[T_N(\hat{\sigma}) + \frac{1}{N}\right]. \quad (7.79)$$

This equation relates states of the system separated by time increment $1/N$, after which a component $\sigma(q)$ of the state vector $\hat{\sigma}$ may change by $\pm 1/[P(q)]$, which is the difference produced by changing opinion of one vertex. By 1_q we denote the vector with one unit element for degree q and the remaining elements equal to 0. In the heterogeneous mean-field approximation, the transition probabilities take the form:

$$P\left(\hat{\sigma} \to \hat{\sigma} + \frac{1_q}{NP(q)}\right) = P(q)[1 - \sigma(q)]d,$$

$$P\left(\hat{\sigma} \to \hat{\sigma} - \frac{1_q}{NP(q)}\right) = P(q)\sigma(q)(1 - d). \quad (7.80)$$

Expanding Eq. (7.79) with these probabilities up to the second order in powers of $1/[NP(q)]$ leads to the famous Kolmogorov equation,

$$\sum_q [d - \sigma(q)]\frac{\partial T_N(\hat{\sigma})}{\partial \sigma(q)} + \sum_q \frac{d + \sigma(q) - 2d\sigma(q)}{2NP(q)}\frac{\partial^2 T_N(\hat{\sigma})}{\partial[\sigma(q)]^2} = -1. \quad (7.81)$$

The drift term in this equation ensures that all components $\sigma(q)$ of the state vector quickly (in time of the order of 1) approach the constant common value $d = \sum_q qP(q)\sigma(q)/\langle q \rangle$, after which Eq. (7.81) greatly simplifies. Taking into account the equality $\partial T_N/\partial \sigma(q) = (qP(q)/\langle q \rangle)\partial T_N/\partial d$ for the derivatives, one gets the equation

$$\frac{\langle q^2 \rangle}{N\langle q \rangle^2}d(1 - d)\frac{\partial^2 T_N(d)}{\partial d^2} = -1. \quad (7.82)$$

The solution of this equation readily provides the average consensus time for the voter model,

$$T_N^{(\mathrm{VM})}(d) = N \frac{\langle q \rangle^2}{\langle q^2 \rangle}[(1-d)\ln(1-d)^{-1} + d\ln d^{-1}], \qquad (7.83)$$

compare with the SIS model, where the average time to reach the absorbing time exponentially diverges with N, Eq. (7.11). Thus the consensus time $T_N \sim N$ if the second moment $\langle q^2 \rangle$ is finite in the limit $N \to \infty$. Otherwise T_N increases slower than N. In this case, a functional form T_N vs. N is determined by the size-dependent cutoff $q_{\mathrm{cut}}(N)$ of the degree distribution, which, in turn, strongly depends on details of a network model. In a similar way one can get $T_N^{(\mathrm{EU})} \propto N$ for the edge update voter model and $T_N^{(\mathrm{IP})} \propto N\langle q \rangle \langle 1/q \rangle$ for the invasion process (Sood, Antal, and Redner, 2008).

7.10   Propagation of memes

Among diverse processes involving social interactions (Zhang, Liu, Zhan, Lu, Zhang, and Zhang, 2016), an important set of processes involves competition between a large number of distinct transmitted items in contrast to standard epidemic models, considering one kind of infection. The propagation of memes in Twitter is a widely known example of such a process. A meme means a distinct piece of information that can be copied and transmitted, for example, a hashtag within a tweet. It is common knowledge that the propagation of memes in social networks by retweeting is cascading ('viral') (Leskovec, Adamic, and Huberman, 2007a). Furthermore, the sizes of such cascades (avalanches) for different memes are distributed according (approximately) to a power law. Here the cascade's size or the popularity of a meme, n, is the number of its retweets increased by one, accounting for the original tweet.

Gleeson, O'Sullivan, Baños, and Moreno (2016) proposed a null model for spreading of memes in complex networks explaining these observations for Twitter. In their model, users, inventing and tweeting new memes and retweeting some of the received ones, are the vertices of an uncorrelated (and hence locally tree-like) directed network with an in-, out-degree distribution $P(q_i, q_o)$. Here the out-degree q_o of a vertex is the number of followers of this user and its in-degree q_i is the number of the users whom the user follows. Users send or resend memes (to all their followers) with the activity rate $\beta(q_i, q_o)$ which depends on their in-, out-degree. This rate is the average number of tweets that a user sends or resends per unit time. It is convenient to normalize the activity rates by choosing time units,

$$\sum_{q_i, q_o} \beta(q_i, q_o)P(q_i, q_o) = 1, \qquad (7.84)$$

that is, on average, one user sends or resends one meme per time unit. A user, receiving a meme, keeps it in memory, as interesting, with probability λ, stamping the meme by the time of arrival. A user, who invents and immediately sends a new meme, also puts it in his or her stream (memory) stamping by the time of sending. The tweeting-retweeting activity is assumed to be the Poisson process characterized by exponentially distributed time intervals between events of activity. Each such event is either (i) inventing and immediately sending a new meme with probability μ (tweeting) or (ii) choosing one of memes from the user's memory and sending it with the complementary probability $1 - \mu$ (retweeting).[26] In the latter case, a meme for retweeting is selected from the memory in the following way. Let the current time be t. Draw t_m (memory time) from a given memory-time distribution $\Phi(t_m)$, whose shape appears to be not very important, and choose the first meme that arrived prior to the time $t - t_m$.[27]

Due to a locally tree-like structure of a network in this model, the propagation of a meme is a branching process, treatable analytically. Each meme in a network at a given instant has its age a (time passed after its creation). The issue of interest is the statistics of popularity of memes of a given age, $n(a)$ (number of times a meme was retweeted by age a increased by one accounting for the original tweeting), especially, a popularity distribution $P(n; a)$.

The key ingredient of the model is the memory of a user, from where memes are selected for retweeting. The memes appearing in the user's memory depend on many choices of other users sitting upstream in the network, which produces the mechanism of competition for memes. Memes enter the stream of a vertex of in-, out-degree (q_i, q_o) as a Poisson process at the rate

$$r(q_i, q_o) = \lambda q_i \tilde{\beta} + \mu \beta(q_i, q_o). \tag{7.85}$$

Here $\tilde{\beta}$ is the average rate of tweeting and retweeting of the users whom our user follows,[28]

$$\tilde{\beta} = \sum_{q_i, q_o} \frac{q_o P(q_i, q_o)}{\langle q_o \rangle} \beta(q_i, q_o). \tag{7.86}$$

The number of these users is q_i, and the memes sent from them are memorized with probability λ, which explains the first term on the right-hand

---

[26] For the sake of simplicity, we consider only tweets with memes.

[27] Note that in this model, a user can retweet the same meme several times.

[28] Moreover, this is the average rate of tweeting and retweeting for all users.

side of Eq. (7.85). The second term is due to the memes invented by the user themself, which are also memorized. Therefore the average occupation time of a meme in the user's stream (mean time interval between two consecutive memes in the memory) is

$$\langle t_{oc}(q_i, q_o)\rangle = \frac{1}{r(q_i, q_o)} = \frac{1}{\lambda q_i \tilde{\beta} + \mu \beta(q_i, q_o)}. \tag{7.87}$$

This expression enables one to arrive at the key conclusion about the meme propagation branching process. The average number of children of a vertex (user) of in-, out-degree (q_i, q_o) in a given meme's propagation tree can be written as

$$\langle b(q_i, q_o)\rangle = (1 - \mu)\beta(q_i, q_o)\frac{1}{\lambda q_i \tilde{\beta} + \mu \beta(q_i, q_o)} q_o \lambda, \tag{7.88}$$

which is the product of the rate of retweets of this user, the average occupation time of a meme in the user's memory, the number of followers of this user, and, finally, the probability that the follower keeps a meme in his or her memory. Then the average branching of this tree equals 1 in the limit of vanishing innovation, $\mu \to 0$,

$$\langle b\rangle = \sum_{q_i,q_o} \frac{q_i P(q_i, q_o)}{\langle q_i\rangle}\langle b(q_i, q_o)\rangle \xrightarrow{\mu\to 0} \sum_{q_i,q_o} \frac{q_i P(q_i, q_o)}{\langle q_i\rangle}\beta(q_i, q_o)\frac{1}{\lambda q_i \tilde{\beta}} q_o \lambda = 1.$$
$$\tag{7.89}$$

This means that meme propagation in this model is a critical branching process,[29] which allows one to easily guess the statistics of meme's popularity. Note that the competition of memes in this model results in a phenomenon which is known as *self-organized criticality*, where a system enters the critical state spontaneously, without adjusting a control parameter to a critical value.[30]

[29] A more rigorous derivation of this result is in Gleeson, Cellai, Onnela, Porter, and Reed-Tsochas (2014) and Gleeson, O'Sullivan, Baños, and Moreno (2016).

[30] Rigorously speaking, innovation rate μ can be treated as a control parameter in this problem, and then $\mu_c = 0$ plays the role of the critical point.
    Immediately before complex networks, self-organized criticality was the hottest topic in non-equilibrium statistical mechanics. Most of the numerous researchers who studied self-organized criticality in the 1980s–1990s, later switched over to complex networks. A simple example (forest-fire model) illustrates this important notion (Schenk, Drossel, Clar, and Schwabl, 2000). Let the vertices of an infinite lattice be occupied or empty, so that we have a set of clusters on a lattice like those in percolation problems. Consider the following process. In parallel, (i) infinitely slowly, fill new and new uniformly randomly

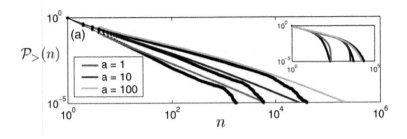

Fig. 7.13 Cumulative popularity distributions $\mathcal{P}_>(n;a)$ obtained for the uncorrelated network with the out-degree distribution exponent $\gamma = 2.5$ and $\langle q_0 \rangle = 11$. The innovation rate $\mu = 0.02$. The points are found by simulations for memes of three different ages and the curves are theoretical. The parameters of the gamma memory-time distribution $\Phi(t_m) \propto t_m^{k-1} e^{-t_m/\theta}$ are $k = 0.1$ and $\theta = 50$. Inset: The same as in the main figure now for the Poisson out-degree distribution with $\langle q_0 \rangle = 11$ and the parameters $k = 0.1$ and $\theta = 0.5$ of the gamma memory-time distribution. Adapted from Gleeson, O'Sullivan, Baños, and Moreno (2016).

For the sake of convenience, in their analysis, Gleeson, O'Sullivan, Baños,

selected vertices, and (ii) also infinitely slowly, choose at random vertices and empty them together with the clusters to which these vertices belong.

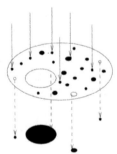

Here the solid arrows show the slow filling of randomly selected vertices (growing trees). The dashed arrows indicate the parallel infinitely slow process: random selection of nodes and elimination clusters (forests) to which these vertices belong (forest fires). Each removed cluster can be treated as an avalanche initiated by the removal of a single vertex.

If we start from an empty lattice, the concentration of occupied vertices will slowly grow until it reaches some value. The maximal size of clusters will grow but will never approach infinity. The point is that a giant connected component cannot emerge in this system since process (ii) efficiently eliminates any cluster containing a finite fraction of vertices. Therefore, the combination of slow processes (i) and (ii) drives a system exactly to the point of the emergence of a giant component, that is, to a critical state. The supposed infinite slowness of the processes is crucially important. If we added vertices and removed clusters at a finite rate, the system would stay away from criticality.

and Moreno (2016) used the following in-, out-degree distribution: $P(q_i, q_o) = \delta_{q_i,q} P(q_o)$, so $\langle q_o \rangle = q$, although real Twitter has a more complex degree distribution. Figure 7.13 shows the cumulative popularity distributions $\mathcal{P}_>(n; a) \equiv \sum_{n' \geq n} \mathcal{P}(n'; a)$ for three different ages a and two distinct out-degree distributions $P(q)$. The points obtained in simulations and theoretical curves indicate skewed popularity distributions. Particularly clear asymptotics of $\mathcal{P}(n; a)$ can be obtained in the limit of infinite age, $a \to \infty$, assuming $\beta(q_i, q_o) = 1$. The following results are independent of the shape of the memory-time distribution $\Phi(t_m)$. Let the second moment of the out-degree distribution be finite, $\langle q_o^2 \rangle < \infty$. Then

$$\mathcal{P}(n; a \to \infty) \sim n^{-3/2} e^{-n/n_{\text{cut}}(\mu)}, \tag{7.90}$$

where the cut-off $n_{\text{cut}} \propto 1/\mu^2$ for small innovation rate μ. If $P(q_o) \sim q_o^{-\gamma}$, $2 < \gamma < 3$, then

$$\mathcal{P}(n; a \to \infty) \overset{\mu \to 0}{\sim} n^{-[1+1/(\gamma-1)]}. \tag{7.91}$$

These power-law distributions are generic for critical branching processes on directed networks.

Generally, the notion 'self-organized criticality' implies self-organization of an open non-equilibrium cooperative system into a critical state. This state features power-law decaying correlations and power-law distributed avalanches, and other critical phenomena.

8
Networks of Networks

8.1 Networks with vertices and edges of different types

Let the vertices of a graph be of M different types, $\alpha = 1, 2, \ldots, M$, forming M subsets (layers) of distinct vertices. Then we can naturally divide the edges of this graph into, at most, $M(M+1)/2$ classes: edges connecting vertices within individual layers (at most, M different types of edges), and edges interconnecting vertices from different layers $\alpha \neq \beta$ (at most, $M/(M-1)/2$ types of such edges). In particular, if all connections are only between layers, then we get a multipartite graph, which we already discussed. In this chapter we proceed further and consider more interesting and more complicated networks having connections both within and between the layers. Distinct types of edges result in different kinds of paths between vertices, and hence to a more complex connectivity. This raises new questions, impossible for ordinary graphs. For example, we can ask about a pair of vertices: how are they interconnected by paths running within particular layers? This rich connectivity enables one to introduce new kinds of connected components playing specific roles in the integrity, robustness, and function of a network.

In this scheme, some layers may be interconnected, while other layers are not. Treating each individual layer as a super-node and all connections between two layers as a super-link, it is natural to call this multilayer structure a *network of networks* (Gao, Buldyrev, Havlin, and Stanley, 2011; Gao, Buldyrev, Stanley, and Havlin, 2012b; Gao, Buldyrev, Stanley, Xu, and Havlin, 2013) (Figure 8.1).

The simplest architecture of this kind arises when any vertex in every layer is interconnected with exactly one vertex in each other layer. Then the interlinked vertices from different layers can be treated as the replicas of a single vertex, forming a fully connected subgraph (M-clique for an M-layer network) with edges—interlinks between layers. In terms of super-nodes and super-links, this leads to the structure shown in Figure 8.1d. This enables one to represent this M-layer network as a graph with vertices of only one kind and edges of M different types ('colours'), as Figure 8.2

demonstrates. Both these representations are called a *multiplex network* or a multiplex graph, if you wish. Using the slang of this science, we simply say: 'a multiplex'. The multiplex networks allow an easier analytical treatment than more general multilayer networks with multiple interconnections between vertices from different layers (Figure 8.3). These more general networks still demonstrate key features and effects similar to their particular case, the multiplex networks, on which we mostly focus.

8.2 Mutually connected components

We shall not discuss here the ordinary giant connected component problem (ordinary percolation) for a multilayer network (Leicht and D'Souza, 2009), where all edges are treated as if they were of the same type. It doesn't differ much from this problem for single networks considered in Sections 6.1 and 6.2. A somewhat exotic phenomena associated with the birth of the giant connected component may occur only in multilayer networks with weakly interconnected layers having specific structural correlations (Hackett, Cellai, Gómez, Arenas, and Gleeson, 2016). In such networks, the giant connected component emerges in a way similar to the lowest curve $S(p)$ in Figure 6.10. In particular, in 2-layer networks, first it emerges in one layer (real percolation threshold, p_c), and, only after some delay, it penetrates into the second layer.

A more interesting concept of specific percolation, feasible only for multilayer networks, was put forward by Buldyrev, Parshani, Paul, Stanley, and Havlin (2010). The famous figure (Figure 8.4) taken from their paper, illustrates the rationale behind this kind of a percolation phenomenon. The figure shows the hypothetic cascade of failures leading to a blackout in two coupled networks in Italy: (i) the power grid (first layer) and (ii) the

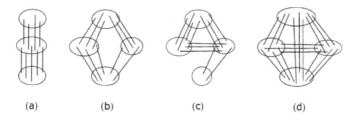

(a)   (b)   (c)   (d)

Fig. 8.1 Examples of the networks of networks. In this scheme, the layers are super-nodes, and the sets of interlinks connecting vertices from two layers are super-links.

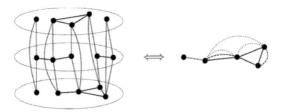

Fig. 8.2 Two representations of a multiplex network. The edges of the bottom, middle, and top layers map to the solid, dashed, and dotted edges of the right graph, respectively. Thin lines show interlinks connecting replica vertices.

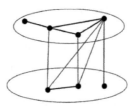

Fig. 8.3 A simple 2-layer graph which is not a multiplex.

network of Internet servers (second layer). Each Internet server is interconnected with a geographically closest power station assuming that just this server probably is used to control the station. On the other hand, when the power station stops, this server (or a number of such servers) assumingly also stops working. For functioning the entire system, both the layers must function. Let one station be removed from the system by a failure—the dark vertex in the first layer in Figure 8.4a. Then the stations separated from the first layer of the system by this removal fail and, together with them, fail all the servers that are close only to these stations. These failed servers switch off a number of other servers from the Internet layer, and these also stop functioning—the dark vertices of the Internet layer in Figure 8.4b. In their turn, the removed servers induce failing stations appearing without controlling servers in the neighbourhood, and so on.

Based on this process one can naturally introduce a *mutually connected component*.[1] This is a subgraph of a given M-layer graph, in which any two vertices of every layer are connected by at least one continuous path within each of the M layers, which passes only edges of this layer and necessary interlinks between layers (two chains of interlinks—starting and finishing).

---

[1] Another name is a *viable cluster* (Baxter, Dorogovtsev, Goltsev, and Mendes, 2012).

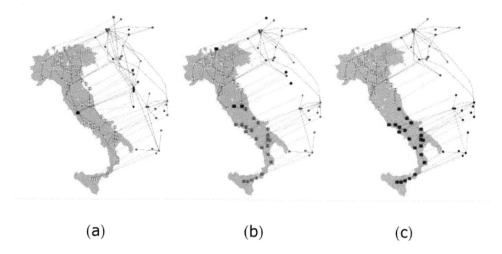

(a)   (b)   (c)

Fig. 8.4 The hypothetic cascade of failures leading to a blackout in two coupled Italian networks (Buldyrev, Parshani, Paul, Stanley, and Havlin, 2010): the net of power stations, attached to the map, and the Internet network shown above the map. The failure starts from the removal of the station represented by the dark vertex in (a) and progressively spreads over the networks in the way explained in the text. The failed stations and controlling Internet servers are shown by dark vertices.

Figure 8.5a explains a mutually connected component in general multilayer networks. In this example, one path between vertices i and j in layer a uses only edges of this layer. The second path passes an interlink between layers a and b, runs through edges within layer b, passes an interlink between layers b and a, and reaches vertex j. The third path, passes an interlink between layers a and b, then immediately passes an adjacent interlink between layers b and c, runs through edges within layer c, passes an interlink between layers c and a, and reaches vertex j. Note that the lengths of the starting and finishing chains of interlinks in these paths can range from 0 (in a path running within the layer to which these two vertices belong) to $M - 1$.[2] Figure 8.5b explains a mutually connected component for multi-

---

[2]Compare the effect of removal of a single interlink in two distinct networks of networks: (a) with a column structure like in Figure 8.1a, assuming in addition that any vertex has exactly $M - 1$ its counterparts in the other layers, forming an M-chain with $M - 1$ interlinks, and (b) with a ring structure like in Figure 8.1b, assuming that any vertex with $M - 1$ its counterparts in the other layers form a ring of M interlinks. If we remove a single interlink from an M-chain belonging to a mutually connected component in the network of networks (a), then these M vertices will drop out from this component. On the

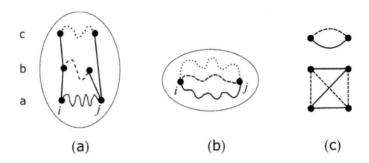

Fig. 8.5 Explaining a mutually connected component. (a) Vertices i and j in layer a in a 3-layer graph belong to a mutually connected component if at least one continuous path is running between them within each of the layers a, b, and c (solid, dashed, and dotted paths, respectively), passing only vertices of this component. The interlinks between the layers, included in two of these paths are also shown. Notice that one of the shown paths appears to use more than two interlinks. (b) Vertices i and j in a multiplex graph with three types of edges (solid, dashed, and dotted) belong to a mutually connected component if at least one continuous path of each of the three 'colours' (solid, dashed, and dotted) runs between i and j, passing only vertices of this component. (c) Two small mutually connected components in a multiplex graph with two types of edges, solid and dashed.

plex networks, and Figure 8.5c shows examples of small mutually connected components in a multiplex graph with edges of two colours. An infinite multilayer network can contain finite mutually connected components and a *giant mutually connected component* including a finite fraction of vertices. This giant mutually connected component, if it exists, is just the final result of the cascading process illustrated above by the developing blackout event in the Italian network (Figure 8.4). The complete blackout in this scheme corresponds to the absence of a giant mutually connected component in a multilayer network. The commonly used term *interdependent networks* implicates the combination of a given multilayer network and of a specific percolation problem for it (or, equivalently, of the process providing connected components for this problem). As is more customary among researchers in this field, here we reserve this term for a multilayer network with the giant mutually connected component problem for it (Gao, Buldyrev, Havlin,

other hand, if we remove one interlink from an M-ring belonging to a mutually connected component in the network of networks (b), then these M vertices will stay in the mutually connected component.

and Stanley, 2012a; Hu, Ksherim, Cohen, and Havlin, 2011; Huang, Gao, Buldyrev, Havlin, and Stanley, 2011).

Let us formulate the algorithm extracting a giant mutually connected component from general infinite interdependent networks (Figure 8.3) more strictly than it was done above for the Italian network:

(i) Delete all finite connected components in the networks formed by vertices and edges of each individual layer. This removal eliminates a number of interlinks between layers.

(ii) In each layer, remove the vertices appearing without at least one interlink to each of the other layers.

(iii) Repeat steps (i) and (ii) until the further process becomes impossible.

The result of this process is a giant mutually connected component or null.[3] The algorithm defines a pruning processes where pruning can be called 'global', since for the detection of all finite components within layers, we need global information about the network. Compare this to local pruning providing a k-core.

We can say this in another way:

(i) Find a giant connected component in each of the networks formed by vertices and edges of individual layers and remove all remaining vertices.

(ii) From each of these giant connected components, remove all vertices appearing without at least one interlink to each of the remaining giant connected components.

(iii) Repeat steps (i) and (ii) until the further process becomes impossible.

In the particular case of infinite multiplex networks, this algorithms takes the following form (we interpret a multiplex as a graph with edges of different colours):

(i) Find and delete all finite connected components (their vertices and edges) in each of the layers (network containing edges of one individual colour).

(ii) Repeat step (i) until the further process becomes impossible.

This can be formulated in yet another, equivalent way:

---

[3]In step (i) we remove finite connected components from all layers and only after that move to step (ii). Alternatively, we can remove finite connected components from one layer, then move to step (ii), then remove finite connected components from another layer, then pass to step (ii), and so on (Buldyrev, Parshani, Paul, Stanley, and Havlin, 2010). The result is the same.

(i) Find a giant connected component for each of the colours of the multiplex.

(ii) Find the overlap of all these giant connected components (their common vertices) and remove the remaining vertices from the multiplex.

(iii) Repeat steps (i) and (ii) until the further process becomes impossible.

The giant mutually connected component contains only a subset of joint vertices of the giant connected components of different colours. The algorithm stops when these giant connected components appear with the same set of vertices, or when they disappear.

Furthermore, one can detect every mutually connected component in interdependent networks, not only the largest components. Here we show only the algorithm for a finite multiplex network with edges of M colours (Baxter, Dorogovtsev, Goltsev, and Mendes, 2012):

(i) Choose a test vertex i at random from the network.

(ii) For each colour of edge s, compile a list of vertices reachable from i by following only edges of colour s.

(iii) The intersection of these M lists forms a new candidate set for the mutually connected component containing vertex i.

(iv) Repeat steps (ii) and (iii) but traversing only the current candidate set. When the candidate set no longer changes, it is either a mutually connected component, or contains only vertex i.

(v) To find further mutually connected components, remove the mutually connected component of vertex i from the network (cutting any edges) and repeat steps (i)–(iv) on the remaining network beginning from a new test vertex.

Repeated application of this procedure identifies every mutually connected component in the multiplex. The reader will find a more refined algorithm for this purpose in Hwang, Choi, Lee, and Kahng (2015).

Let each layer (colour) in a random multiplex be uncorrelated. Then the multiplex is completely characterized by the joint probability distribution $P(q_a, q_b, \ldots, q_M)$ for vertex degrees of all colours and by the number of vertices N, which we tend to infinity. If there is no other correlations between layers, a multiplex of this sort is locally tree-like in the sense that it has a negligible number of finite cycles. Here cycles can also contain edges of different colours. The local tree-likeness guarantees that finite mutually connected components are almost surely absent and only a giant mutually connected component is possible. The local tree-likeness enables one

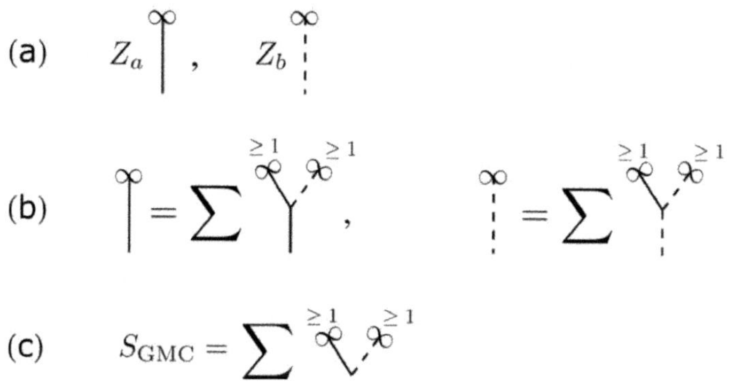

Fig. 8.6 (a) Notations for the probabilities Z_a and Z_b for a 2-layer multiplex (giant mutually connected component problem). (b) and (c) Diagrammatic representation of the self-consistency equations for these probabilities, Eq. (8.1), and the expression of S_{GMC} (relative size of a giant mutually connected component) in terms of Z_a and Z_b, Eq. (8.2), respectively.

to develop the theory of these networks and to find the characteristics of a giant connected component by employing the same techniques as for one-layer random networks in Sections 6.1, 6.2, and 6.7 (Son, Bizhani, Christensen, Grassberger, and Paczuski, 2012; Baxter, Dorogovtsev, Goltsev, and Mendes, 2012). Here we shall use this techniques. A more powerful approach based on the message-passing techniques is described in detail in Bianconi (2018a). For locally tree-like networks these techniques are equivalent.

For the sake of clarity, we first consider a 2-layer infinite multiplex with a given joint distribution $P(q_a, q_b)$ of two degrees of a vertex in layers a and b. Proceeding in the way described in Sections 6.1, 6.2, and 6.7, we introduce two probabilities, Z_a and Z_b, for edges of layers a and b, respectively (see graphical notations in Figure 8.6a). These are the probabilities that if we choose uniformly at random an edge in layer a or b, respectively, and follow it in one of two directions, then we appear in an infinite tree with a structure explained in Figure 8.6b. This diagrammatic representation of the self-consistency equations for the probabilities Z_a and Z_b,

$$Z_a = \sum_{q_a, q_b} \frac{q_a P(q_a, q_b)}{\langle q_a \rangle} [1 - (1 - Z_a)^{q_a - 1}][1 - (1 - Z_b)^{q_b}],$$

$$Z_b = \sum_{q_a, q_b} \frac{q_b P(q_a, q_b)}{\langle q_b \rangle} [1 - (1 - Z_a)^{q_a}][1 - (1 - Z_b)^{q_b - 1}], \qquad (8.1)$$

is also the definition of these infinite trees. Figure 8.6c explains the expression of the relative size of a giant mutually connected component S_{GMC} in terms of Z_a and Z_b,

$$S_{\text{GMC}} = \sum_{q_a,q_b} P(q_a, q_b)[1 - (1 - Z_a)^{q_a}][1 - (1 - Z_b)^{q_b}]. \qquad (8.2)$$

For the site problem, one must minimally modify Eqs. (8.1) and (8.2) by inserting the factor p (occupation probability of vertices) in front of their right-hand sides.

If the degrees q_a and q_b of a vertex are independent, that is, $P(q_a, q_b) = P_a(q_a)P_b(q_b)$, then we can introduce generating functions for each of the distributions $P_a(q_a)$ and $P_b(q_b)$,

$$G_{a,b}(z) \equiv \sum_q P_{a,b}(q)z^q,$$

$$G_{1a,b}(z) \equiv \frac{G'_{a,b}(z)}{\langle q_{a,b} \rangle}, \qquad (8.3)$$

which enable us to represent Eqs. (8.1) and (8.2) in a convenient form,

$$Z_a = [1 - G_{1a}(1 - Z_a)][1 - G_b(1 - Z_b)],$$
$$Z_b = [1 - G_a(1 - Z_a)][1 - G_{1b}(1 - Z_b)], \qquad (8.4)$$

and

$$S_{\text{GMC}} = [1 - G_a(1 - Z_a)][1 - G_b(1 - Z_b)], \qquad (8.5)$$

respectively.

In the simplest symmetric situation, where $P(q_a, q_b) = P(q_a)P(q_b)$, with a Poisson degree distribution $P(q)$, Eqs. (8.4) and (8.5) are reduced to the following single equation for $S_{\text{GMC}} \equiv S$:

$$S = p\left(1 - e^{-\langle q \rangle S}\right)^2, \qquad (8.6)$$

where for the sake of completeness we inserted the factor p (occupation probability for a vertex) on the right-hand side. Figure 6.16 explaining the graphical solution of the k-core problem in Section 6.7 is applicable as well to Eq. (8.6). Consequently the phase transition associated with the emergence of a giant mutually connected component is hybrid, like in the k-core

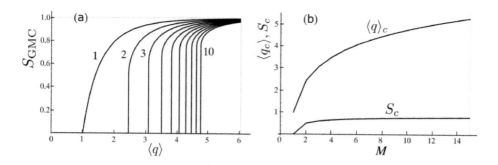

Fig. 8.7 (a) Relative size S_{GMC} of a giant mutually connected component vs. the average intra-layer degree $\langle q \rangle$ of a vertex in a symmetric M-layer multiplex with the same Poisson degree distribution for each of the layers. $M = 1, 2, \ldots, 10$. (b) The critical values $\langle q \rangle_c$ and S_c vs. M.

problem, with the combination of a discontinuity of S and the square root singularity

$$S(p) \cong (S_c + C\sqrt{p - p_c})\theta(p - p_c), \tag{8.7}$$

with the control parameter p.[4]

In a similar way we can derive the set of equations for an M-layer multiplex with uncorrelated layers, $M \geq 1$, directly generalizing Eqs. (8.1), (8.2) and Eqs. (8.4), (8.5). For $M = 1$, these equations reduce to the one for an ordinary giant connected component in a one-layer uncorrelated network. In particular, in the case of the same Poisson degree distribution for all layers, we arrive at the direct generalization of Eq. (8.6),

$$S = p\left(1 - e^{-\langle q \rangle S}\right)^M. \tag{8.8}$$

Figure 8.7 shows the results of the solution of this equation for different M with the occupation probability p set to 1. The asymptotics of the critical average degree $\langle q \rangle_c$ for large M equals

$$\langle q \rangle_c \cong \ln M, \tag{8.9}$$

while the discontinuity (relative size S_c of a giant mutually connected component at $\langle q \rangle_c + 0$) approaches 1 asymptotically,

$$1 - S_c \cong \frac{1}{\ln M}. \tag{8.10}$$

With increasing M, the number of conditions that must satisfy vertices in a giant mutually connected component (existence of at least M paths of

---

[4]One can set $p = 1$ and use $\langle q \rangle$ as a control parameter in this expression.

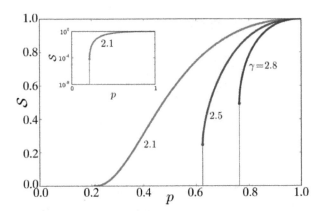

Fig. 8.8 Relative size of the giant mutually connected component cluster S as a function of the occupation probability p of vertices for two symmetric scale-free networks with $\gamma = 2.1$, 2.5, and 2.8. The height of the jump becomes very small as γ approaches 2, but it is not 0, as seen in the inset, which is S vs. p on a logarithmic vertical scale for $\gamma = 2.1$. Adapted from Baxter, Dorogovtsev, Goltsev, and Mendes (2012).

all M colours between two vertices) increases, and hence $\langle q \rangle_c(M)$ increases, while the dependence $S(\langle q \rangle)$ approaches a step function form for large M. The functional form of $\langle q \rangle_c(M)$ is determined by the tail of the degree distribution.[5]

One can explore Eqs. (8.4) and (8.5) for symmetric multiplex networks with a power-law degree distribution $P(q) \sim q^{-\gamma}$ for each of the layers. The verdict is that a hybrid transition happens if $\gamma > 2$, although, when exponent γ is close to 2, the discontinuity is so tiny, that it is hardly observable (Figure 8.8) (Baxter, Dorogovtsev, Goltsev, and Mendes, 2012). As γ approaches 2, the threshold value p_c tends to 0 (Buldyrev, Parshani, Paul, Stanley, and Havlin, 2010), and the relative size of a giant mutually connected component shows the same singularity $S \sim e^{-\text{const}/p}$ as in the giant connected component problem in a one-layer network with $\gamma = 3$, Eq. (6.26). In other words, these multiplex networks are hyper-resilient against random damage when $\gamma \leq 2$ (in the sense of the indestructibility of a giant mutually connected component) in contrast to uncorrelated one-layer networks,

---

[5]Compare the critical value of the total average degree of a vertex (total number of the edges in all layers, adjacent to a vertex), $M\langle q \rangle_c \cong M \ln M$, with the much lower point of the emergence of a giant connected component in the projection of this multiplex to a one-layer network, which is close to 2.

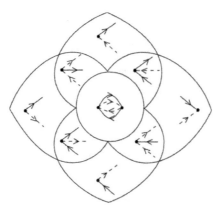

Fig. 8.9 Set of giant components in a 2-layer directed multiplex, generalizing the original bow-tie diagram for a one-layer networks (Figure 6.11a). Solid and dashed lines represent directed paths within the first and second layers. The circle in the centre shows a giant strongly mutually connected component. The directed paths, indicated in eight in-, and out-components, connect their vertices with a giant strongly mutually connected component. Adapted from Azimi-Tafreshi, Dorogovtsev, and Mendes (2014a).

hyper-resilient if $\gamma \leq 3$.

A similar hybrid phase transition occurs in general interdependent networks with a more complex set of interlinks between vertices in different layers than for multiplexes, like in Figure 8.3. Calculations for these networks are more cumbersome than above for multiplexes, but main conclusions stay valid, inspect Shao, Buldyrev, Havlin, and Stanley (2011) and Appendix G.

8.2.1 Directed multiplex networks

Similarly to one-layer directed networks, one can consider multiplexes with directed edges in all layers or in some of them. For such directed multiplex networks, we introduce a *giant strongly mutually connected component* (Azimi-Tafreshi, Dorogovtsev, and Mendes, 2014a). Any two vertices in this component are reachable from each other by directed paths running within each of the layers (directed paths of all colours). The vertices of this giant component are a subset of common vertices of the giant strongly connected components of all layers (colours) in the multiplex. The phase transition associated with the emergence of this component is hybrid. Notably, the set of distinct giant in- and out-components in multiplexes is much richer than for one-layer directed networks. There are up to 3^M distinct giant components

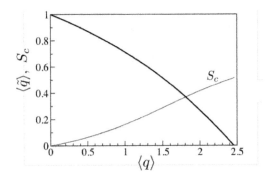

Fig. 8.10 Phase diagrams in the $\langle q \rangle - \langle \tilde{q} \rangle$ plane for a 2-level symmetric multiplex with edge overlaps. Here $\langle q \rangle$ is the average number of non-overlapping edges of a vertex within each of the levels, and $\langle \tilde{q} \rangle$ is the average number of overlapping connections of a vertex. The degree distributions are Poissonian. A giant component is present above the phase boundary (thick line). The thin line shows the value of the relative size of a giant mutually connected component, S_c, immediately above the discontinuous transition, S_c vs. $\langle q \rangle$. Adapted from Baxter, Bianconi, da Costa, Dorogovtsev, and Mendes (2016).

in an M-layer directed multiplex, including a giant strongly mutually connected component. Figure 8.9 explains these components for $M = 2$. Liu, Stanley, and Gao (2016) took into account in- and out-degree correlations in this problem, which enabled them to describe the robustness of real-world interdependent networks with directed connections, including international trade networks.

8.2.2 Overlapping edges

The locally tree-like interdependent networks allow an easy analytical treatment, however neglecting correlations between layers, typical in real network. One simple type of such correlations, overlapping edges from different layers, still does not spoil the local tree-likeness and so it is treatable analytically (Cellai, López, Zhou, Gleeson, and Bianconi, 2013b; Hu, Zhou, Zhang, Han, Rozenblat, and Havlin, 2013). Let us, for instance, consider a 2-layer multiplex with some of edges in one layer overlapping with the edges in the other layer. In other words, two edges from different layers, can have the same pair of end vertices. We can treat the overlapping edges as a special type of edges in a such a multiplex. Hu, *et al.* noticed that the clusters of overlapping edges can be treated as supernodes connected with each other by non-overlapping edges. Therefore the presence of any concentration of overlapping edges, smaller than 1, does not change qualitatively the hybrid phase transition associated with the emergence of the giant mutually connected component, as was confirmed by Min, Lee, Lee, and Goh (2015). In a multiplex of this kind, each vertex has three degrees, q_a, q_b, and $\tilde{q}$, being, respectively, the number of connections only in layer a, only in layer b, and the number of overlapping connections. If other correlations are absent, then the multiplex is defined by the joint degree distribution $P(q_a, q_b, \tilde{q})$ and it is locally tree-like. Figure 8.10 shows the phase diagram for a symmetric 2-layer multiplex with the joint degree distribution $P(q_a, q_b, \tilde{q}) = P(q_a)P(q_b)\widetilde{P}(\tilde{q})$, where $P(q)$ and $\widetilde{P}(\tilde{q})$ are Poisson distributions with the first moments $\langle q \rangle$ and $\langle \tilde{q} \rangle$, respectively. In the case of $\langle \tilde{q} \rangle = 0$, we get an ordinary Erdős–Rényi graph whose edges are overlapping edges of the multiplex, and a giant mutually connected component emerges at $\langle q \rangle_c = 1$ without a discontinuity. For $\langle q \rangle > 0$, the phase transition is hybrid.[6]

In M-layer multiplexes, the number of the possible types of overlaps equals $2^M - M - 1$ (in particular, in a 3-layer multiplex with layers a, b, c, there are 4 combinations of overlapping edges: ab, bc, ac, and abc), and for $M > 2$ the situation becomes more complicated than for a 2-layer multiplex. Nonetheless, it was found by the message passing techniques that the qualitative conclusions are similar for all $M \geq 2$ (Cellai, Dorogovtsev, and Bianconi, 2016).

8.2.3 Finite multiplexes

In finite random multiplex networks, the size of the largest mutually connected component fluctuates in different members of a statistical ensemble of multiplexes. Coghi, Radicchi, and Bianconi (2018) observed that in finite multiplexes the size of this component in a single realization can strongly deviate from the average over the entire ensemble. These deviations are particularly strong near p_c, the critical point of the corresponding infinite multiplex, and below it. Notably, the distribution of the relative size of the largest mutually connected component, $\mathcal{P}(S)$, has two peaks near the critical point and in some subcritical region. One of the peaks is at small S and

---

[6]A more general joint degree distribution $P(q_a, q_b, \tilde{q})$ with correlations between degrees provides a spectrum of scenarios. Depending on specific correlations, one can observe multiple hybrid transitions (giant mutually connected component, while monotonously increasing with the control parameter, experiences a chain of discontinuous transitions—jumps) and even recurrent hybrid transitions (with the increasing control parameter, a giant mutually connected component first emerges, then vanishes, and, after a while, reemerges) (Baxter, Bianconi, da Costa, Dorogovtsev, and Mendes, 2016).

the second one is near the value of S immediately after the hybrid phase transition in the corresponding infinite multiplex. This bimodal distribution indicates the mixture of two phases: normal phase and the phase with a 'giant' component typical for finite systems with a hybrid phase transition. Compare this with a one-peak distribution $\mathcal{P}(S)$ for a continuous transition (recall Figure 6.33). We refer the reader to Lee, Choi, Stippinger, Kertész, and Kahng (2016a) for theory and measurements of finite-size scaling of a 'giant' component in multiplex networks and in other problems with a hybrid phase transition.

8.2.4 Relation between multiplexes and general interdependent networks

Up to now we mostly focused on multiplex networks and that was not only for demonstration purposes. Already Gao, Buldyrev, Stanley, and Havlin (2012b) noticed that, with respect to a giant mutually connected component, a large class of networks of networks is equivalent to multiplexes, and hence the case of multiplexes is particularly important among interdependent networks. Moreover, even a wider class of interdependent networks was found to have the same sets of mutually connected components as in corresponding multiplexes (Bianconi, Dorogovtsev, and Mendes, 2015). Imagine a multiplex in the multilayer representation. Each vertex in the multiplex together with the replicas of this vertex in other layers form a fully connected subgraph—M-clique—linked together by interlinks between layers, and no interlinks connect vertices from different M-cliques. Let us remove some of the interlinks in such a way that each of these M-graphs remain connected. This removal does not change any of paths within layers and keeps intact the connectivity of each of the M-graphs. This guarantees that any network generated in this way has the same set of mutually connected components as the original multiplex. On the other hand, interdependent networks with vertices non-separable into sets of replicas can show some peculiarities unseen in usual multiplexes, for example, multiple hybrid transitions (Bianconi and Dorogovtsev, 2014).

8.3 Avalanches and cascading failures

When considering a hybrid transition associated with the emergence of a k-core in one-layer networks in Section 6.7, we paid particular attention to its weakest, critical part, namely its corona, easily collapsing and producing large avalanches near the transition. A giant mutually connected component

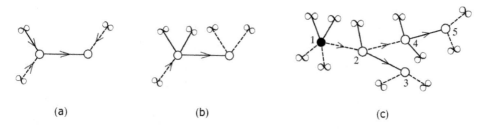

(a)   (b)   (c)

Fig. 8.11 Examples of critical clusters in a 2-layer multiplex, where we use the graphical notations from Figure 8.6b. In (a) and (b), all vertices are critical (shown by open circles). Arrows mark single edges of critical vertices leading to infinite trees introduced in Figure 8.6(b). Deleting the left vertex in (a) or (b) immediately removes the right vertex from a giant mutually connected component. Deleting the right vertex in (a) or (b) keeps the left vertex in a giant mutually connected component. (c) Removal of (non-critical) vertex 1 will result in all of the shown vertices being removed from a giant mutually connected component, while removal of vertex 4 results only in vertex 5 also being removed from this component. Removal of vertices 3 and 5 has no further effect. Adapted from Baxter, Dorogovtsev, Goltsev, and Mendes (2012).

in interdependent networks contains a similar critical subgraph. For the sake of clarity, we focus on locally tree-like multiplex networks. Let us introduce *critical vertices* in a giant mutually connected component. By definition, these vertices have exactly one edge in one of the layers (colours) of the multiplex leading to an infinite tree of the kind introduced in Figure 8.6b. In addition, they must have at least one edge in each of the remaining layers (colours), leading to the infinite trees of this kind. One can indicate such single edges connecting the infinite trees of Figure 8.6b to critical vertices in a multiplex by arrows as is shown in Figure 8.11. This figure explains that the direction of each of such edges is unique. Note that the source vertex of a 'directed' edge can be either critical or non-critical, belonging to a giant mutually connected component. The set of these 'directed' edges with their end vertices, critical and non-critical, form the critical subgraph of a giant mutually connected component. This subgraph in its turn contains numerous separate finite critical clusters. Removal of a vertex in a critical cluster results in the removal of all downstream critical vertices, that is, in the avalanche transmitted in the direction of the arrows. These avalanches can produce new critical vertices, creating a substrate for next avalanches.

The critical subgraph of a giant mutually connected component and the associated avalanches are similar to the corona of a k-core and the avalanches on it, except the directedness, which turns out to be not of primary importance. The theory describing the critical subgraph and the statistics of these avalanches was developed in Baxter, Dorogovtsev, Goltsev, and Mendes (2012). The main conclusions are similar to the k-cores. The susceptibility is given by the average size of a critical cluster to which a uniformly randomly chosen vertex belongs—the average size of the avalanche induced by the removal of a random vertex. This susceptibility has an inverse square root singularity at the critical point. Fluctuations of the size of a giant mutually connected component in the infinite network limit produce another 'susceptibility' showing the Curie–Weiss singularity (Lee, Choi, Stippinger, Kertész, and Kahng, 2016a). The size distribution $\mathcal{P}(s)$ of the avalanches induced by the removal of a random vertex decays as $s^{-3/2}$.

Moreover, despite the global nature of pruning resulting in a giant mutually connected component in interdependent networks in contrast to the local pruning algorithm leading to a k-core, the cascading failure of interdependent networks was found to be qualitatively close to the k-core pruning (Section 6.8). For a detailed discussion of this process in interdependent networks, we refer the reader to Zhou, Bashan, Cohen, Berezin, Shnerb, and Havlin (2014).

8.4 Other percolation problems

8.4.1 Redundant mutually connected component

The definition of a mutually connected component containing the requirement that its vertices must be connected by paths within each of M layers of an interdependent network implies the increasing with M number of conditions, which leads to a threshold diverging as M approaches infinity. Radicchi and Bianconi (2017) introduced a percolation problem for multilayer networks, without such a divergence. The idea was to reduce from M to 2 the number of required paths running within different layers between each two vertices in a component. Here we indicate two options, for the sake of brevity focusing on M-*layer* multiplex networks. (i) In a 2-*layer connected component* each two vertices are connected by continuous paths within at least two layers (at least one path in one layer and at least one path in another layer). (ii) In a *redundant mutually connected component* each two vertices are connected by at least one path within a given layer (say, layer 1) and at least one path running within any other layer. Radicchi and Bianconi (2017) explored option (ii). Let us analyse equations for the

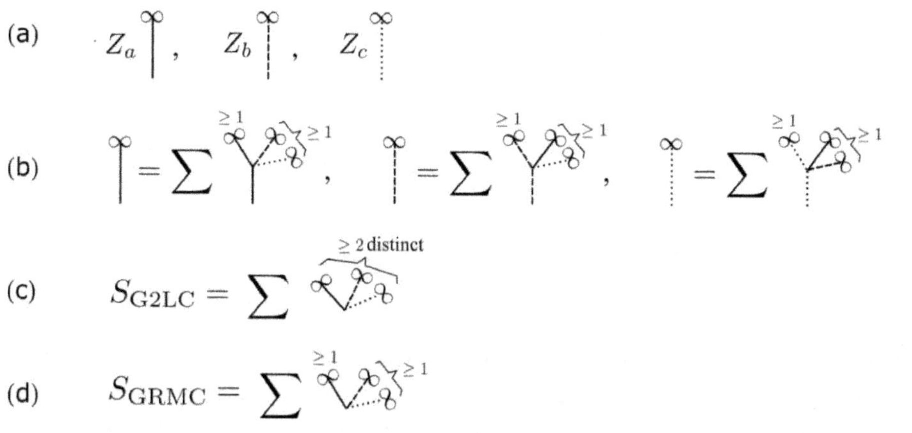

(a) Z_a, Z_b, Z_c

(b)

≥ 2 distinct

(c) $S_{\text{G2LC}} = \sum$

(d) $S_{\text{GRMC}} = \sum$

Fig. 8.12 (a) Notations for the probabilities Z_a, Z_b, and Z_c for a 3-layer multiplex (problems of giant 2-layer connected and redundant mutually connected components). (b), (c), and (d) Diagrammatic representation of the self-consistency equations for these probabilities, Eq. (8.11), and the expressions of S_{G2LC} (relative size of a giant 2-layer connected component), Eq. (8.12), and of S_{GRMC} (relative size of a giant redundant mutually connected component), Eq. (8.13), in terms of Z_a, Z_b, and Z_c.

giant components in these problems. We consider an infinite uncorrelated M-layer multiplex described by its joint degree distribution $P(q_1, \ldots, q_M)$ and use the local tree-likeness of this network. Figure 8.12 explains the equations in the case of $M = 3$. Note that the equations for the probabilities Z_α, $\alpha = 1, 2, \ldots, M$ coincide for problems (i) and (ii). Consequently the thresholds in these problems coincide. The equations for Z_α have the following form:

$$Z_\alpha = \sum_{q_1, \ldots, q_M} \frac{q_\alpha P(q_1, \ldots, q_M)}{\langle q \rangle} \left[1 - (1 - Z_\alpha)^{q_\alpha - 1} \right]$$

$$\times \left[1 - \frac{1}{(1 - Z_\alpha)^{q_\alpha}} \prod_{\beta=1}^{M} (1 - Z_\beta)^{q_\beta} \right], \tag{8.11}$$

while the expressions of the relative sizes S_{G2LC} and S_{GRMC} of a giant 2-layer connected component and a giant redundant mutually connected component, respectively, in terms of the probabilities Z_α (Figure 8.12c, d), can be written as

$$S_{\text{G2LC}} = 1 + \sum_{q_1,\ldots,q_M} P(q_1,\ldots,q_M)$$

$$\times \prod_{\alpha=1}^{M} (1 - Z_\alpha)^{q_\alpha} \left[M - 1 - \sum_{\beta=1}^{M} (1 - Z_\beta)^{-q_\beta} \right], \tag{8.12}$$

$$S_{\text{GRMC}} = \sum_{q_1,\ldots,q_M} P(q_1,\ldots,q_M) \left[1 - (1 - Z_1)^{q_1}\right] \left[1 - \prod_{\beta=2}^{M} (1 - Z_\beta)^{q_\beta} \right]. \tag{8.13}$$

Clearly, in the limit $M \to \infty$, S_{G2LC} exactly equals 1 above the critical point, and S_{GRMC} coincides with the relative size of the giant connected component of a single-layer network with a degree distribution $\widetilde{P}(q) = \sum_{q_2,\ldots,q_M} P(q, q_2, \ldots, q_M)$. On the other hand, in the case of $M = 2$, two-layer connected components and redundant mutually connected components coincide with mutually connected components. Assuming that the layers are symmetric and degrees in different layers are independent, that is $P(q_1, \ldots, q_M) = \prod_{\alpha=1}^{M} P(q_\alpha)$ and $Z_\alpha \equiv Z$, and using generating functions of the distribution $P(q)$, one can write Eqs. (8.11)–(8.13) in the form:

$$Z = [1 - G_1(1 - Z)]\{1 - [G(1 - Z)]^{M-1}\}, \tag{8.14}$$

$$S_{\text{G2LC}} = 1 + (M - 1)[G(1 - Z)]^M - M[G(1 - Z)]^{M-1}, \tag{8.15}$$

$$S_{\text{GRMC}} = [1 - G(1 - Z)]\{1 - [G(1 - Z)]^{M-1}\}. \tag{8.16}$$

Additionally assuming that the degree distribution $P(q)$ is Poisson with the average degree $\langle q \rangle$ readily gives

$$Z = \left[1 - e^{-\langle q \rangle Z}\right] \left[1 - e^{-(M-1)\langle q \rangle Z}\right], \tag{8.17}$$

$$S_{\text{G2LC}} = 1 + (M - 1)e^{-M\langle q \rangle Z} - Me^{-(M-1)\langle q \rangle Z}, \tag{8.18}$$

$$S_{\text{GRMC}} = Z. \tag{8.19}$$

Figure 8.13 shows the resulting relative sizes of a giant 2-layer connected component and a giant redundant mutually connected component versus $\langle q \rangle$ for different M and the dependences of the critical value $\langle q \rangle_c$ and the sizes of the discontinuities at the critical point, $S_{\text{G2LC}}^{(c)}$ and $S_{\text{GRMC}}^{(c)}$, on M. One can see that $\langle q \rangle_c$ decreases to 1 as M goes to infinity. Compare this figure to its contrasting counterpart (Figure 8.7) for a giant mutually connected component, where $\langle q \rangle_c$ diverges with M. Compare also the asymptotics

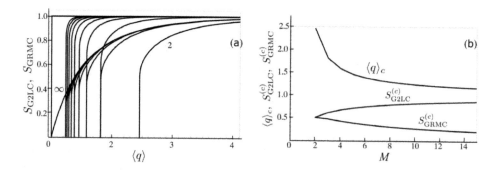

Fig. 8.13 (a) Relative sizes S_{G2LC} and S_{GRMC} of, respectively, a giant 2-layer connected component and a giant redundant mutually connected component vs. the average intra-layer degree $\langle q \rangle$ of a vertex in a symmetric M-layer multiplex with the same Poisson degree distribution for each of the layers (upper and lower sets of curves). $M = 2, \ldots, 10, \infty$. (b) The critical values $\langle q \rangle_c$ and $S_{\text{G2LC}}^{(c)}$, $S_{\text{GRMC}}^{(c)}$ (relative sizes of the giant components immediately above the discontinuous transition) vs. M. Compare to Figure 8.7 for a giant mutually connected component.

$$\langle q \rangle_c - 1 \cong \frac{\ln(2M) + 1}{2M}, \tag{8.20}$$

$$1 - S_{\text{G2LC}}^{(c)} \cong \frac{\ln(2M) + 1}{2M}, \tag{8.21}$$

$$S_{\text{GRMC}}^{(c)} \cong \frac{\ln(2M)}{M}, \tag{8.22}$$

with the corresponding asymptotics for a giant mutually connected component in Eqs. (8.9) and (8.10).

8.4.2   Weak percolation

The definition of a giant mutually connected component and its versions and variations is about continuous paths between vertices, running within layers, and so it is non-local. In contrast to that, in a *weak percolation problem*, relevant components are defined by a strictly local rule (Baxter, Dorogovtsev, Mendes, and Cellai, 2014). By definition, every vertex of a *weakly mutually connected component* in a multilayer network has at least one nearest neighbour belonging to this component in each of the layers. In particular, for multiplexes, every vertex of a weakly mutually connected component shares at least one edge of each colour with other vertices of this component (Figure 8.14). This condition is weaker than the one for a

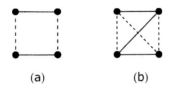

(a)   (b)

Fig. 8.14 The difference between weakly mutually connected (a) and mutually connected (b) components. In (a) all nodes have connections of both solid and dashed types and belong to one weakly mutually connected component. This small graph contains no mutually connected components. (b) Two paths—solid and dashed—run between each two vertices of this graph, and so this is a mutually connected component (coinciding in this case with a weakly mutually connected component). Adapted from Baxter, da Costa, Dorogovtsev, and Mendes (2020).

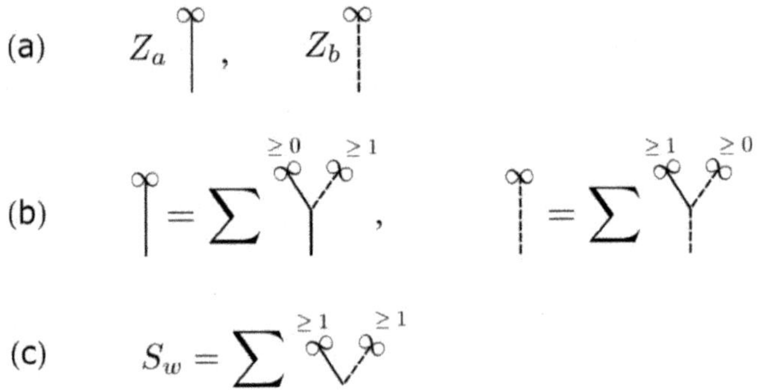

Fig. 8.15 (a) Notations for the probabilities Z_a and Z_b in the weak percolation problem on a 2-layer multiplex. (b) Diagrammatic representation of the self-consistency equations for these probabilities, Eq. (8.23), and, simultaneously, the definition of the relevant infinite trees. (c) The expression of S_w (relative size of a giant weakly mutually connected component) in terms of Z_a and Z_b, Eq. (8.24).

mutually connected component, and hence a mutually connected component is a subgraph of a weakly mutually connected component. Clearly, a network can have one weakly mutually connected component or null, and this component can have separate parts, including a giant one. Similarly to a k-core in single-layer networks, a weakly mutually connected component can be easily obtained by a local pruning algorithm, that is, by progressing removal of all vertices having no nearest neighbours in at least one of the layers of a network.

Let us obtain the relative size S_w of a giant weakly mutually connected component in uncorrelated multiplexes, easily treatable due to their local tree-likeness (Baxter, da Costa, Dorogovtsev, and Mendes, 2020). Figure 8.15 explains the self-consistency equations for the relevant probabilities Z_a and Z_b in this problem for a 2-layer uncorrelated multiplex with a joint degree distribution $P(q_a, q_b)$. These equations look as follows:

$$Z_a = \sum_{q_a, q_b} \frac{q_a P(q_a, q_b)}{\langle q_a \rangle}[1 - (1 - Z_b)^{q_b}],$$

$$Z_b = \sum_{q_a, q_b} \frac{q_b P(q_a, q_b)}{\langle q_b \rangle}[1 - (1 - Z_a)^{q_a}], \qquad (8.23)$$

while the expression of the relative size of a giant weakly mutually connected component S_w in terms of the probabilities Z_a and Z_b has the form

$$S_w = \sum_{q_a, q_b} P(q_a, q_b)[1 - (1 - Z_a)^{q_a}][1 - (1 - Z_b)^{q_b}]. \qquad (8.24)$$

Equations for multiplexes with a higher number M of layers can be written straightforwardly. For the sake of brevity, we touch upon here only the case of a symmetric multiplex with independent degrees, $P(q_a, q_b) = P(q_a)P(q_b)$, which can be analysed by using generating functions. Find a complete consideration in Baxter, da Costa, Dorogovtsev, and Mendes (2020). For a 2-layer multiplex, the conclusion is that

$$S_w \sim S^2, \qquad (8.25)$$

where S is the relative size of a giant connected component in one layer. Therefore, if $\langle q^2 \rangle < \infty$, which corresponds to exponent $\gamma > 3$ for power-law degree distributions, then the phase transition is continuous with the critical exponent β of S_w equal to 2. On the other hand, if the second moment of the degree distribution $\langle q^2 \rangle$ diverges, that is, $\gamma \leq 3$, then the giant weakly mutually connected component is hyper-resilient to a random damage, similarly to a giant connected component.

For $(M \geq 3)$-layer multiplexes, the results markedly differ from the 2-layer case. See Figure 8.16 for a 3-layer symmetric uncorrelated multiplex, covering a wide range of power-law degree distributions with $1 < \gamma < \infty$, including the region $1 < \gamma \leq 2$, where the first moment of a degree distribution, $\langle q \rangle$, diverges in the infinite networks. The occupation probability p of edges is used as a control parameter (bond problem), although the results for the site problem are qualitatively similar. Notice that this theory, exploiting local tree-likeness, works surprisingly well in the region $1 < \gamma \leq 2$ where one

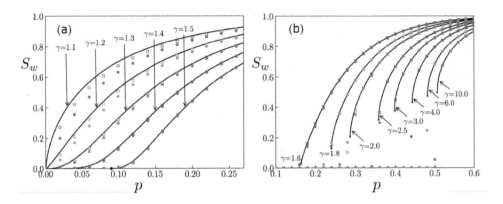

Fig. 8.16 Relative size S_w of the giant weakly mutually connected component for a 3-layer multiplex with identically power-law distributed layers ($P(q) \propto q^{-\gamma}$, $q \geq 4$), vs. occupation probability p of edges, for various values of γ. For $\gamma \leq 2$, relevant size-dependent cut-offs are introduced in the degree distributions to avoid 'finite' cycles (see the text). (a) For $\gamma \leq 1 + 1/(M-1) = 3/2$ ($M = 3$), the giant component exists for any $p > 0$, where $p_c = 0$ is the critical point of a continuous transition. Symbols show measurements averaged over 100 networks of $N = 10^4$ vertices (circles) and $N = 10^7$ vertices (squares). (b) For γ exceeding this limit, the giant component appears with a discontinuous hybrid transition at a finite critical point. Symbols show measurements in networks of $N = 10^5$ vertices (circles, 100 realizations) and $N = 10^4$ vertices (squares, one realization). Curves are theoretical calculations. Adapted from Baxter, da Costa, Dorogovtsev, and Mendes (2020).

could expect numerous finite cycles for degree distributions without cut-offs. To avoid such cycles in simulations, one has to introduce size-dependent cut-offs $q_{\text{cut}}(N)$ in the degree distributions. It turns out sufficient to set $q_{\text{cut}}(N) \sim 1/N$. The phase transition associated with the emergence of a giant weakly mutually connected component is hybrid for $\gamma > 1 + 1/(M-1)$. As we approach this limit from above, the discontinuity decreases to zero, and for $\gamma = 1 + 1/(M-1) + 0$ the transition appears to be continuous with a finite critical point $p_c^*(M) > 0$ and the following critical singularity:

$$S_w \propto (p - p_c^*)^{M/(M-2)}. \tag{8.26}$$

Below $\gamma = 1 + 1/(M-1)$, the critical point is zero, a giant weakly mutually connected component is present at any non-zero p, and the hyper-resilience regime is realized. This scenario markedly differs from that for the giant mutually connected component in 2-layer multiplex networks (Figure 8.8) where a hybrid transition is present for $\gamma > 2$, the hyper-resilience phenomenon takes place for $\gamma < 2$, and at the border, $\gamma = 2$, between these regimes, an infinite order continuous transition occurs with $p_c = 0$.

The weak percolation problem for multiplex networks is a particular case of the **k**-core problem for them (Azimi-Tafreshi, Gómez-Gardeñes, and Dorogovtsev, 2014*b*; Di Muro, Valdez, Rêgo, Buldyrev, Stanley, and Braunstein, 2017). By definition, every vertex in the **k**-core has at least k_α nearest neighbours in it connected to the vertex by edges of colour α, $\alpha = 1, 2, \ldots, M$. Here k_α are the components of the vector $\mathbf{k} = (k_1, k_2, \ldots, k_M)$. In the weak percolation problem, $\mathbf{k} = (1, 1, \ldots, 1)$.

We touched upon only a few basic percolation problems for multilayer networks, demonstrating main effects. In fact, the range of options is very wide. For example, we can define a new component by demanding that any two its vertices are connected by at least a single path within each of a given set of layers, within each of the second set of layers, they must be connected by at least two independent paths, for the third set of layers some local conditions are imposed, and so on.

8.5 Dynamical systems on multilayer networks

Introducing each of the percolation problems for multilayer networks, we first defined specific components, accounting for connections between layers, and then explored the emergence of the giant one. For cooperative systems or dynamical models placed on such networks, one has to proceed in a different manner and introduce into models a special coupling between layers, distinct from coupling between agents or spins within individual layers. This special interlayer coupling corresponds to what is called 'interdependence' in percolation problems, qualitatively changing critical phenomena in such systems compared to their single-layer counterparts. Danziger, Bonamassa, Boccaletti, and Havlin (2019) indicated a way to get discontinuous critical effects, similar to the birth of a giant mutually connected component, in dynamical systems placed on multilayer networks. They considered two dynamical models: Kuramoto oscillators (Section 12.4), and the quenched mean-field theory for the SIS model (Section 7.4) Here we touch upon only the second as it is actually about final stationary states.

Recall the quenched mean field theory equations for the probability $x_i(t) \equiv \rho_i(t)$ that vertex i infective in the SIS model for a single-layer network:

$$\frac{dx_i}{dt} = -\mu x_i + \beta(1 - x_i) \sum_{j=1}^{N} A_{ij} x_j, \tag{8.27}$$

where the recovery rate μ can be set to 1, and $\lambda = \beta/\mu$ is the infection rate.[7] The stationary solution of this equation demonstrates a continuous phase transition at the epidemic threshold. Consider a 2-layer multiplex network, that is, each vertex in one layer (adjacency matrix $A_{ij}^{(a)}$) is interconnected with its counterpart in the second layer (adjacency matrix $A_{ij}^{(b)}$). For each vertex i introduce two dynamic variables $x_i^{(a)}$ and $x_i^{(b)}$, whose evolution is described by the coupled equations:

$$\frac{dx_i^{(a)}}{dt} = -x_i^{(a)} + \lambda_a x_i^{(b)}(1 - x_i^{(a)}) \sum_{j=1}^{N} A_{ij}^{(a)} x_j^{(a)},$$

$$\frac{dx_i^{(b)}}{dt} = -x_i^{(b)} + \lambda_b x_i^{(a)}(1 - x_i^{(b)}) \sum_{j=1}^{N} A_{ij}^{(b)} x_j^{(b)}. \tag{8.28}$$

The coupling between layers in these equations effectively modifies the infection rates for each of the layers,

$$\lambda_a \to \lambda_a x_i^{(b)}, \quad \lambda_b \to \lambda_b x_i^{(a)}, \tag{8.29}$$

resulting in the interdependence.[8] Indeed, due to this coupling, an infective vertex in one layer stimulates the spreading process in the neighbourhood

---

[7]More general equations of this sort for dynamical variables $x_i(t)$ on vertices and $y_{ij}(t)$ on edges have the form:

$$\frac{d\mathbf{x}}{dt} = F(\mathbf{x}, \mathbf{y}),$$

$$\frac{d\mathbf{y}}{dt} = G(\mathbf{x}, \mathbf{y}),$$

where the vectors $\mathbf{x}(t)$ and $\mathbf{y}(t)$ have the components $x_i(t)$, $i = 1, 2, \ldots, N$, and $y_{ij}(t)$, $(ij) = 1, 2, \ldots, E$, respectively. For these dynamical systems, one can organize coupling between layers in a way similar to Eq. (8.28).

[8]For dynamical systems with variables having varying signs, this coupling should be exchanged with the following:

$$\lambda_a \to \lambda_a |x_i^{(b)}|, \quad \lambda_b \to \lambda_b |x_i^{(a)}|.$$

If $0 \leq |x_i| \leq 1$, one can also introduce a competitive coupling,

$$\lambda_a \to \lambda_a(1 - |x_i^{(b)}|), \quad \lambda_b \to \lambda_b(1 - |x_i^{(a)}|),$$

having the opposite effect. That is, a vertex with a large value of $|x_i|$ in one layer hinders the spreading process in the neighbourhood of this vertex within the other layer.

of this vertex within the other layer and, vice versa, a recovered vertex in one layer hinders spreading within the second layer. This resembles the rule defining a mutually connected component. The analysis of the stationary solution of Eq. (8.28) showed that the phase transition becomes discontinuous, of the first order.[9] This first-order transition corresponds to the hybrid transition in the problem of a giant mutually connected component.

Another way of treating interdependence for coupled dynamical systems is based on the exploration of the response of these systems to local perturbations. The steady state of a system is perturbed by applying a small permanent perturbation Δx_i to a single vertex i: $x_i \rightarrow x_i + \Delta x_i$. This induces changes in the states of other vertices, $x_j \rightarrow x_j + \Delta x'_j$, which become steady in the limit $t \rightarrow \infty$. The number of vertices j whose relative response exceed some threshold δ, that is, $\sum_{j \neq i} \theta(|\Delta x'_j/x_j| - \delta)$, where $\theta(x)$ is the step function, is interpreted as the size of the 'cascade of failures' induced by the perturbation of vertex i (Barzel and Barabási, 2013). In coupled dynamical systems, such a response, mimicking cascades of failures, demonstrates discontinuous phase transition phenomena, which resemble the critical effects observed in interdependent networks (Duan, Lv, Si, Wang, Li, Gao, Havlin, Stanley, and Boccaletti, 2019).

---

[9]This result can be easily obtained by applying the heterogeneous mean-field approximation.

9

Spectra and Communities

9.1 Adjacency matrix spectra

Spectral properties of networks essentially determine the key features of processes and cooperative phenomena taking place in them. For example, the largest eigenvalue of the adjacency matrix of a network determines the epidemic threshold of the SIS model in the quenched mean-field approximation, while the corresponding eigenvector indicates the probability that a vertex is infective near this threshold (Section 7.4). To gain some intuition about how the spectra of graphs are organized, the reader should review the chain of examples for small sample graphs in Section 2.9. In the present chapter, we focus on the spectra of undirected random networks. For an adjacency matrix, the key quantity of interest is the spectral density

$$\rho(\lambda) = \left\langle \frac{1}{N} \sum_{\alpha=1}^{N} \delta(\lambda - \lambda_\alpha) \right\rangle, \tag{9.1}$$

where λ_α are the eigenvalues of the adjacency matrix A of a given realization of a random network, $\delta(x)$ is the Dirac delta function, and the averaging is over the statistical ensemble providing this random network.

The spectrum of the adjacency matrix of the simplest random network, which is an infinite random k-regular graph (equivalently, a Bethe lattice), was obtained by McKay (1981). Apart from two eigenvalues, $\lambda_{1,N} = \pm k$ (eigenvector, corresponding to the largest eigenvalue $\lambda_1 = k$, has all components equal), the spectrum is continuous with the spectral density

$$\rho(\lambda) = \frac{k\sqrt{4(k-1) - \lambda^2}}{2\pi(k^2 - \lambda^2)} \theta(2\sqrt{k-1} - |\lambda|). \tag{9.2}$$

In particular, for the infinite ring (or chain), that is, with vertex degree $k = 2$, this gives the well-known spectral density

$$\rho_{\text{ring}}(\lambda) = \frac{1}{\pi\sqrt{2^2 - \lambda^2}} \theta(2 - |\lambda|), \tag{9.3}$$

divergent at the edges, $\lambda = \pm 2$. For any k, the statistics of the components of all eigenvectors in this spectrum is known (Elon, 2008), in contrast to the other random graphs. The components of each eigenvector in the continuous spectrum are distributed according to the symmetric normal law. For each eigenvector, the pair correlation function for components at vertices separated by distance ℓ decay as $(k-1)^{-\ell/2}$, where the exponential decay is corrected by an oscillating factor. The period of these oscillations is the shortest, 4, in the centre of the spectrum, $\lambda = 0$, diverging as λ approaches the edges of the continuous spectrum. Since the number of vertices at distance not exceeding ℓ from a given vertex is about $(k-1)^{\ell}$, this exponential decay of correlations, $(k-1)^{-\ell/2}$, still indicates that the eigenvectors of the continuous spectrum are delocalized, spreading over the entire network.[1]

Notice two eigenvalues $\lambda = \pm k$, sitting outside of the continuous spectrum. In general, the largest eigenvalue of the adjacency matrix of a simple undirected graph is bound by two numbers, namely,

$$\frac{\sum_i q_i}{N} \leq \lambda_1 \leq q_{\max} \tag{9.4}$$

(Krivelevich and Sudakov, 2003), where $\sum_i q_i / N = 2E/N$ is the mean degree of vertices in the graph (E is the number of edges), and $q_{\max}$ is its maximum degree. In the case of a random regular graph, these two bounds coincide resulting in $\lambda_1 = k$.

If $k \gg 1$, Eq. (9.1) results in the semicircle distribution,

$$\rho(\lambda) \overset{k \gg 1}{\cong} \frac{\sqrt{4k - \lambda^2}}{2\pi k} \theta(2\sqrt{k} - |\lambda|), \tag{9.5}$$

which is the famous Wigner law (Wigner, 1955, 1957a; Mehta, 2004; Tao, 2012).[2]

---

[1]See Section 9.3 for a more detailed discussion of localized and delocalized eigenvectors.

[2]The semicircle distribution of eigenvalues takes place in a rather broad class of random matrices. In particular, this class includes: (i) infinite real symmetric matrices with independent elements, each of which has a normal distribution, $\langle A_{ij}^2 \rangle = 1 + \delta_{ij}$; (ii) infinite real symmetric matrices, whose diagonal elements are zero while the non-diagonal elements are ± 1 with equal probabilities (Wigner, 1957b); (iii) infinite real matrices with independent, almost surely symmetric, elements, $\langle A_{ij} \rangle = 0$, $\langle A_{i \neq j}^2 \rangle = \sigma^2 < \infty$, $\langle A_{ij}^k \rangle < M_k < \infty$ (Olson and Uppuluri, 1972). On the other hand, for the great majority of random matrices, Wigner's distribution fails. For example, the spectrum of a sparse infinite symmetric random matrix with $O(N)$ random entries ± 1 and the remaining entries 0 has the tails extending beyond the edges of a semicircle distribution (Rodgers and Bray, 1988). Hence the semicircle law is used as a landmark for the sake of comparison.

For an undirected simple graph of N vertices, the number of closed walks of length n equals

$$\mathrm{Tr}\,A^n = \sum_{\alpha=1}^{N} \lambda_\alpha^n. \tag{9.6}$$

This equality provides the full set of the moments of the distribution of eigenvalues. The problem is to obtain the distribution from this set of its moments. Applying the following equality for the delta function,

$$\delta(x) = -\frac{1}{\pi} \lim_{\varepsilon \downarrow 0} \mathrm{Im}\frac{1}{x+i\varepsilon}, \tag{9.7}$$

where $\varepsilon \downarrow 0$ means that ε tends to zero from above, to Eq. (9.1) we arrive at the well-known expression of the spectral density,

$$\rho(\lambda) = -\left\langle \frac{1}{\pi N} \lim_{\varepsilon \downarrow 0} \mathrm{Im} \sum_{\alpha=1}^{N} \frac{1}{\lambda+i\varepsilon-\lambda_\alpha} \right\rangle$$

$$= -\left\langle \frac{1}{\pi N} \lim_{\varepsilon \downarrow 0} \mathrm{Im}\mathrm{Tr}[(\lambda+i\varepsilon)I - A]^{-1} \right\rangle, \tag{9.8}$$

in which I is the $N \times N$ identity matrix.[3] This equality formally returns the spectral density from the set of moments, Eq. (9.6). Indeed, if ε is treated as arbitrary, then Eq. (9.8) provides the following series for the spectral density on the complex plain, $\rho(z)$, with $z = \lambda + i\varepsilon$,

$$\rho(z) = -\left\langle \frac{1}{\pi N z} \sum_{n=0}^{\infty} \frac{1}{z^n} \mathrm{Tr}\,A^n \right\rangle. \tag{9.9}$$

This expresses the spectral density in terms of the number of closed walks of length n. Equations for this number (more strictly, for generating functions related to this number) can be derived explicitly in the case of locally tree-like networks, enabling us to obtain $\rho(\lambda)$ numerically (Dorogovtsev, Goltsev, Mendes, and Samukhin, 2003a). We can greatly simplify the solution of these equations by applying the effective-medium approximation. Equations for generating functions related to the number of closed walks of length n can be also obtained approximately for arbitrary simple graphs by a message passing algorithm approximating a graph by its infinite tree-like

---

[3]Compare the function $[(\lambda+i\varepsilon)I - A]^{-1}$ appearing on the right-hand side of Eq. (9.8) with the generating function of walks between two vertices, Eq. (2.33).

counterpart (Newman, Zhang, and Nadakuditi, 2019).[4] In this technique, a single (that is, non-random) graph is considered, and hence the expressions for the spectral density, Eqs. (9.1), (9.8), and (9.9), do not contain averaging. The approximate solution of these equations provides the spectral density.[5] We do not describe these involved calculations in detail and only discuss their results.[6]

For a dense Erdős–Rényi random graph, $\langle q \rangle \gg 1$, the spectral density approaches the semicircle distribution, Eq. (9.5) with $\langle q \rangle$ substituted for k. For a finite $\langle q \rangle > 1$, this density has rapidly decaying tails

$$\rho(\lambda) \overset{|\lambda| \gg \sqrt{\langle q \rangle}}{\sim} e^{-\lambda^2 \ln(\lambda^2/\langle q \rangle)} \tag{9.10}$$

(Semerjian and Cugliandolo, 2002), significantly deviating from the semicircle form (Figure 9.1). Similarly to random regular graphs, the largest eigenvalue of an Erdős–Rényi graph can occur separated from the main body of the spectrum. Asymptotically,

$$\lambda_1 = \begin{cases} [1 + o(1)]\langle q \rangle & \text{if } \langle q \rangle \gg \ln N, \\ [1 + o(1)] \max(\sqrt{q_{max}}, \langle q \rangle) & \text{if } \langle q \rangle = O(\ln N) \end{cases} \tag{9.11}$$

(Krivelevich and Sudakov, 2003), where $o(1)$ in the second line tends to zero as $\max(\sqrt{q_{max}}, \langle q \rangle)$ tends to infinity.

---

[4]Newman (2019) proposed a way to overpass the locally tree-likeness ansatz and compute spectra of loopy networks (see the discussion of techniques of this sort in Section 6.4).

[5]For calculations of spectra of random networks with hidden variables, see Nadakuditi and Newman (2013) and references therein.

[6]An alternative method for getting the spectral density is based on the classical approach to the spectra of random matrices (Edwards and Jones, 1976). Applying the equality $\operatorname{Tr} \ln K = \ln \det K$ to Eq. (9.8) we have

$$\rho(\lambda) = \frac{2}{\pi N} \operatorname{Im} \frac{\partial}{\partial \lambda} \langle \ln Z(\lambda + i\varepsilon) \rangle,$$

where ε is a positive infinitesimal and

$$Z(\lambda + i\varepsilon) = \int_{-\infty}^{\infty} \left(\prod_k d\phi_k \right) \exp\left[-\frac{i}{2} \left((\lambda + i\varepsilon) \sum_k \phi_k^2 - \sum_{k,j} A_{kj} \phi_k \phi_j \right) \right] \propto \det^{-1/2}[(\lambda + i\varepsilon) - A].$$

This reduces the problem to a Gaussian theory, and the calculations can be straightforwardly (though cumbersomely) performed by using the replica trick, $\ln Z = \lim_{n \to 0}[(Z^n - 1)/n]$ (Rodgers and Bray, 1988; Semerjian and Cugliandolo, 2002; Rodgers, Austin, Kahng, and Kim, 2005; Susca, Vivo, and Kühn, 2021).

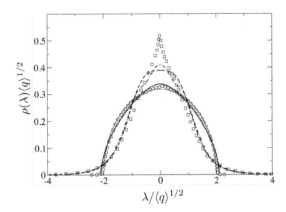

Fig. 9.1 Density of eigenvalues of the adjacency matrices of two networks. (i) The Erdős–Rényi random graph with the average degree $\langle q \rangle = 10$: the effective-medium approximation (solid line) and numerical calculations for the graphs of 20,000 vertices (Farkas, Derényi, Barabási, and Vicsek, 2001) (open circles). Notice the tiny rapidly decaying tails of the distribution including the principal eigenvalue $\lambda_1 \approx \langle q \rangle$. (ii) The uncorrelated network with the degree distribution $P(q) \propto q^{-\gamma} \theta(q - k_0)$, exponent $\gamma = 3$, and the smallest degree $k_0 = 5$: the effective-medium approximation (dashed line), the improved effective-medium approximation of Dorogovtsev, Goltsev, Mendes, and Samukhin (2003a) (dashed-dotted line). The results of the simulations of the Barabási–Albert model of 7,000 vertices with the same k_0 (Farkas, Derényi, Barabási, and Vicsek, 2001) (open squares). The semicircle law is shown by the thin solid line. Adapted from Dorogovtsev, Goltsev, Mendes, and Samukhin (2003a).

For uncorrelated networks with a power-law degree distribution $P(q) \cong Aq^{-\gamma}$, the spectral density deviates even stronger from the semicircle law, and it has a power-law asymptotics,

$$\rho(\lambda) \overset{|\lambda| \gg 1}{\cong} 2A|\lambda|^{-(2\gamma-1)} \tag{9.12}$$

(Dorogovtsev, Goltsev, Mendes, and Samukhin, 2003a).

Notably, apart from the continuum part produced by the delocalized eigenvectors, the spectrum also contains delta peaks due to localized states inevitable if the network has finite connected components (Bauer and Golinelli, 2001b). If a giant connected component is absent, the spectrum completely consists of the delta peaks. The localized states can occur even within a giant connected component (Figure 9.2b). The central delta peak in this distribution is due to the localized states on the small subgraphs with leaves

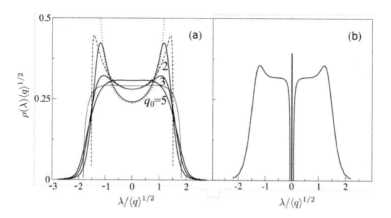

Fig. 9.2 Spectral density for the adjacency matrices of uncorrelated networks with the degree distribution $P(q) \propto q^{-\gamma}\theta(q - k_0)$, $\gamma = 5$, for different smallest degrees k_0. (a) The spectra of these networks with $k_0 = 2$, 3, and 5, which have no finite components. The dotted line corresponds to the spectral density of an infinite chain. The dashed and thin solid lines present the spectrum of the $(k = 3, 6)$-regular Bethe lattices. (b) The spectrum of the network with $k_0 = 1$ having the fraction of dead $P(1) = 0.3$. The central peak is produced by the states localized on the subgraphs with leaves shown in Figure 9.3. Adapted from Dorogovtsev, Goltsev, Mendes, and Samukhin (2003a).

shown in Figure 9.3.[7] Localized states can be also on hubs and on dense finite subgraphs, resulting in delta peaks at large λ. Recall that an isolated q-star has two of its eigenvalues equal to $\pm\sqrt{q}$, Eq. (2.56).[8] Then, in particular, a strongly connected hub with the largest degree in the network, $\sqrt{q_{max}}$, has a chance to produce the localized states and the delta peaks at $\lambda \sim \pm\sqrt{q_{max}}$ in the spectrum (see Section 9.3 for more detail).

Furthermore, Castellano and Pastor-Satorras (2017) conjectured that the largest eigenvalue of an adjacency matrix for general networks can be expressed as

$$\lambda_1 \approx \max(\sqrt{q_{max}}, \langle q \rangle k_{\text{highest}}), \qquad (9.13)$$

[7]This statement can be checked straight-forwardly. Multiply the adjacency matrix of a graph containing one of this subgraphs, say, of size n, by a vector with n arbitrary components at the vertices of the subgraph and the remaining components equal to zero. Equating this product to the zero vector, leads to n linear equations that have at least one solution, which is a localized state.

[8]This also hints at the reason for the term $\sqrt{q_{max}}$ in Eq. (9.11).

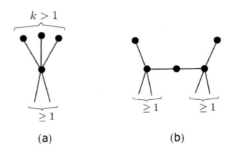

Fig. 9.3 Subgraphs producing localized states with $\lambda = 0$; see the central peak in Figure 9.2b. If the vertex in configuration (a) has $k > 1$ leaves, then there are $k - 1$ states with eigenvectors localized on this subgraph. Configuration (b) has one such an eigenvector. Adapted from Dorogovtsev, Goltsev, Mendes, and Samukhin (2003a).

where $\langle q \rangle_{k_{\text{highest}}}$ is the average degree of vertices in the highest k-core subgraph, while the label k_{highest} indicates the threshold degree of this k-core. Since the degree distribution of this subgraph is narrow, the largest eigenvalue of its adjacency matrix is close to $\langle q \rangle_{k_{\text{highest}}}$. This approximate formula was found to be correct for a large set of real-world and synthetic networks. In particular, for uncorrelated networks, $\langle q \rangle_{k_{\text{highest}}} \approx k_{\text{highest}} \approx \langle q^2 \rangle / \langle q \rangle$, and hence

$$\lambda_1 \approx \max(\sqrt{q_{\text{max}}}, \langle q^2 \rangle / \langle q \rangle). \tag{9.14}$$

Thus, for uncorrelated scale-free networks with a degree distribution $P(q) \sim q^{-\gamma}$, the largest eigenvalue of an adjacency matrix can be approximated as

$$\lambda_1 \approx \begin{cases} \sqrt{q_{\text{max}}} & \text{if } \gamma > 5/2, \\ \dfrac{\langle q^2 \rangle}{\langle q \rangle} & \text{if } \gamma < 5/2. \end{cases} \tag{9.15}$$

9.2 Laplacian matrix spectra

This section considers the spectra of a Laplacian matrix $L = D - A$, where D is the degree matrix, and of its two variations, namely, a symmetric normalized Laplacian matrix $\mathcal{L} = D^{-1/2} L D^{-1/2}$ and a normalized Laplacian matrix $\tilde{\mathcal{L}} = D^{-1} L$. The matrices $\mathcal{L}$ and $\tilde{\mathcal{L}}$ have the same spectrum, since $\tilde{\mathcal{L}} = D^{-1/2} \mathcal{L} D^{1/2}$. We denote the spectral density of the Laplacian matrix L by $\rho_L(\lambda)$ and of the matrices $\mathcal{L}$ and $\tilde{\mathcal{L}}$ by $\rho_{\mathcal{L}}(\lambda)$. If a graph is connected, that is it consists of a single connected component, then in both spectra, of L and $\tilde{\mathcal{L}}$, the first eigenvalue $\lambda_1 = 0$ is non-degenerate, and the second eigenvalue λ_2 plays a key role in dynamical processes associated with these linear operators, which are, in particular, diffusion and random walks.

9.2.1 Diffusion

Diffusion on a graph of N vertices is described by the linear equation

$$\frac{d}{dt}\mathbf{x}(t) = -L\mathbf{x}(t), \tag{9.16}$$

where each entry of the N-vector $\mathbf{x}(t)$ describes the current state of a particular vertex. Equation (9.16) guarantees the conservation of the sum $\sum_i x_i(t) = \text{const} = \sum_i x_i(0)$. If we know this vector at the initial moment, $\mathbf{x}(0)$, then the solution of this equation is

$$\mathbf{x}(t) = e^{-Lt}\mathbf{x}(0). \tag{9.17}$$

Since the set of orthogonal eigenvectors $\mathbf{v}$ of L, where $\alpha = 1, 2, \ldots, N$, form a basis, we can express $\mathbf{x}(0)$ as a linear combination of the eigenvectors,

$$\mathbf{x}(0) = \sum_{\alpha=1}^{N}(\mathbf{x}(0), \mathbf{v}_\alpha)\,\mathbf{v}_\alpha = \sum_{\alpha=1}^{N} c_\alpha \mathbf{v}_\alpha. \tag{9.18}$$

Then Eq. (9.18) gives

$$\mathbf{x}(t) = c_1 \mathbf{v}_1 + c_2 e^{-\lambda_2 t}\mathbf{v}_2 + \ldots, \tag{9.19}$$

where the eigenvector $\mathbf{v}_1 = (1, 1, \ldots, 1)$ (Section 2.9). Hence, if there is a spectral gap (interval between $\lambda_1 = 0$ and λ_2), then at long times the relaxation of the system to the final uniform state $\mathbf{x}(t{\to}\infty) = c_1 \mathbf{v}_1$ is exponential,

$$\mathbf{x}(t) - \mathbf{x}(\infty) \cong c_2 e^{-t/\tau_{\text{diffusion}}}\mathbf{v}_2 \tag{9.20}$$

with $\tau_{\text{diffusion}} = 1/\lambda_2$ defining the longest time scale of the diffusion process.[9]

---

[9]The spectrum simplifies when the Laplacian matrix of a network has a block structure, which occurs in multiplex networks. Gomez, Diaz-Guilera, Gómez-Gardeñes, Perez-Vicente, Moreno, and Arenas (2013) considered the case of a 2-layer weighted multiplex network, and the results of their work were generalized to an arbitrary number M of layers in Solé-Ribalta, De Domenico, Kouvaris, Díaz-Guilera, Gómez, and Arenas (2013). In particular, for a 2-layer weighted multiplex network with the weights $w_{ij}^{(\alpha)}$ of edges (ij) within the layers $\alpha = 1, 2$ and the weight v of each of N edges between the layers, the Laplacian matrix has the following block form:

SPECTRA AND COMMUNITIES

9.2.2 Random walk

Another basic process on a graph, a *simple random walk* is similarly described by a linear equation, this time containing the normalized Laplacian matrix $\tilde{\mathcal{L}} = LD^{-1} = I - \mathcal{P}$, where $\mathcal{P} = AD^{-1}$ is the transition matrix. In the simple random walk, at each step, a walker moves from its current vertex to one of the nearest-neighbouring vertices chosen uniformly randomly. The element $\mathcal{P}_{ij} = A_{ij}/q_j$ of the transition matrix is the probability that the walker will move from vertex j to vertex i in one time step. The random

$$L = \begin{pmatrix} L^{(1)} + vI & -vI \\ -vI & L^{(2)} + vI \end{pmatrix},$$

where $L^{(1)}$ and $L^{(2)}$ are the Laplacian matrices of layers 1 and 2, respectively, $L_{ij}^{(\alpha)} = \delta_{ij} \sum_k w_{ik}^{(\alpha)} - [1 - \delta_{ij}] w_{ij}^{(\alpha)}$, $\alpha = 1, 2$, and I is the $N \times N$ identity matrix. For an M-layer multiplex network, the edges interconnecting the set of vertices with the same label i in different layers can have $M(M-1)/2$ different weights, $v_{\alpha\beta}$, where $\alpha = 1, 2, \ldots, M$. These weights are assumed to be independent on i. Let the Laplacian matrix of this weighted network of interconnections between replica vertices with a fixed label be L_I. Solé-Ribalta, *et al.* proved that the spectrum of the Laplacian matrix L of the entire multiplex network with $N \times M$ vertices contains all M eigenvalues of the Laplacian matrix L_I of interconnections, including an eigenvalue zero.

If the weights $v_{\alpha\beta}$ of the edges between the layers are sufficiently small, smaller than some number w^* comparable to a typical edge weight for the layers, then the smallest non-zero eigenvalue λ_{2I} of L_I appears to be the smallest nonzero eigenvalue λ_2 of the Laplacian of the multiplex,

$$\lambda_2 = \lambda_{2I}.$$

In this situation, the relaxation time $\tau_{\text{diffusion}}$ of the diffusion process on the multiplex network can be even longer than for each of its layers. In particular, if all the weights of edges between the layers are equal, $v_{\alpha\beta} = v$, then the spectrum of the multiplex includes $M - 1$ degenerate eigenvalues $\lambda = Mv$, and hence $\lambda_2 = Mv$ for sufficiently small v. If $v \ll w^*$, then the relaxation time $\tau_{\text{diffusion}} = 1/(Mv)$ for the multiplex is longer than for each of its layers.

On the other hand, if $v_{\alpha\beta} \gg w^*$, then

$$\lambda_2 \cong \frac{\lambda_{2s}}{M},$$

where λ_{2s} is the smallest non-zero eigenvalue of the Laplacian spectrum for the superposition of the layers, which is the network of N vertices with the weights $\sum_{\alpha=1}^M w_{ij}^{(\alpha)}$ of its edges. In this regime, according to inequalities from Mohar (1991), the eigenvalue λ_2 is bounded from below,

$$\lambda_2 \geq \frac{1}{M} \sum_{\alpha=1}^M \lambda_2^{(\alpha)} \geq \min_\alpha \lambda_2^{(\alpha)},$$

where $\lambda_2^{(\alpha)}$ is the smallest non-zero eigenvalue of the Laplacian of layer α. Hence in this situation the diffusion relaxation time of the multiplex is shorter than in its 'slowest' layer,

$$\tau_{\text{diffusion}} < \max_\alpha \tau_{\text{diffusion}}^{(\alpha)}.$$

292

walk is 'simple', that is, the walker always changes its place, if all diagonal elements of the adjacency matrix are zero. If $x_i(t)$ is the probability that a walker occupies vertex i at time t, $i = 1, 2, \ldots, N$, then we have the following probabilities after one step:

$$x_i(t+1) = \sum_j \mathcal{P}_{ij} x_j(t), \tag{9.21}$$

or, in matrix form, $\mathbf{x}(t+1) = \mathcal{P}\mathbf{x}(t)$. In the continuous version of this process, the state vector $\mathbf{x}(t)$, whose elements are the probabilities $x_i(t)$, evolves according to the following equation:

$$\frac{d}{dt}\mathbf{x}(t) = -\mathbf{x}(t) + \mathcal{P}\mathbf{x}(t) = -\widetilde{\mathcal{L}}\mathbf{x}(t), \tag{9.22}$$

which guarantees that $\sum_i x_i(t) = \sum_i x_i(0) = 1$. If there is a spectral gap, then we can reproduce the derivation of Eq. (9.18), keeping in mind that $\widetilde{\mathcal{L}}$ is non-symmetric and hence the right and left eigenvectors have different components, $\mathbf{v}_\alpha^{(\text{left})} \neq (\mathbf{v}_\alpha^{(\text{right})})^T$. In this way, we arrive at the exponential relaxation to the final stationary set of probabilities,

$$\mathbf{x}(t) - \mathbf{x}(\infty) \cong c_2 e^{-t/\tau_{\text{mixing}}} \mathbf{v}_2^{(\text{right})}, \tag{9.23}$$

where $\tau_{\text{mixing}} = 1/\lambda_2$ defines the *mixing time* of the random walk process on a graph, λ_2 is the second eigenvalue of the matrix $\widetilde{\mathcal{L}}$ and of $\mathcal{L}$, and $\mathbf{v}_2$ is the eigenvector corresponding to this eigenvalue. The constant c_2 is determined by initial conditions. The final stationary state $\mathbf{x}(t \to \infty)$ equals the right eigenvector $\mathbf{v}_1^{(\text{right})}$ of $\widetilde{\mathcal{L}}$, normalized to get $\sum_i x_i = 1$. Hence

$$\mathbf{x}(\infty) = \frac{1}{\sum_i q_i}(q_1, q_2, \ldots, q_N)^T \tag{9.24}$$

(Section 2.9), that is, in the steady state of the process, the probability that the walker uses any given edge of a connected graph to make a step is $1/E$.

9.2.3   Spectral gap and diameter

In addition to the mixing time, the eigenvalue λ_2 of the normalized Laplacian matrix of a graph of N vertices enables one to estimate the diameter d of the earlier graph,

$$d \leq \frac{\ln(N-1)}{\ln[(\lambda_N + \lambda_2)/(\lambda_N - \lambda_2)]} \tag{9.25}$$

(Chung, 1997).[10] Equations (9.18), (9.23), and (9.25) assume the existence of the spectral gap (whose width is substantially determined by the relative size of a bottleneck in the graph, Eq. (2.84) in Section 2.9). If this is not the case, then often $\rho^{(L)}(\lambda) \sim \rho^{(\mathcal{L})}(\lambda) \sim \lambda^{D_s/2-1}$ for small λ, where D_s is the spectral dimension. In this situation, occurring, in particular, in finite-dimensional lattices, long-time relaxation in diffusion and random walk processes become power-law. In this case, for example, the probability to find a walker on its initial vertex decays as $\int d\lambda\, \rho^{(\mathcal{L})}(\lambda) e^{-\lambda t} \sim t^{-D_s/2}$. A finite spectral dimension and a power-law relaxation can happen even in small worlds as is allowed by the inequality in Eq. (2.81) relating D_s and the Hausdorff dimension. For a few examples of this combination, namely $D_s < \infty$ while $D_H = \infty$, in simplicial complexes, see Bianconi and Dorogovstev (2020) and da Silva, Bianconi, da Costa, Dorogovtsev, and Mendes (2018).

9.2.4 The largest eigenvalue of Laplacian matrix

While all eigenvalues of the spectrum of the matrix $\widetilde{\mathcal{L}}$ (and matrix $\mathcal{L}$) are within the interval $[0, 2]$ (Section 2.9), the largest eigenvalue λ_N of the Laplacian matrix can be large. In uncorrelated networks with fat-tailed degree distributions, λ_N can be estimated by using perturbation theory if we treat the degree matrix D in the Laplacian $L = D - A$ as an unperturbed matrix and the adjacency matrix as a perturbation (Kim and Motter, 2007). Labelling the vertices of an uncorrelated network in ascending order of their

---

[10] Adding an extra edge to an undirected graph, linking together any pair of its non-adjacent vertices, improves the connectivity of the graph and diminishes its diameter. Hence one could expect that this addition decreases the mixing time of a random walk and increases the spectral gap in the spectrum of the normalized Laplacian matrix of a graph. Counter-intuitively, this is not the case for particular extra edges in some graphs, as is demonstrated by the following figure.

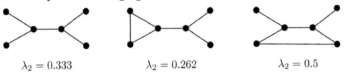

$\lambda_2 = 0.333 \qquad\qquad \lambda_2 = 0.262 \qquad\qquad \lambda_2 = 0.5$

Indeed, the middle graph has a smaller spectral gap than the left one. This phenomenon can be called Braess's paradox for the spectral gap in graphs. Chung conjectured (2014) and Eldan, Rácz, and Schramm (2017) proved for the $G(N,p)$ model that the addition of a random edge decreases the spectral gap of the normalized Laplacian with a non-zero probability. A similar effect was observed in the spectrum of a Laplacian matrix for directed graphs, where an addition of an extra edge has a chance to reduce synchronizability (Pade and Pereira, 2015).

degrees, $q_1 \leq q_2 \leq \ldots \leq q_N = q_{max}$, one can get the perturbation expansion of λ_N,

$$\lambda_N = q_N - A_{NN} + \sum_{j \neq N} \frac{(A_{Nj})^2}{q_N - q_k} + \ldots \approx q_N + \sum_{j \neq N} A_{Nj} \left(\frac{1}{q_N} + \frac{q_j}{q_N} + \ldots \right)$$

$$= q_N + 1 + \frac{1}{q_N} \frac{\langle q^2 \rangle}{\langle q \rangle} + \ldots \approx q_N. \tag{9.26}$$

The degree distribution here is assumed to be power law, $P(q) \sim q^{-\gamma}$, with $\gamma > 2$. Hence the average degree of a nearest neighbour of a vertex (including vertex N) is $\langle q^2 \rangle / \langle q \rangle \sim q_N^{3-\gamma} \ll q_N$, which explains the last equality. Thus the average maximal eigenvalue over the statistical ensemble of such a random network is estimated as $\lambda_N \approx q_{max}$. Importantly, both λ_2 and λ_N play a special role in synchronization phenomena. It is the smallness of the ratio λ_N / λ_2 that determines the synchronizability of linearly coupled dynamical systems (Barahona and Pecora, 2002).[11] The heterogeneity of a

---

[11]Consider N linearly coupled dynamical systems ('elements'), assuming that each of them is described by an m-dimensional vector dynamical variable $\vec{z}_i(t)$, $i = 1, 2, \ldots, N$. (We use here this notation for m-vectors to distinguish them from N-dimensional vectors, like the eigenvectors $\mathbf{v}_\alpha$ of the Laplacian matrix, with components $v_{i\alpha}$.) Let the dynamics of the entire system be governed by the following set of equations

$$\frac{d}{dt}\vec{z}_i = \vec{F}(\vec{z}_i) - J \sum_j L_{ij} \vec{H}(\vec{z}_j),$$

where J is the coupling strength, $L_{ij} = q_i \delta_{ij} - A_{ij}$ are the entries of the Laplacian matrix L, and $\vec{F}(\vec{z}_i)$ and $\vec{H}(\vec{z}_j)$ are m-dimensional vector functions, continuous and differentiable. Since $\sum_j L_{ij} = 0$, any solution of the equation $\frac{d}{dt}\vec{s} = \vec{F}(\vec{s})$ satisfies all N equations previous, that is, all N systems evolve coherently, $\vec{z}_i(t) = \vec{s}(t)$.

The synchonizability of this set of systems is determined by the stability of the fully synchronized state $s(t)$ against small perturbations $\vec{u}_i(t)$, $\vec{z}_i(t) = \vec{s}(t) + \vec{u}_i(t)$. The eigenvalues of the Laplacian correspond to N ortogonal eigenvectors, including the uniform eigenvector $\mathbf{v}_1 = \mathbf{1}$ (fully synchronized or coherent state) corresponding to $\lambda_1 = 0$. The perturbation of the coherent state is a linear combination of the $N - 1$ transverse modes $\mathbf{v}_2, \ldots, \mathbf{v}_N$, namely, $\vec{u}_i = \sum_{\alpha=2}^N \vec{u}_\alpha v_{i\alpha}$, whose coefficients $\vec{u}_\alpha$ satisfy the equations

$$\frac{d}{dt}\vec{u}_\alpha = [\mathcal{D}\vec{F}(\vec{s}) - g\mathcal{D}\vec{H}(\vec{s})]\vec{u}_\alpha.$$

Here $g = J\lambda$ and $\mathcal{D}\vec{F}$ and $\mathcal{D}\vec{H}$ are the Jacobian matrices. The fully synchronized state is stable if the largest eigenvalue $\Lambda(g)$ of the matrix $\mathcal{D}\vec{F}(\vec{s}) - g\mathcal{D}\vec{H}(\vec{s})$ (Lyapunov exponent of this equation), called the *master stability function*, is negative (Pecora and Carroll, 1998). Usually, the function $\Lambda(g)$ is negative in a bound region $g_l < g < g_u$. Hence, a network is synchronizable if

$$\frac{g_l}{\lambda_2} < J < \frac{g_u}{\lambda_N}.$$

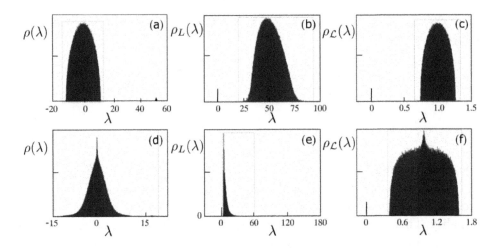

Fig. 9.4 Adjacency matrix, $\rho(\lambda)$, (a) and (d), Laplacian, $\rho_L(\lambda)$, (b) and (e), and normalized Laplacian, $\rho_{\mathcal{L}}(\lambda)$, (c) and (f), spectra of the Erdős–Rényi random graph and the Barabási–Albert network, obtained from 100 realizations of the networks of 1000 vertices. The average degree of vertices in the Erdős–Rényi graph, (a), (b), and (c), is $\langle q \rangle = 50$. In the Barabási–Albert network, (d), (e), and (f), each new vertex is attached to $m = 4$ existing ones. The not-shown y-axes' scales are different in different panels. Adapted from Mirchev (2017).

network, which assumes a large $q_{\max}$ and hence a large λ_N, may suppress synchronization (Motter, Zhou, and Kurths, 2005).

9.2.5 Spectral densities

Figure 9.4 compares different spectral densities—of an adjacency matrix, Laplacian, and normalized Laplacian—for two contrasting networks, namely, for an Erdős–Rényi random graph and for the Barabási–Albert network.[12] Notice the peak in the spectrum of the normalized Laplacian at $\lambda = 1$ visible for the Barabási–Albert network (Figure 9.4f). This peak is determined by vertices of small degrees and their neighbours (Vukadinović, Huang, and

This is equivalent to the following condition:

$$\frac{\lambda_N}{\lambda_2} < \frac{g_u}{g_1}$$

(Barahona and Pecora, 2002). Therefore, for small λ_2, for example, in infinite lattices, where $\lambda_2 \to 0$, synchronization is impossible. Furthermore, the increase of λ_N suppresses synchronization.

[12]Note that the spectral densities $\rho_{\mathcal{L}}(\lambda)$ and of the transition matrix, $\rho_{\mathcal{P}}(\lambda)$, are related as $\rho_{\mathcal{L}}(\lambda) = \rho_{\mathcal{P}}(1 - \lambda)$, since $\mathcal{L} = I - D^{1/2}\mathcal{P}D^{-1/2}$.

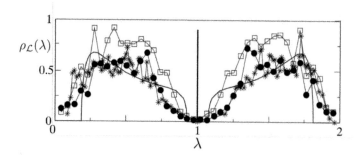

Fig. 9.5 Eigenvalue density $\rho_{\mathcal{L}}(\lambda)$ of the normalized Laplacian matrix for the Internet graph compared with the effective-medium calculations for its uncorrelated counterpart with the same degree distribution. (i) Measurements for the Internet graph—Autonomous Systems of 2000 (the filled circles) and for its randomized counterpart (the open squares) from Eriksen, Simonsen, Maslov, and Sneppen (2003). (ii) Measurements for the Internet graph extracted from Vukadinović, Huang, and Erlebach (2002) (the stars). (iii) Results of the effective-medium calculations (the solid line) for the infinite uncorrelated network with the degree distribution $P(1) = 0.358$, $P(2) = 0.4$, $P(3) = 0.12$, and $P(q \geq 4) \propto q^{-2.1}$, taken from Vukadinović, Huang, and Erlebach (2002). Adapted from Dorogovtsev, Goltsev, Mendes, and Samukhin (2003a).

Erlebach, 2002). If a network contains leaves, then the spectral density contains a degenerate delta peak at $\lambda = 1$.[13] The spectral densities of the Laplacian and normalized Laplacian matrices of various networks can be obtained in a way outlined for the adjacency matrix spectra in Section 9.1. Figure 9.5 compares the measurements for the Internet graph (Autonomous Systems level) with the results of the effective-medium calculations for the uncorrelated network with a close degree distribution (Dorogovtsev, Goltsev, Mendes, and Samukhin, 2003a). Notice a delta peak at $\lambda = 1$ and a wide dip in the spectral density.

A spectral gap in Laplacian spectra of infinite uncorrelated random

---

[13]The multiplicity μ of this peak is determined by three numbers: (i) number l of leaves in a graph, (ii) number n of the nearest neighbours of leaves, and (iii) number r of isolated vertices in the graph after removal of its leaves together with their nearest neighbours. The multiplicity satisfies the following inequality

$$\mu \geq l - n + r,$$

where the lower bound turned out to be close to the observed multiplicity for the Internet graph (Vukadinović, Huang, and Erlebach, 2002).

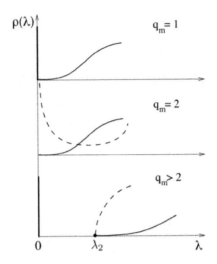

Fig. 9.6 Spectral density $\rho(\lambda)$ near the lower edge of a Laplacian spectrum of infinite uncorrelated random networks with the minimum degree q_m (there should be a finite fraction of vertices with the minimum degree). The dashed lines show the Laplacian spectral densities of k-regular Bethe lattices with $k = q_m$. For $q_m > 2$, the spectral gap is $\lambda_2 = q_m - 2\sqrt{q_m - 1}$. Adapted from Samukhin, Dorogovtsev, and Mendes (2008).

networks is determined by the minimum degree q_m of vertices in a network (excluding degree 0). The width of this gap coincides with that of a $(k=q_m)$-regular Bethe lattice (infinite random regular graph), namely $\lambda_2 = q_m - 2\sqrt{q_m - 1}$. There is no spectral gap in infinite uncorrelated networks with a finite fraction of vertices of degree 1 or 2. The spectral gap disappears due to long chains existing in these networks.[14] Note that a single chain of length ℓ in a network guarantees that $\lambda_2 \lesssim \ell^{-2}$. Figure 9.6 shows schematically the Laplacian spectral densities of infinite uncorrelated networks with different q_m. The singularity at the edge of the spectrum in these networks is essentially determined by the minimum vertex degree

---

[14]Here it is an example of such a chain in a network containing a finite fraction of vertices of degree 1:

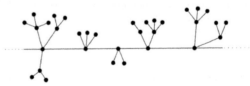

and their randomness, while the high-degree part of the degree distribution is not that important. These singularities of the BKT type and the corresponding long-time asymptotics for the diffusion process were obtained explicitly (Samukhin, Dorogovtsev, and Mendes, 2008). The singularities of this type in the Laplacian spectrum lead to a very slow approach of the spectral gap $\lambda_2(N)$ of a network of N vertices to its value for the infinite network. For $q_m = 1$ and 2,

$$\lambda_2(N) \sim (\ln N)^{-2}, \tag{9.27}$$

and for $q_m > 2$,

$$\lambda_2(N) - \lambda_2(\infty) \sim (\ln \ln N)^{-2}. \tag{9.28}$$

9.3   Localization

We explained that at least some of eigenvectors of adjacency, Laplacian, and other matrices are closely related with key features of processes and cooperative phenomena in networks. For each eigenvector, two distinct situations are possible, assuming that a network is infinite or very large: (i) an eigenvector is spread over the network (*delocalized eigenvector*), and (ii) an eigenvector is concentrated on a very small part of a network, namely, on a finite number of vertices or, at least, on a vanishing fraction of an infinite network (*localized eigenvector*). For the sake of simplicity, let a matrix be symmetric. Then its eigenvectors are real. For any eigenvector $\mathbf{v}_\alpha$, where $\alpha = 1, 2, \ldots, N$, we can introduce the following normalization $(\mathbf{v}_\alpha, \mathbf{v}_\alpha) = \mathbf{v}_\alpha^T \mathbf{v}_\alpha = \sum_{i=1}^{N} v_{i\alpha}^2 = 1$. With this normalization, if a given eigenvector is delocalized, then all its components $|v_{i\alpha}| \sim 1/\sqrt{N}$ approaching zero as $N \to \infty$. If a vector is localized on, say, n vertices of an infinite network, where n is finite or increases with N slower than N, e.g., $n \sim N^\varsigma$, where $\varsigma < 1$, then $|v_{i\alpha}| \sim 1/\sqrt{n}$ for the vertices in this set. Localization can be quantified by the *inverse participation ratio* of an eigenvector,

$$\text{IPR}_\alpha \equiv \sum_{i=1}^{N} |v_{i\alpha}|^4, \tag{9.29}$$

whose sufficiently large value serves as a universal criterion of localization. Indeed, for a delocalized eigenvector $\mathbf{v}_\alpha$, the estimate above gives $\text{IPR}_\alpha \sim 1/N$ vanishing as $N \to \infty$. On the other hand, for a localized eigenvector, this estimate provides a much larger inverse participation ratio $\text{IPR}_\alpha \sim 1/n$, which is finite for a finite n.

Let us focus on the eigenvectors of an adjacency matrix determining the quenched mean-field solution of the SIS model, Eqs. (7.27) and (7.28), and providing the critical point $\lambda_c = \lambda_1^{-1}$. Near this point, the prevalence increases linearly with λ, $\rho(\lambda) \cong C_1 \lambda_1 (\lambda - \lambda_1^{-1})$, Eq. (7.29), where the coefficient C_1 is expressed in terms of the components of the principal eigenvector $\mathbf{v}_1$, which is the eigenvector centrality (Section 2.9). Substituting the estimate given above into the expression for C_1, namely $C_1 = \sum_{i=1}^{N} v_{i1}/(N\sum_{i=1}^{N} v_{i1}^3)$, Eq. (7.30), we readily get $C_1 = O(1)$ if the principal eigenvector is delocalized, and $C_1 \sim n/N$ for the localized eigenvector. Thus, in the quenched mean-field approximation, two situations are possible. First, if the principal eigenvector is delocalized, then λ_c is the endemic epidemic threshold above which a finite fraction of vertices in a large network are infective. Second, if the principal eigenvector is localized, then in the neighbourhood of λ_c, a finite number of vertices are infective, and only at significantly greater infection rates λ, a finite fraction of vertices appears to be infective, and a real epidemic happens. Figure 9.7 demonstrates these two contrasting situations for the quenched mean-field solutions of the SIS model on two real networks.[15]

It is easy to obtain an exact localized solution of the equation $\lambda \mathbf{v} = A\mathbf{v}$ for a k-regular Bethe lattice (branching $B = k - 1 \geq 1$) with a single hub of degree q (Figure 9.8). This simple network provides a representative null model for localization. The components of the resulting symmetric principal eigenvector depend only on the distance ℓ from a vertex to the hub, v_ℓ, and exponentially decay with ℓ,

$$v_\ell = \sqrt{\frac{q - 2B}{2(q - B)}} (q - B)^{-\ell/2}. \tag{9.30}$$

Clearly, the solution is localized if $(q - B)^{-\ell/2} B^\ell \to 0$ as $\ell \to \infty$, that is, if $q > B(B + 1) \geq 2B$. The eigenvalue associated with this localized eigenvector and the inverse participation ratio take the form:

---

[15] Note that the quenched mean-field approximation neglects the absorbing state in the SIS model and fluctuations, due to which a finite number of infected vertices all will finally become susceptible due to fluctuations. Hence this solution with a finite number of infective vertices has sense only in the quasi-stationary, metastable state, and so localization on a hub in the SIS model is actually metastable (Ferreira, Da Costa, Dorogovtsev, and Mendes, 2016). This trouble disappears for localization on a large cluster still containing a vanishingly small fraction of an infinite network.

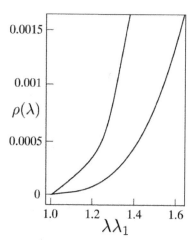

Fig. 9.7 Dependence of prevalence ρ on infection rate λ for the SIS model on two real networks of the arXiv co-authorships (quenched mean-field approximation): astro-phys-1999 (upper line), where the principal eigenvector is delocalized, and cond-mat-2005 (lower line), where the principal eigenvector is localized. Adapted from Goltsev, Dorogovtsev, Oliveira, and Mendes (2012).

Fig. 9.8 Regular Bethe lattice (vertex degree $k = B - 1$, where B is a branching) with a single hub of degree $q > k$.

$$\lambda_1 = \frac{q}{\sqrt{q - B}},$$

$$\text{IPR}_1 = \frac{1}{4}\left(\frac{q - 2B}{q - B}\right)^2\left[1 + \frac{q}{(q - B)^2 - B}\right] \qquad (9.31)$$

(Goltsev, Dorogovtsev, Oliveira, and Mendes, 2012). As is natural, this eigenvalue must be the largest one, which gives the condition $\lambda_1 > k = B - 1$ of the existence of this localized state (Section 9.1) leading to the same condition for the degree q of the hub,

$$q > B(B + 1), \qquad (9.32)$$

as shown, and the value $q_{loc} = B(B+1)$ is the point of the *delocalization* (or, equivalently, *localization*) *transition*. If this condition is fulfilled, the second, delocalized eigenvector, has the associated eigenvalue $\lambda_2 = k$. For large q, the eigenvalue of the localized eigenvector has the asymptotics $\lambda_1 \cong \sqrt{q}$, coinciding with the largest eigenvalue of the isolated q-star (Section 2.9) and $IPR_1 \cong (1 + 1/q)/4$. Similarly, localization can occur on a cluster of hubs in a network.

One can also get a localized principal eigenvector in another way, by ascribing a sufficiently large weight w to, say, a single edge of the k-regular Bethe lattice. It is easy to check that this localized eigenvector exists, if $w^2 > B = k - 1$, and its eigenvalue and the inverse participation ratio are

$$\lambda_1 = w^2 + \frac{B}{w^2} > B - 1 = k,$$

$$IPR_1 = \frac{1}{2}\frac{(1 - B/w^2)^2}{1 - B/w^4}. \tag{9.33}$$

Approximate formulas for localization on a vertex with the largest degree q_{max} in a random uncorrelated network follow from Eq. (9.31) after replacing q by q_{max} and branching B by the average branching $\langle q^2\rangle/\langle q\rangle - 1$ of an uncorrelated network.[16] Localization on this vertex happens when the value λ_1 obtained in this way exceeds the largest eigenvalue λ_d for the delocalized eigenvectors,

$$\lambda_1 = \frac{q_{max}}{\sqrt{q_{max} - \langle q^2\rangle/\langle q\rangle + 1}} > \lambda_d. \tag{9.34}$$

In turn, the value of λ_d typically slightly exceeds $\lambda_a = \langle q^2\rangle/\langle q\rangle$ (eigenvalue of the weighted adjacency matrix of the annealed counterpart of our network) as one can check numerically.[17] Comparing $\lambda_1 \approx \sqrt{q_{max}}$ and

[16]See Martin, Zhang, and Newman (2014) for the description of localization of the adjacency matrix eigenvector on a hub attached to a set of uniformly randomly selected vertices of a classical random graph.

[17]Recall that the entries of the adjacency matrix of an annealed network are expressed in terms of the sequence of degrees of the original network, $A_{ij} = q_i q_j/(N\langle q\rangle)$. Then, for the spectrum of this matrix, the following relations hold:

$$\sum_\alpha \lambda_\alpha = \mathrm{Tr}A = \frac{\langle q^2\rangle}{\langle q\rangle}$$

and

$$\sum_\alpha \lambda_\alpha^2 = \mathrm{Tr}A^2 = \frac{\langle q^2\rangle^2}{\langle q\rangle^2}.$$

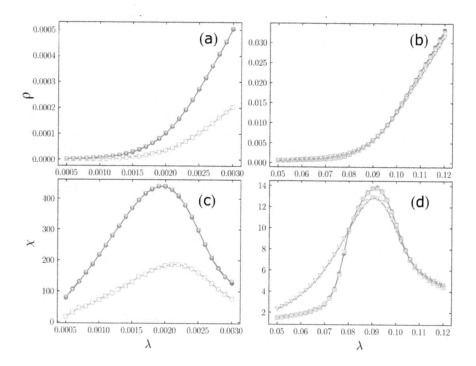

Fig. 9.9 Effects of removal of the hub (vertex with the highest degree) and of the highest k-core on characteristics of a quasi-stationary state of the SIS model on the network of DBpedia—3.92×10^6 vertices, $q_{max} = 4.70 \times 10^5$ (!), the highest k-core of 70 vertices, (a) and (c), and on the Hep-Th,1995-1999 network of co-authorships—5,835 vertices, $q_{max} = 50$, the highest k-core of 19 vertices, (b) and (d). (a) and (b) Prevalence $\rho = \langle \sigma \rangle$ (average fraction of infective vertices) vs. infection rate λ. (c) and (d) Susceptibility $\chi = N(\langle \sigma^2 \rangle - \langle \sigma \rangle^2)/\langle \sigma \rangle$ vs. λ. Filled circles—an original network, open squares—the hub is removed, triangles—the highest k-core is removed. Adapted from Pastor-Satorras and Castellano (2018).

These equalities suggest that the spectrum of the annealed network has a unique non-zero eigenvalue,

$$\lambda_a = \frac{\langle q^2 \rangle}{\langle q \rangle}.$$

One can directly check that the components of the eigenvector associated with this eigenvalue are proportional to the degrees of the original network,

$$v_i^{(a)} = \frac{q_i}{\sqrt{N \langle q^2 \rangle}},$$

where the normalization $\sum_i v_i^2 = 1$ is taken into account. Hence the inverse participation ratio of this eigenvector is given by

$\lambda_d \sim \langle\langle q^2 \rangle\rangle / \langle q \rangle \sim (q_{max})^{3-\gamma}$ in uncorrelated scale-free networks with a degree distribution exponent $\gamma < 3$, we see that the condition for localization on a single hub can be fulfilled only if $\gamma > 5/2$. On the other hand, even if localization on a single vertex is impossible, there may be localization on a cluster of vertices with large degrees in a network, in particular, on a rich club occurring in networks with the divergent second moment of the degree distribution (Section 4.8).[18] Pastor-Satorras and Castellano (2016) inspecting a number of synthetic and real networks concluded that, while for $\gamma > 5/2$, localization is on single hubs, for $\gamma < 5/2$ localization takes place, approximately, on the highest k-core. Note that in this situation, the highest k-core is assumed to contain a vanishingly small fraction of vertices in an infinite network. Formulas from Section 6.7 enable us to estimate this fraction. To ensure that a scale-free network with $2 < \gamma < 3$ is uncorrelated, the cutoff degree $q_{cut}(N)$ must grow with N not faster than $\sqrt{N}$, say, $q_{cut}(N) \sim N^a$ where $0 \leq a < 1/2$ (Section 4.8). Then the relative size of the highest k-core $S_{k_{highest}} \sim (q_{cut})^{-(\gamma-1)} \sim N^{-(\gamma-1)a}$, which indeed guarantees the smallness of $S_{k_{highest}}$ for any $0 < a < 1/2$. Pastor-Satorras and Castellano (2018) observed these two contrasting situations by simulating the SIS model on two real networks with and without removed the highest degree vertices and the highest k-cores near the endemic epidemic threshold. Figure 9.9 highlights the effects of these removals on the prevalence and susceptibility of the quasi-stationary states of these systems.[19] Localization on the highest k-cores conforms with two earlier observations of Kitsak, Gallos, Havlin, Liljeros, Muchnik, Stanley, and Makse (2010) for the SIS and SIR models, namely, (i) the most efficient spreaders of diseases are located within the high-k-cores, and (ii) infections persist in these k-cores. Further developing these ideas, Morone, Del Ferraro, and Makse (2019) proposed to monitor the highest k-cores for anticipating catastrophic events in dynamical systems on networks.

Localization of eigenvectors also happens in other matrices of networks, in particular, in a Laplacian matrix, in a normalized Laplacian, etc. In widely used spectral clustering techniques, where one needs the full set of compo-

$$\text{IPR}_a = \frac{1}{N} \frac{\langle q^4 \rangle}{\langle q^2 \rangle^2}.$$

---

[18] Recall that in infinite networks, a rich club includes infinite number of vertices, though still a vanishing fraction of a network, and hence it can be a centre of localization.

[19] Notice that the huge hub in the DBpedia network is outside of its highest k-core (Figure 9.9a, c).

nents of appropriate eigenvectors for a thorough analysis, this phenomenon can introduce a serious hardship, effectively shading the major part of a network in favour of a hub and its neighbourhood. The same can be said about centratity measures, which fail when localization concentrates most of the weight of a centrality on a hub and its neighbours. One might think that, if different matrices are prone to this effect to a different extent, then it should be easy to choose the optimal one for a given technique. However, it turns out that the reason for localization of eigenvectors is actually the same for the majority of matrix representations of networks, including the adjacency, Laplacian, and normalized Laplacian matrices, and so for a given network, localization may happen in each of these matrices. Speaking figuratively in terms of eigenvector centrality, 'the localization effect arises because a hub with high eigenvector centrality gives high centrality to its neighbours, which in turn reflect it back again and inflate the hub's centrality' (Martin, Zhang, and Newman, 2014). In tree-like networks, these 'reflections' are only possible if a matrix allows backtracking paths.

9.3.1 Non-backtracking centrality

A non-backtracking matrix (Section 6.3 and Appendix F) was invented just to exclude such paths, and so one might hope that its eigenvectors and related quantities avoid localization, or at least display it in a less-wide set of situations than the eigenvectors of an adjacency and Laplacian matrices. Focusing on centrality measures, Martin, Zhang, and Newman (2014) defined a novel, *non-backtracking centrality*

$$x_i \equiv \sum_{j \in \partial i} v_{i \leftarrow j} = \sum_j A_{ij} v_{i \leftarrow j}, \tag{9.35}$$

where $v_{i \leftarrow j}$ are the components of the principal eigenvector of an $E \times E$ non-backtracking matrix, where E is the number of edges in the network. The non-backtracking centralities for all vertices, x_i, $i = 1, 2, \ldots, N$ and the associated eigenvalue can be actually obtained from a smaller $2N \times 2N$ matrix

$$M = \begin{pmatrix} A & I - D \\ I & 0 \end{pmatrix}, \tag{9.36}$$

where A, I, and D are the adjacency, identity, and degree matrices, respectively. The full set of non-backtracking centralities, x_i, are the first N components of the principal eigenvector of the matrix M, and its largest eigenvalue coincides with that of the non-backtracking matrix. For more details on non-backtracking centrality and the properties of these matrices, see Appendix F. Recalling Section 6.3, one can easily understand the meaning of non-backtracking centrality. Remove at random a fraction of edges from the network almost down to the percolation threshold (birth point of a giant connected component) indicated by the message-passing algorithm. Then the component x_i of the non-backtracking centrality is proportional to the probability that vertex i belongs to a giant connected component, found in this approximation. Similarly, for the SIR process with an infection rate slightly exceeding the epidemic threshold for this network, x_i is proportional to the probability that vertex i is removed. In uncorrelated networks this probability is proportional to the degree of a vertex (Section 6.1) and one can estimate the non-backtracking centrality for vertices of these networks as

$$x_i \sim q_i \tag{9.37}$$

(Krzakala, Moore, Mossel, Neeman, Sly, Zdeborová, and Zhang, 2013; Martin, Zhang, and Newman, 2014).[20]

Martin, Zhang, and Newman (2014) compared eigenvector and non-backtracking centralities for a number of networks and observed that in most of the cases considered, non-backtracking centrality avoids localization on hubs, unlike eigenvector ones, even in networks with many short cycles (Figure 9.10). Thus non-backtracking centrality provides a good remedy against localization on hubs, and this is why it is currently one of most efficient and widely used instruments in spectral clustering. On the other hand, non-backtracking centrality still demonstrates localization effects in networks with sufficiently large cliques, with 'overlapping hubs' (hubs having many joint neighbours) and other small strongly connected complexes (Martin, Zhang, and Newman, 2014; Kawamoto, 2016; Pastor-Satorras and Castellano, 2020).[21]

---

[20] A better approximation can be obtained by substituting the estimate $v_{i\leftarrow j} \sim q_j - 1$ for uncorrelated networks, Eq. (F.5), into Eq. (9.35), which gives

$$x_i \approx \frac{\sum_j A_{ij}(q_j - 1)}{\sum_j q_j(q_j - 1)}$$

(Pastor-Satorras and Castellano, 2020). The sum $\sum_j A_{ij}(q_j - 1)$ is the number of the second-nearest neighbours of vertex i in a locally tree-like network.

[21] Apart from direct computing non-backtracking centrality, one can estimate it, even accounting for localization effects (Pastor-Satorras and Castellano, 2020). A good approximation of non-backtracking centrality was obtained by taking into account degree–

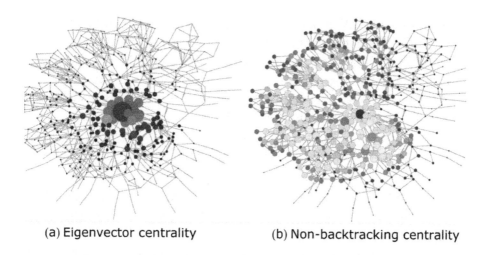

(a) Eigenvector centrality   (b) Non-backtracking centrality

Fig. 9.10 Eigenvector and non-backtracking centralities for the electronic circuit network of 512 vertices. Vertex sizes are proportional to centrality. Inverse participation ratios for eigenvector and non-backtracking centralities equal, respectively, 0.179 and 0.0056. Adapted from Martin, Zhang, and Newman (2014).

9.4 Stochastic block model

Among the key issues of interest in network science is a modular structure of a complex network, in particular, the possibility and ways to divide a network into the set of parts with either (i) denser connections within them then connections between these parts or, (ii) vice versa, with sparser connections within the parts and denser connections between them. This criterion of division into blocks, according to density of inter- and intra-connections, is the simplest one. The *stochastic block model* (Holland, Laskey, and Leinhardt, 1983) is the basic model of random networks with a modular structure just of this kind, directly generalizing the $G(N,p)$ random graph. Let N vertices of an undirected random network, $i = 1, 2, \ldots, N$, be parted into m blocks (parts, groups, modules, communities) of sizes N_a, $a = 1, 2, \ldots, m$. To each vertex i, one can ascribe a label a_i indicating block a to which it belongs. Any two vertices, i and j, in this network are connected with probability depending only on their labels—their blocks, that is, the edge (ij) is present with probability $p_{a_i a_j} = p_{a_j a_i}$ and absent with probability $1 - p_{a_i a_j}$. In particular, for $m = 1$, there is only one such probability $p_{1_i 1_j} \equiv p_{11}$, and we

---

degree correlations between the nearest-neighbouring vertices and using a branching matrix (Timár, da Costa, Dorogovtsev, and Mendes, 2021).

get the ordinary $G(N, p)$ model. For a general m, there can be at the most $m(m + 1)/2$ different probabilities p_{ab}.

Introducing the fraction n_a of vertices of vertices in block a, $n_a = N_a/N$ and assuming that the network is large and sparse, so the probabilities should scale as $1/N$, $p_{ab} = c_{ab}/N$, where c_{ab} is a constant, we have the average degree of a vertex in this model

$$c \equiv \langle q \rangle = \frac{1}{N} \sum_a p_{aa} N_a (N_a - 1) + \frac{2}{N} \sum_{a<b} p_{ab} N_a N_b \cong \sum_a c_{aa} n_a^2 + 2 \sum_{a<b} c_{ab} n_a n_b.$$

$$(9.38)$$

The terms $\sum_a p_{aa} N_a (N_a - 1)/2$ and $\sum_{a<b} p_{ab} N_a N_b$ are, respectively, the average numbers of edges within blocks and between them. In the simplest situation, $c_{aa} \equiv c_{in}$ for connections in every block and $c_{a \neq b} \equiv c_{out}$ for connections between any two blocks (Newman and Girvan, 2004). When $c_{in} > c_{out}$, the structure is assortative, that is, with dense intraconnections and sparse interconnections, when $c_{in} < c_{out}$, the structure is disassortative. Furthermore, if all m blocks have equal sizes, that is, $n_a = 1/m$, then

$$c = \frac{1}{m} c_{in} + \frac{m-1}{m} c_{out}.$$

$$(9.39)$$

Importantly, in the sparse, infinite network regime, the stochastic block model provides locally tree-like networks, for which message-passing and belief-propagation algorithms work exactly. This model was used to demonstrate and explain main effects in networks with modular structure, and it is also one of standard benchmarks for community detection algorithms, which we will discuss later (Lee and Wilkinson, 2019). More complex, still locally tree-like, models of modular networks can be constructed by straightforwardly generalizing the configuration model (microcanonical statistical ensemble) and a random network with hidden variables (grand canonical ensemble). In particular, in the microcanonical version, this is a uniformly random graph with a given sequence of lists of intra- and inter-degrees for vertices. An i-th member of this sequence is the list $(q_i^{(1)}, \ldots, q_i^{(a)}, \ldots, q_i^{(m)})$, where $q_i^{(a)}$ is a given number of connections of vertex i to vertices in block a (Karrer and Newman, 2011).

9.5 Modularity

Let a network with an arbitrary modular structure be divided into m subgraphs by edge cuts, N_a vertices in subgraph a, where $a = 1, 2, \ldots, m$. These subgraphs not necessarily coincide with authentic blocks, modules, or communities in a network, if they exist at all. We can still suggest that our partition sets some modular structure of a network. How can one quantify the deviation of such an induced modular structure from a uniform one? Furthermore, how can one figure out that this partition well fits the genuine modular (or community) structure of a network? An integral measure defined for a given partition of a network, *modularity* Q, helps to answer these questions (Newman, 2003; Newman and Girvan, 2004). The modularity compares the total number of edges within the subgraphs—members of a given partition—with the average total number of edges within the corresponding subgraphs of a uniform random counterpart of the original network—its 'null model'.

$$Q(\text{partition of network})$$

$$\equiv \frac{1}{E}\left[\begin{pmatrix}\text{total number of edges}\\ \text{within members}\\ \text{of network's partition}\end{pmatrix} - \begin{pmatrix}\text{expected number of edges}\\ \text{within members of partition}\\ \text{of uniform counterpart}\end{pmatrix}\right], \quad (9.40)$$

where E is the total number of edges in the network. A partition with a single member, the network itself, has $Q = 0$. Denoting the number of edges within member a of a given partition by E_a, one can write

$$Q = \frac{1}{E}\left[\sum_{a=1}^{m} E_a\big|_{\text{given network}} - \sum_{a=1}^{m} \langle E_a\rangle\big|_{\text{uniform counterpart}}\right]. \quad (9.41)$$

As for the uniform counterpart, the typical choice is the Chung–Lu model with the desired degrees of vertices coinciding with their degrees in the original network (Section 4.4). In the uniformly random counterpart, the ratio $\sum_a \langle E_a\rangle / E$ can easily be expressed in terms of the total degrees $q_a \geq 2E_a$ of the vertices belonging to individual modules of the original network, where $a = 1, 2, \ldots, m$ and $\sum_a q_a = 2E$. We must find the probability $p_a = (\langle E_a\rangle / E)|_{\text{uniform counterpart}}$ that a randomly chosen edge in the uniformly random network is in module a. Similarly to the configuration model, the probability that a given end of an edge is in module a is proportional to q_a. Consequently, the probability p_a is proportional to q_a^2. Then $p_a = (q_a/2E)^2$. In particular, if the entire network consists of a single module, we have the probability $(2E/2E)^2 = 1$, as is should be. Then modularity takes the form

$$Q = \sum_{a=1}^{m}\left[\frac{E_a}{E} - \left(\frac{q_a}{2E}\right)^2\right]. \quad (9.42)$$

Denoting the fraction of edges that connect vertices in module a to those in module b by $\tilde{e}_{ab}$ and introducing $\tilde{a}_a \equiv \sum_b \tilde{e}_{ab}$, one can write

$$Q = \sum_{a=1}^{m} (\tilde{e}_{aa} - \tilde{a}_a^2). \tag{9.43}$$

In terms of the matrix $\tilde{e}$ with elements $\tilde{e}_{ab}$, the sum $\sum_{a=1}^{m} \tilde{a}_a^2$ in this equality is the sum of all elements of the matrix $\tilde{e}^2$.

If a network has strong community structure, that is, well-distinguished modules, and the chosen partition of the network fits these modules well, then we get a large modularity for this partition. Clearly, Q cannot be greater than 1. On the other hand, let a network consist of a set of connected components where each component is a member of the optimal partition providing the maximal modularity. In this situation, $Q = 1 - \sum_a (E_a/E)^2 < 1$.

Ascribing the label a_i to each vertex in group a and again assuming that the uniform counterpart of a network is generated by the Chung–Lu model, one can represent modularity, Eq. (9.40), in the following form:

$$Q = \frac{1}{2E} \sum_{ij} \left(A_{ij} - \frac{q_i q_j}{2E} \right) \delta_{a_i, a_j}, \tag{9.44}$$

where the Kronecker symbol removes edges between different groups from the sum. In the particular case of partition into two parts, $a = 1$ and 2, for each vertex i in a network, one can introduce a spin $s_i = \pm 1$, where $s_i = 1$ if vertex belongs to group 1 and $s_i = -1$ if vertex belongs to group 2. Taking into account in Eq. (9.44) that $\delta_{a_i, a_j} = (s_i s_j + 1)/2$ leads to the form

$$Q = \frac{1}{4E} \sum_{ij} \left(A_{ij} - \frac{q_i q_j}{2E} \right) (s_i s_j + 1) = \frac{1}{4E} \sum_{ij} \left(A_{ij} - \frac{q_i q_j}{2E} \right) s_i s_j \tag{9.45}$$

(Newman, 2006).[22] Let partition in Eq. (9.45) be not fixed. Then this expression in terms of spins enables one to treat modularity as a benefit function

---

[22]In term of these spins, the edge cut size, that is, the number of edges between the two groups in the partition, can be written as

$$F = \frac{1}{2} \sum_{ij} A_{ij} \frac{1 - s_i s_j}{2} = \frac{1}{4} \sum_{ij} (q_i \delta_{ij} - A_{ij}) s_i s_j$$

since $\sum_{ij} A_{ij} = \sum_i q_i s_i^2 = \sum_{ij} q_i \delta_{ij} s_i s_j$. Noticing that $q_i \delta_{ij} - A_{ij} = L_{ij}$ are the elements of the Laplacian matrix, one can represent F in matrix form

$$F = \frac{1}{4} \mathbf{s}^T L \mathbf{s}.$$

Using Eq. (9.45), we derive the relation between Q and F, whose largest value is hit for the optimal partition—the optimal configuration of spins. This defines the optimization problem for modularity.[23] Introducing the *modularity matrix* $\mathcal{M}$

$$\mathcal{M} = A_{ij} - \frac{q_i q_j}{2E},$$
(9.46)

and the vector **s** with components s_i for vertices $i = 1, 2, \ldots, N$, one can rewrite Eq. (9.45) in matrix form

$$Q = \frac{1}{4E} \mathbf{s}^T \mathcal{M} \mathbf{s}.$$
(9.47)

Typical modularity for the optimal partition of real-world networks falls in the range of 0.3–0.8. For demonstration purposes, we compile some of modularity values, obtained for diverse networks of different sizes and architectures (Newman, 2006):

- For the karate club network of Zachary, $Q = 0.419$.[24]
- For a metabolic network for the nematode *C. elegans* of 453 vertices, $Q = 0.435$.
- For a co-authorship network of scientists working on condensed matter physics of 27,519 vertices, $Q = 0.723$.

Notably, it is hard to find a real network with low modularity. Furthermore, the definition of modularity, Eq. (9.40), with the Chung–Lu random graph used as a uniform counterpart of the original network often leads to rather large values of modularity for networks clearly having no modular structure, in particular, for uniform trees, for chains, and so on.[25] To avoid such artefacts, one can impose some of the structural constraints of an original network on its zero-modularity counterpart. Still, the standard

$$Q + \frac{1}{E}F = \frac{1}{8}\left[1 - \frac{1}{E^2}\left(\sum_{i \in \text{group 1}} q_i - \sum_{i \in \text{group 2}} q_i\right)^2\right].$$

[23]For a partition into $m \geq 2$ communities, this optimization problem can be formulated in terms of the m-state Potts model (Reichardt and Bornholdt 2004, 2006).

[24]The small Zachary karate club graph (Zachary, 1977), consisting of 34 vertices and 78 edges, was a reference network in numerous studies of modularity and communities. This is an undirected one-partite network of friendships between members of a karate club at a US university. A social conflict in this group split the network into two factions centered around the club president and the instructor. A sharp border between these factions was absent, which made community indexing non-trivial.

[25]Check that for the partition of the chain ●—●—●—● into two equal groups, ●—● + ●—●, the modularity is

uniform counterpart, Eq. (9.42) is commonly used, for which, as a rule of thumb, a modularity exceeding, say, 0.3 indicates that a network indeed has a modular structure.

Equation (9.42) suggests the following necessary condition that a subgraph—a member of a given partition—is a module in the network:

$$\frac{E_a}{E} - \left(\frac{2E_a + E_a^{(\text{out})}}{2E}\right)^2 > 0, \tag{9.48}$$

where $E_a^{(\text{out})}$ is the number of edges connecting this subgraph with the rest of the network. In turn, this criterion leads to the upper bound for the number of such edges of a module

$$\frac{E_a^{(\text{out})}}{E_a} < 2\left(\sqrt{\frac{E}{E_a}} - 1\right). \tag{9.49}$$

Notice that this upper bond depends on the total number of edges in a network.

9.5.1 Resolution limit

Fortunato and Barthélemy (2007) (see also Lancichinetti and Fortunato, 2011) showed that optimizing modularity, defined by Eq. (9.42), cannot detect communities smaller than some *resolution limit*. More precisely, no optimization techniques based on this form of modularity can resolve communities with E_a smaller than $O(\sqrt{E})$. To obtain this resolution limit, let us consider two modules, *1* and *2*, in a network, which are weakly interconnected with each other and with the rest of the network. We explore the situation where these modules are most sharply distinguished, which should favour their detection, and hence we interconnect them together and with the rest of the network by single edges (Figure 9.11). The idea is to compare the modularities of two partitions, namely, (i) the partition of the network into three parts—module *1*, module *2*, and the remaining network, modularity $Q_{1,2}$, and (ii) the partition of the network into two parts—the

$$Q = \frac{1}{3}\left[(1+1) - \frac{1}{2}\left(\frac{1^2 + 2^2 + 2 \times 1 \times 2}{6} + \frac{2^2 + 1^2 + 2 \times 2 \times 1}{6}\right)\right]$$

$$= \left(\frac{1}{3} - \frac{(1+2)^2}{6^2}\right) + \left(\frac{1}{3} - \frac{(2+1)^2}{6^2}\right) = \frac{1}{6}.$$

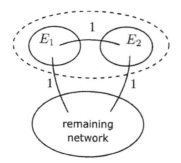

Fig. 9.11 Comparison of two possible partitions of partitions of a network: (i) module *1*, module *2*, and the rest of the network and (ii) the union of modules *1* and *2* with their interlink (dashed line) and the rest of the network. E_1 and E_2 are the numbers of internal edges in modules *1* and *2*.

union of the modules *1* and *2* with their interlink and the rest of the network, modularity $Q_{1\cup 2}$. Clearly, any optimization of modularity can detect modules *1* and *2* only if $Q_{1,2} - Q_{1\cup 2} > 0$. Denoting the numbers of internal edges in modules *1* and *2* by E_1 and E_2, respectively, and using Eq. (9.42), we get

$$Q_{1,2} - Q_{1\cup 2} = \frac{E_1}{E} - \left(\frac{2E_1 + 2}{E}\right)^2 + \frac{E_2}{E} - \left(\frac{2E_2 + 2}{E}\right)^2$$
$$- \left[\frac{E_1 + E_2 + 1}{E} - \left(\frac{2E_1 + 2E_2 + 4}{E}\right)^2\right] > 0. \quad (9.50)$$

Assuming $E_1 E_2 \gg 1$, this leads to the equality

$$E_1 E_2 > \frac{E}{2}, \quad (9.51)$$

which suggests the resolution limit for the number of edges within a module

$$E_{\text{resolution limit}} \sim \sqrt{E}. \quad (9.52)$$

To overpass this limit, one can try to progressively split the network into disconnected parts thus reducing E in Eq. (9.52) (Granell, Gomez, and Arenas, 2012). Alternatively, one can modify the definition of modularity (Li, Zhang, Wang, Zhang, and Chen, 2008; Chen, Kuzmin, and Szymanski, 2014). The point is, however, that a similar resolution limit was found for the stochastic block model by entropy based reasoning, irrespective of modularity optimization (Peixoto, 2013*b*), which suggests a more general nature of this limit than one would expect.

9.6   Detection of communities

One major difficulty in community detection is the absence of a unique definition of a community. Modularity provides only one of options for treating and distinguishing communities in networks.[26] Furthermore, the direct optimization of modularity is very costly and usually infeasible, and hence extra ideas, a heuristic, or approximations have to be applied.

9.6.1   Optimization of modularity

Section 4.12 outlined Granovetter's hypothesis stating that in social networks main informations flows run through links between different 'densely knit clumps of close friends'. In fact, the same should be true for general networks with modular structure. This observation was exploited in the Girvan–Newman algorithm for community detection (Girvan and Newman, 2002; Newman and Girvan, 2004):

(i) Compute the betweenness for each edge of the network.

(ii) Remove the edge with the largest betweenness.

(iii) Recalculate betweennesses of all edges affected by the removal.

(iv) Repeat (ii)–(iv) until no edges remain.

Along this process, the network progressively splits into a growing set of diminishing disconnected clusters, with N bare vertices in the finite state. The process can be depicted as a hierarchical tree of partitions—a dendrogram— shown in Figure 9.12. The evolution proceeds from the left to the right. A

---

[26]In particular, the following two definitions of a community in a strong and a weak sense are among other options (Radicchi, Castellano, Cecconi, Loreto, and Parisi, 2004). Let C be a subgraph of a graph G, and q_i, $i = 1, 2, \ldots, |C|$, be the degrees of vertices in this subgraph. Each of these degrees is the sum of two numbers, $q_i = q_i^{(\mathrm{in})} + q_i^{(\mathrm{out})}$, namely the number of connections $q_i^{(\mathrm{in})}$ of this vertex to vertices in the community and the number of connections $q_i^{(\mathrm{out})}$ to other vertices within the graph.

- The subgraph C is a *community in a strong sense* if

$$q_i^{(\mathrm{in})} > q_i^{(\mathrm{out})} \quad \text{for any vertex } i \text{ in } C,$$

  see also Flake, Lawrence, Giles, and Coetzee (2002).

- The subgraph C is a *community in a weak sense* if

$$\sum_{i \in C} q_i^{(\mathrm{in})} > \sum_{i \in C} q_i^{(\mathrm{out})}.$$

One can easily check that for a weak community, the inequality in Eq. (9.48) is surely satisfied, while Eq. (9.48) does not guarantee that a community is weak.

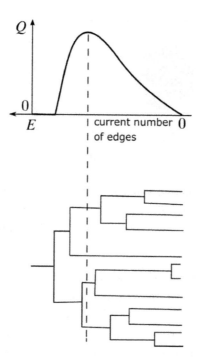

Fig. 9.12 Illustration of the Girvan–Newman algorithm for community detection. Dendrogram on the bottom depicts the progressive splitting of a network into smaller and smaller clusters along the process of edge removal. A cross-section of the dendrogram provides the set of these clusters at a given stage of the process characterizing by the number of remaining edges. Evolution of modularity during this process is shown on the top. The dashed line indicates the cross-section corresponding to the highest modularity and reveals, in this case, four communities present in the network.

cross-section of the dendrogram provides the full set of disconnected clusters remaining of a network at a given step of the algorithm. After each removal of an edge, modularity is calculated, and the cross-section corresponding to the maximum of modularity indicates all communities in the network. The fast algorithm for computing edge betweenness for all E edges in a network of N vertices does it in time $O(NE)$ (Newman, 2001b; Brandes, 2001). This task should be repeated after the removal of each edge with the largest betweenness, and so, at the worst, the entire Girvan–Newman algorithm runs in time $O(NE^2)$. The quality and performance of this algorithm was tested on a stochastic block model, the Zachary karate club graph, and many other networks. It worked well in most of cases.

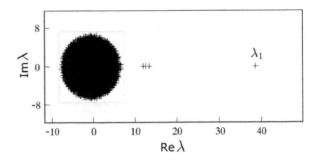

Fig. 9.13 Real and imaginary parts of the eigenvalues of the adjacency matrix of a directed scale-free network with four sufficiently well distinguished equal-sized communities. Degree distribution exponent $\gamma = 2.5$. Adapted from Chauhan, Girvan, and Ott (2009).

The more rapid, though less precise, greedy algorithm of Newman (2004b) constructs a resembling dendrogram in the opposite direction, from the right (N bare vertices) to the left (original network). Clusters are repeatedly joined together in pairs by choosing at each step the interlink—a relevant edge of the original network—that results in the largest increase (or smallest decrease) in modularity. One step of the algorithm takes time $O(E + N)$, since, at the worst, E edges should be tried and, at the worst, N vertices should be reassigned to merging clusters. The maximum number of join operations equals $N - 1$, which gives time $O(N(E + N))$ for the entire algorithm. The cross-section corresponding to the maximum of modularity indicates all communities in a network, similarly to the Girvan–Newman algorithm. A more refined version of this algorithm runs even faster in time $O(Ed \ln N)$, where d is the depth of the resulting dendrogram (Clauset, Newman, and Moore, 2004). In the sparse networks that have communities at many scales, $d \sim \ln N$, this algorithm runs in time $O(N \ln^2 N)$.

9.6.2 Spectral clustering

Most of community detection algorithms employ various spectral clustering techniques. The core idea of this approach is generally the same for all matrices characterizing a network, including the adjacency matrix, the Laplacian, the normalized Laplacians, the random walk matrix, the modularity matrix, the non-backtracking matrix, etc. For instance, let us first focus on an adjacency matrix. If a network has no modular structure, then the spectrum of its adjacency matrix consists of the largest eigenvalue λ_1

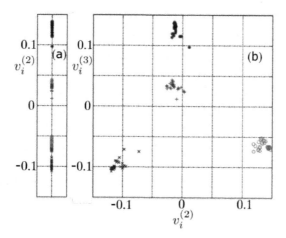

Fig. 9.14 (a) Components of the first non-trivial eigenvector $v_i^{(2)}$ of the Laplacian spectrum of a stochastic block model network of 128 vertices containing four well distinguished communities of 32 edges each. (b) Components of the second non-trivial eigenvector $v_i^{(3)}$ vs. corresponding components of the first non-trivial eigenvector $v_i^{(2)}$. Adapted from Donetti and Munoz (2004).

separated by a gap from the main body of the spectrum (Section 9.1).[27] Imagine m disconnected copies of this network. Their spectrum contains the m-degenerate largest eigenvalues λ_1 separated by the same gap from the $(m+1)$-th largest eigenvalue. Clearly, if we weakly interconnect these copies, then the degeneracy disappears, and the spectrum appear with m close eigenvalues still separated from the bulk of the spectrum, if the interconnection is sufficiently weak. One can suggest that the spectrum of a network with m well-distinguished modules should be orgainized similarly.[28] Figure 9.13 shows an example of such a spectrum for a directed network containing four modules (Chauhan, Girvan, and Ott, 2009).[29] Spectral clustering techniques detects communities by analysing the eigenvectors corresponding to the m eigenvalues separated from the bulk of the spectrum.

Importantly, to ensure detecting each of, say, m communities in a network, all these m eigenvectors should be inspected. The following example is for the Laplacian matrix spectrum (Donetti and Munoz, 2004). The dif-

[27]This separation is particularly conspicuous in sufficiently large dense networks where the continuous bulk of the spectrum has sharp boundaries.

[28]See Peixoto (2013a) for more details on the spectra of the adjacency matrix, Laplacian, and normalized Laplacian for modular networks.

[29]For an undirected network, all eigenvalues of the spectrum would be real.

ference from the adjacency matrix is that for the Laplacian, one should look at the smallest positive eigenvalues. For the spectrum of the Laplacian matrix of a network with four communities, Figure 9.14a demonstrates that the components of the first non-trivial eigenvector $v_i^{(2)}$ of the matrix do not allow us to find the community structure of the network (one could mistakenly conclude that there are only three communities). On the other hand, the inspection of the components of the first two non-trivial eigenvectors, $v_i^{(2)}$ and $v_i^{(3)}$, $i = 1, 2, \ldots, N = 128$, plotted in the 2-dimensional space $(\mathbf{v}^{(2)}, \mathbf{v}^{(3)})$ already allows us to distinguish four communities (Figure 9.14b). Four separate sets of points in this plane correspond to four communities. Even better separation of these clusters occurs in the 4-dimensional vector space $(\mathbf{v}^{(2)}, \mathbf{v}^{(3)}, \mathbf{v}^{(4)}, \mathbf{v}^{(5)})$. Diverse spectral clustering algorithms perform an automated assortment of such sets in various ways for different matrices. For details of these techniques, we direct the readers to Von Luxburg (2007), Newman (2006), Capocci, Servedio, Caldarelli, and Colaiori (2005), and Moore (2017).

9.6.3 Detectability threshold

A spectral algorithm enables one to detect communities only if the community-correlated eigenvectors of the exploited matrix are not drown in the sea of non-informative eigenvectors. The eigenvalues of these m eigenvectors must be out of the bulk of the spectrum (if there are m communities). When is this the case? Following Nadakuditi and Newman (2012), let us consider the spectrum of the adjacency matrix of a large network with two equal-sized blocks, $m = 2$, provided by the stochastic block model. Let vertices in each of the blocks be linked with probability $p_{in} = c_{in}/N$, and vertices in different blocks be interlinked with probability $p_{out} = c_{out}/N$. The average degree of a vertex is

$$c = \frac{c_{in} + c_{out}}{2}. \tag{9.53}$$

If the network is large and sufficiently dense, the spectral density for the bulk of the adjacency matrix spectrum asymptotically follows the Wigner semicircle law, Eq. (9.5),

$$\rho(\lambda) = \frac{1}{2\pi} \frac{\sqrt{4c - \lambda^2}}{c} \theta(2\sqrt{c} - |\lambda|). \tag{9.54}$$

In addition, one or two eigenvalues can occur above the upper edge, $2\sqrt{c}$, of this continuous spectrum, namely,

$$\lambda_1 = c + 1 \tag{9.55}$$

and

$$\lambda_2 = \frac{1}{2}(c_{in} - c_{out}) + \frac{2c}{c_{in} - c_{out}}. \tag{9.56}$$

Clearly, $\lambda_1 > 2\sqrt{c}$. Furthermore, λ_1 exceeds λ_2 when $c_{in} - c_{out} > 2$ and therefore $c > 1 + c_{out} > 1$, which is the range of values interesting for us.

The components of the eigenvector associated with the largest eigenvalue (spectral centrality) are essentially related to local properties of a network (degrees of vertices), while the eigenvector associated with λ_2 is correlated with the blocks. To identify the blocks, one needs the fulfilment of the inequality $\lambda_2 > 2\sqrt{c}$ resulting in the following condition of the *detectability threshold* (sometimes, detectability limit)[30]

$$c_{in} - c_{out} = 2\sqrt{c}. \tag{9.57}$$

Recall that Eq. (9.57) was derived by assuming that the upper edge of the continuous spectrum of the adjacency matrix is sharp, which is the case only for sufficiently dense networks. In a sparse networks, for example, in the Erdős–Rényi graphs, as discussed in Section 9.1, the spectral density has tails extending far beyond the edges of the semicircle in Eq. (9.54). Hence in the spare regime, λ_2 may be lost within the tail of the spectral density even when the condition $\lambda_2 > 2\sqrt{c}$ is satisfied. The same is true for the eigenvectors of the Laplacian, normalized Laplacian, and modularity matrices. However, it turns out that the detectability threshold can be found exactly even in sparse networks. Decelle, Krzakala, Moore, and Zdeborová (2011b, 2011a) solved this problem by applying believe propagation, which works exactly in locally tree-like networks including a sparse stochastic block model, and proved that the detectability threshold for $m = 2$ is indeed given by Eq. (9.57). Importantly, no other algorithm can perform, in this case, better than belief propagation and reach a lower detectability threshold. In general, for an arbitrary number m of equal-sized communities, the detectability threshold is set by the value

$$c_{in} - c_{out} = m\sqrt{c}, \tag{9.58}$$

where the average degree of a vertex $c = [c_{in} + (m-1)c_{out}]/m$. Equation (9.58) is valid both in the sparse and dense regimes. Speaking more strictly, in the infinite size limit, below the detectability threshold, that is, when $c_{in} - c_{out} < m\sqrt{c}$, no algorithm can classify the vertices better than

[30]That is, the blocks can be detected by the spectral clustering with the adjacency matrix only if $c_{in} - c_{out} > 2\sqrt{c}$.

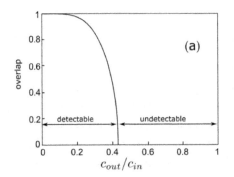
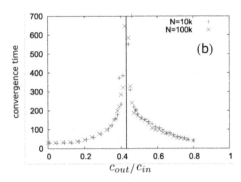

Fig. 9.15 (a) Overlap (fraction of vertices correctly classified by the belief propagation algorithm) vs. the ratio c_{out}/c_{in} for the stochastic block model with $m = 4$ equal-sized communities, having average vertex degree $c = 16$. The detectability threshold value is $c_{out}/c_{in} = 3/7$. (b) For the same parameters of the model, the number of iterations needed for convergence of the belief propagation algorithm for two different sizes of a network, $N = 10^4$ and 10^5. Adapted from Decelle, Krzakala, Moore, and Zdeborová (2011a).

chance. On the other hand, above the detectability threshold, believe propagation allows one to correctly classify the majority of vertices. Note that spectral algorithms with the adjacency, Laplacian, normalized Laplacian, and modularity matrices are successful all the way down to this detectability threshold only if the network is sufficiently dense, which seriously limits their applicability.

The detectability threshold can be treated as the critical point of a phase transition with the order parameter given by the fraction of vertices correctly classified by an algorithm, that is, by the overlap of the assignment and real blocks. It is convenient to fix the average degree c and choose the ratio c_{in}/c_{out} as a control parameter. Figure 9.15a shows that this transition is continuous for $m = 4$. Furthermore, the number of iterations needed for convergence of the belief propagation algorithm diverges at the detectability threshold demonstrating a critical slow-down typical for continuous phase transitions (Figure 9.15b). Decelle, et al. found that this transition is continuous only when $m \leq 4$. For $m > 4$ the transition is first-order though with the hysteresis occurring within a quite narrow window of the control parameter.

9.6.4 Spectral clustering with a non-backtracking matrix

Krzakala, Moore, Mossel, Neeman, Sly, Zdeborová, and Zhang (2013) proposed using a non-backtracking matrix in spectral algorithms, overpassing the limitations of spectral clustering with the adjacency, Laplacian, and modularity matrices. One can indicate two key advantages of this matrix (Decelle, Hüttel, Saade, and Moore, 2014):

(i) The eigenvectors of the non-backtracking matrix are less prone to localization around vertices with large degrees, and so, in this clustering technique, hubs do not obscure communities.

(ii) For infinite locally tree-like networks, this spectral clustering provides the best possible solution for any algorithm.[31] For such networks, including a sparse stochastic block model, this spectral clustering is successful all the way down to the detectability threshold.

The following example considered by Krzakala, *et al.* highlights the success of this spectral clustering in a situation where the spectral algorithm with the adjacency matrix fails. Consider the network with two equal communities, generated by the stochastic block model. Let the size of the network be $N = 4000$ vertices, $c_{in} = 5$, and $c_{out} = 1$, so the average degree of a vertex is $c = (c_{in} + c_{out})/2 = 3$. Then, according to Eqs. (9.55) and (9.56), two largest eigenvalues of the adjacency matrix are $\lambda_1 = 4$ and $\lambda_2 = 7/2$. Hence the second eigenvalue, λ_2, is outside the semicircle of radius $2\sqrt{c} = 3.4641\ldots$, and the detectability condition is formally satisfied. In reality, the spectral density of the adjacency matrix of this sparse network has tails extending out of the semicircle and overpassing λ_2, and so the second eigenvector vanishes in the bulk. Thus spectral algorithms with the adjacency matrix cannot detect communities in this network. Now look at the spectrum of the non-backtracking matrix for the same network (Figure 9.16). The second eigenvalue of this matrix appears to be well separated from the bulk of the spectrum enabling one to correctly classify the majority of vertices.

Spectral algorithms based on the non-backtracking matrix perform also well for more challenging non-tree-like networks, including Zachary's karate club network.

---

[31]In this sense, the spectral clustering with the non-backtracking matrix is 'exact', similarly to the message-passing, and belief-propagation algorithms involving the non-backtracking matrix.

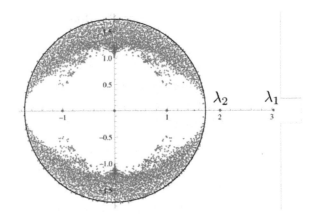

Fig. 9.16 Spectrum of the non-backtracking matrix for the stochastic block model network of $N = 4000$ vertices with two equal-sized blocks, $c_{in} = 5$, and $c_{out} = 1$, and so $c = (c_{in} + c_{out})/2 = 3$. The two largest eigenvalues are $\lambda_1 = c$ and $\lambda_2 = (c_{in} - c_{out})/2$. The bulk of the spectrum is within the circle of radius $\sqrt{c} = \sqrt{3}$. Notice that $\lambda_2 = 2$ is markedly separated from the bulk. Adapted from Krzakala, Moore, Mossel, Neeman, Sly, Zdeborová, and Zhang (2013).

9.6.5 Other algorithms

Spectral clustering algorithms are, unfortunately, too time-consuming to be applicable to very large networks. Most of communit- detection algorithms aiming at this difficult task are various versions of the greedy optimization of modularity (Clauset, Newman, and Moore, 2004; Blondel, Guillaume, Lambiotte, and Lefebvre, 2008).[32] A set of clustering algorithms exploits information flows and random walks on networks (Rosvall and Bergstrom, 2008). A special class of community-detection algorithms account not only for network structure, but also for additional information about the network vertices, for example, the age of individuals in a social network (Newman and Clauset, 2016). This metadata enables one to identify communities more accurately. See Danon, Diaz-Guilera, Duch, and Arenas (2005) for the comparison of various community detection algorithms tested on benchmark networks.

---

[32]In particular, the algorithm of Blondel, Guillaume, Lambiotte, and Lefebvre (2008) reliably identified a hierarchical community structure (with communities embedded within other communities) in a web graph of 118 million vertices and more than one billion edges.

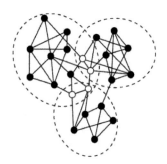

Fig. 9.17 Overlapping communities. Adapted from Lancichinetti, Fortunato, and Kertész (2009).

9.7 Overlapping communities

In many real networks, vertices can belong to more than one community, and communities overlap (Figure 9.17). A few approaches were developed for community structures of this sort. In particular, Palla, Derényi, Farkas, and Vicsek (2005) interpreted overlapping communities in the spirit of k-clique percolation (Section 6.12). By their definition, two k-cliques are adjacent if they share the maximum possible number of vertices, namely, $k - 1$. Using this adjacency, one can naturally introduce the 'connected components' of k-cliques. According to Palla, *et al.*, a community is a subgraph of the original network, consisting of the vertices and edges of such a connected component of k-cliques. Clearly, two k-cliques belonging to different communities of this kind still can overlap by less than $k - 1$ vertices, leading to overlapping communities. Furthermore, more than one k-clique of one community can overlap with k-cliques of another community, and so the total overlap can be large. The only parameter in this scheme, the number k, determines the size of the smallest community which can be resolved.[33]

Lancichinetti, Fortunato, and Kertész (2009) proposed a greedy optimization algorithm for identification of overlapping communities, including hierarchically organized ones. A metric chosen for the optimization was the *fitness* of a subgraph C, defined as

$$f_C \equiv \frac{q_C^{(\text{in})}}{\left[q_C^{(\text{in})} + q_C^{(\text{out})}\right]^{\alpha}}, \tag{9.59}$$

[33]See Adamcsek, Palla, Farkas, Derényi, and Vicsek (2006) for the details and applications of this algorithm.

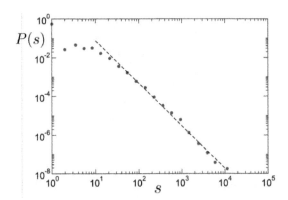

Fig. 9.18 Distribution of community sizes $P(s)$ for the largest connected component of the subset of the domain .gov obtained with the resolution parameter $\alpha = 1$. The graph includes 774,908 vertices and 4,711,340 edges. The slope of the dashed line equals -2.2. Adapted from Lancichinetti, Fortunato, and Kertész (2009).

where $q_C^{(in)}$ is the total internal degree of the vertices in C, that is $q_C^{(in)} \equiv \sum_{i \in C} q_i^{(in)}$, and $q_C^{(out)}$ is the total number of connections between these vertices and the remaining vertices in the network, that is $q_C^{(out)} \equiv \sum_{i \in C} q_i^{(out)}$. The positive parameter α controls the size of the communities and the resolution of the method; large values of α favour small communities. The idea is to search for local maxima of the fitness function $f_{C,i}$ in the space of all subgraphs including a given vertex i. The global maximum of the function $f_{C,i}$ with any vertex i is realized for the whole network G, $f_G = D^{1-\alpha}$, where D is the total degree of all vertices in the network. This searching is performed in a greedy fashion, starting from the subgraph C consisting of vertex i and first adding to C the neighbouring vertex maximally increasing fitness. After that, the fitness of all vertices in the increased subgraph is recalculated and those of them whose exclusion would increase the fitness are removed. The process is repeated until every neighbouring vertex added to C decreases the fitness. The result is the first local maximum of the function $f_{C,i}$, starting from vertex i. The subgraph C_i corresponding to this local maximum is called the 'natural community' of vertex i. From the set of natural communities of all vertices one can extract a 'cover' of the graph G, that is, a subset of this set such that every vertex in G belongs to at least one community.[34]

The resolution parameter α determines the resulting cover of the network, which is interpreted in this approach as the set of overlapping communities. The value $\alpha = 1$ is typically used, but one can proceed in a more sophisticated way. Compare the average fitnesses of covers obtained for different α and select the cover with the maximum average fitness.[35] Figure 9.18 demonstrates the power-law distribution of community sizes for a large Web graph, found by this method with $\alpha = 1$. Note that this approach is also applicable to directed and weighted networks (Lancichinetti, Radicchi, Ramasco, and Fortunato, 2011).

[34]One can actually approximately get the cover in a more economic way, without computing the full set of natural communities for all vertices in G.

[35]Since each cover is realized for some range of α, one has also to take into account the width of this region.

10
Walks and Search

10.1 Diffusion and random walks on networks

10.1.1 Basic times

This section considers simple random walks. Note however, that for regular graphs, there is not much difference between the diffusion and random walk processes. Section 9.2 explained that the time scale of the exponential relaxation to the final state of the random walk process on a graph is determined by the mixing time $\tau_{\text{mixing}} = 1/\lambda_2$, where λ_2 is the first nonzero eigenvalue of the normalized Laplacian matrix. In undirected graphs, the final probability to find a walker on a vertex is proportional to the vertex degree. Knowing λ_2, we readily get the mixing time. For networks with a finite spectra dimension D_s, the second eigenvalue (Fiedler eigenvalue) is $\lambda_2(N) \sim N^{-2/D_s}$ for large N, Eq. (2.83), which gives $\tau_{\text{mixing}} \sim N^{2/D_s}$.

- For example, for a uniformly random tree, $D_s = 4/3$ (Section 2.9) and so $\tau_{\text{mixing}} \sim N^{3/2}$.

In general, one can use the inequalities for λ_2, containing the diameter of a graph, Eq. (9.25), or the Cheeger constant, Eq. (2.84), to get an idea how the mixing time of a random walk on a given graph depends on its size. In particular,

- for a ring of N vertices or a chain, $\tau_{\text{mixing}} \sim N^2$
- for a D-dimensional lattice, $\tau_{\text{mixing}} \sim N^{2/D}$
- for a complete graph, $\tau_{\text{mixing}} = 1$
- for a random regular graph, $\tau_{\text{mixing}} \sim \ln N$, and so on (Levin and Peres, 2017).

To define other key times of a random walk on a graph, we should introduce the *mean first-passage time* $\langle t_{ij} \rangle$ (or *hitting time*) which is the expected time of a random walk started at vertex i and arrived first time at vertex j (Redner, 2001). For coinciding source and target vertices, $i = j$, this is the *mean first-return time* $\langle t_{ii} \rangle$ (or *recurrence time*). In 1947, Mark Kac proved his famous formula, which directly relates the first-return time for a vertex and the final probability $P_i(t \to \infty)$ to find a walker on this vertex:

$$\langle t_{ii} \rangle = \frac{1}{P_i(\infty)}. \tag{10.1}$$

From the Kac formula, accounting for Eq. (9.24), we readily obtain the value of the first-return time to vertex i of degree q_i:

$$\langle t_{ii} \rangle = \frac{\langle q \rangle N}{q_i}. \tag{10.2}$$

Lau and Szeto (2010) showed that if the mixing time of a random walk is sufficiently small, $\tau_{\text{mixing}} \ll N$, and so the information about the start vertex is quickly lost—then the mean first passage time $\langle t_{ij} \rangle$ can be approximated by $\langle t_{jj} \rangle = \langle q \rangle N/q_j$. The hitting time of a graph, τ_{hitting}, is the average of $\langle t_{ij} \rangle$ over over all pairs of vertices i and j. The *cover time* of a graph, τ_{cover}, is the average time needed for a random walk to visit every vertex of the graph at least once. Clearly, $\tau_{\text{mixing}} < \tau_{\text{hitting}} < \tau_{\text{cover}}$. In particular, one can easily estimate the cover time for a complete graph of N vertices in the following way. In this graph, the probability that a random walker moves from a vertex to any other vertex equals $1/(N-1)$ and so the probability that a vertex is visited at least once after t moves is equal to $1 - [1 - 1/(N-1)]^t$. Then for sufficiently large t, the probability $\widetilde{P}(t, N)$ to visit all N vertices at least once can be estimated as $\left[1 - [1 - 1/(N-1)]^t \right]^N$. For large N and $t > N$, this probability takes the form

$$\widetilde{P}(t, N) = \left[1 - (1 - 1/N)^{Nt/N} \right]^N \cong \left[1 - e^{-t/N} \right]^N. \tag{10.3}$$

If we consider $t = cN \ln N$, where c is a positive number, we get $\widetilde{P}(t = N \ln N, N) \to 1/e$, while this probability tends to 0 for $c < 1$ and it tends to 1 for $c > 1$.

- This leads to the cover time $\tau_{\text{cover}} \cong N \ln N$ of a complete graph, which coincides with the leading term of the exact asymptotics. This time should be compared with the short mixing time for this graph, $\tau_{\text{mixing}} = 1$. On the other hand, the hitting time for this graph is $\tau_{\text{hitting}} \sim N$.

See the list of the asymptotics of these three times for a few large basic graphs in Alon, Avin, Koucký, Kozma, Lotker, and Tuttle (2011). In particular,

- The cover and hitting times for a chain or a ring approach $N^2/2$, similarly to the mixing time.

- For a plain lattice, $\tau_{\text{cover}} \sim N(\ln N)^2$ and $\tau_{\text{hitting}} \sim N \ln N$.
- For a $(D{>}2)$-dimensional lattice, $\tau_{\text{cover}} \sim N \ln N$ and $\tau_{\text{hitting}} \sim N$.

In addition, we outline a few asymptotics of the basic times for random graphs.

- For a uniformly random tree, $\tau_{\text{cover}} \sim N^{3/2}$ (Aldous, 1991).
- For a random k-regular graph,

$$\tau_{\text{cover}} \cong \frac{k-1}{k-2} N \ln N \tag{10.4}$$

(Cooper and Frieze, 2005). Tishby, Biham, and Katzav (2021) obtained the distribution of cover times for these graphs and its standard deviation, which scales like N.

- For a sufficiently dense $G(N,p)$ random graph with $p(N)N = \langle q \rangle$ increasing with N as $\ln N$ or faster, $\tau_{\text{cover}} \sim N \ln N$ (Jonasson, 1998; Cooper and Frieze, 2007a), $\tau_{\text{hitting}} \sim N$, and $\tau_{\text{mixing}} \sim \ln N$. For a giant component of a sparse $G(N,p)$ random graph, $pN = \langle q \rangle < \infty$, out of the critical region, $\tau_{\text{mixing}} \sim \ln N$ (Berestycki, Lubetzky, Peres, and Sly, 2018), while $\tau_{\text{cover}} \sim N(\ln N)^2$. At the critical point, the largest connected component is an equilibrium tree containing $N^{2/3}$ vertices, and so $\tau_{\text{mixing}} \sim N$ for the random walk on this component. In this situation, also, $\tau_{\text{cover}} \sim N$ (Barlow, Ding, Nachmias, and Peres, 2011).
- For a Barabási–Albert preferential attachment network,

$$\tau_{\text{cover}} \cong \frac{2m}{m-1} N \ln N, \tag{10.5}$$

where m is the number of connections of each new edge (Cooper and Frieze, 2007b).

Maier and Brockmann (2017) indicated an efficient way to compute numerically the cover time for an arbitrary finite graph.

Another widely studied quantity, the *coverage $C(t)$*, is the expected number of vertices of a graph visited at least once during t moves of a random walk and averaged over starting vertices. As an example, let us obtain the coverage for a complete graph of N vertices,

$$C(t) = N \left[1 - \left(1 - \frac{1}{N-1} \right)^t \right] \cong N(1 - e^{-t/N}), \tag{10.6}$$

for a large N and $t/N \to$ const. The coverage of an arbitrary undirected graph can estimated in the following way (Starnini, Baronchelli, Barrat, and Pastor-Satorras, 2012). In the steady state of a simple random walk, the walker moves to vertex i with probability $P_i(\infty)$ at any step. Let us approximate the corresponding probability for sufficiently large time by this value, $P_i(\infty)$. Then the probability $\mathcal{R}_i(t)$ that vertex i is visited at least one time by a random walk initiated at an arbitrary vertex, at any time less than or equal to t, is given by

$$\mathcal{R}_i(t) \approx 1 - [1 - P_i(\infty)]^t \cong 1 - e^{-P_i(\infty)t} \qquad (10.7)$$

for large t and a finite product $P_i(\infty)t$. This leads to the following expression for the coverage

$$\mathcal{C}(t) \approx N\left(1 - \frac{1}{N}\sum_i e^{-P_i(\infty)t}\right). \qquad (10.8)$$

10.1.2 Random walk centrality

While the mean first-return time $\langle t_{ii}\rangle$ for a vertex is determined only by a local property of a network, namely its degree, Eq. (10.2), the first-passage time $\langle t_{ij}\rangle$ depends on the structure of the entire network. There is a marked difference between random walks on regular and heterogeneous networks. For regular lattices and networks, the first-passage times are symmetric in the sense that $\langle t_{ij}\rangle = \langle t_{ji}\rangle$, while this is not the case for heterogeneous networks. Noh and Rieger (2004) derived the expressions for these mean times, valid for an arbitrary connected simple graph,

$$\langle t_{ij}\rangle = \begin{cases} \dfrac{\langle q\rangle N}{q_i} & \text{if } i = j, \\[2ex] \dfrac{\langle q\rangle N}{q_j}[R_{jj} - R_{ij}] & \text{if } i \neq j, \end{cases} \qquad (10.9)$$

where

$$R_{ij} = \sum_{t=0}^{\infty}[P_{ij}(t) - P_j(\infty)] \qquad (10.10)$$

and $p_{ij}(t)$ is the probability to find the walker at vertex j at time t if it started at vertex i at time $t = 0$ ($P_i(t) \equiv P_{ii}(t)$), for which the following equality is valid,[1]

---

[1]In the limit $t \to \infty$, this implies the equality

$$q_i P_j(\infty) = q_j P_i(\infty),$$

see Eq. (9.24).

$$q_i P_{ij}(t) = q_j P_{ji}(t). \tag{10.11}$$

Equations (10.9)–(10.11) lead to the equality

$$\langle t_{ij} \rangle - \langle t_{ji} \rangle = C_i^{-1} - C_j^{-1}, \tag{10.12}$$

where C_i is the *random walk centrality* (Noh and Rieger, 2004)

$$C_i \equiv \frac{P_i(\infty)}{\tau_i} = \frac{1}{\langle t_{ii} \rangle \tau_i}, \tag{10.13}$$

where the characteristic relaxation time τ_i for vertex i is

$$\tau_i \equiv \sum_{t=0}^{\infty} [P_i(t) - P_i(\infty)]. \tag{10.14}$$

The measure C_i quantifies the centrality of vertex i in respect of the quick access to information randomly spreading over the network. If two vertices, i and j, with, say, $C_i > C_j$, simultaneously launch two random walks, then, on average, the walker from j will reach i before the walker from i will visit j. Figure 10.1 shows the dependence of the characteristic relaxation time and the random walk centrality on the vertex degree in the Barabási–Albert model, demonstrating that, on average, $\langle C(q) \rangle \propto q$.

10.1.3 Return probability

Loosely speaking, in a network with a finite spectral dimension D_s, after a time t, a random walker browses a region of radius $r(t) \sim \sqrt{t}$ around the initial point of the walk, and so the number of accessible vertices $S(t)$ for a random walk of t moves is [2]

$$S(t) \sim t^{D_s/2}. \tag{10.15}$$

The probability of return to the origin at time t in an infinite network is inversely proportional to $S(t)$,

$$P(t) \equiv \frac{1}{N} \sum_i P_i(t) \sim t^{-D_s/2} \sim \frac{1}{S(t)}, \tag{10.16}$$

where $P_i(t)$ is the probability to find a walker at the initial vertex i at time t. According to Eq. (10.16), when $D_s > 2$, a random walk is *transient*,

---

[2] See Condamin, Bénichou, Tejedor, Voituriez, and Klafter (2007) for more detail and rigour.

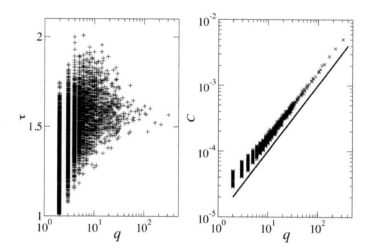

Fig. 10.1 Dependence of the characteristic relaxation time τ and the random walk centrality C on the vertex degree q in the Barabási–Albert network of 10^4 vertices where each new vertex attaches to two existing ones. The line has slope 1. Adapted from Noh and Rieger (2004).

which means that in an infinite network, there is a finite probability that the walk never returns to the initial vertex. If $D_s \leq 2$, then a random walk is *recurrent*, that is, it surely returns to the initial point.

In heterogeneous networks, the functional form of the decay of the probability $p_i(t)$ with time essentially depends on a vertex i having degree q_i. Hwang, Lee, and Kahng (2012a, 2014) studied this probability in scale-free networks (degree-distribution exponent γ) with a finite spectral dimension D_s, where this decay is power law and observed three behaviours:[3]

$$
P_i(t) \sim
\begin{cases}
t^{-D_s(\gamma-2)/[2(\gamma-1)]} & \text{for } t \ll q_i^{2(\gamma-1)/D_s}, \\[2mm]
q_i t^{-D_s/2} & \text{for } q_i^{2(\gamma-1)/D_s} \ll t \ll (\langle q \rangle N)^{2/D_s}, \\[2mm]
\dfrac{q_i}{\langle q \rangle N} & \text{for } t \gg (\langle q \rangle N)^{2/D_s}.
\end{cases}
\tag{10.17}
$$

Within the first region, the decay is slower than within the second. If the maximum vertex degree in the network is $q_{\max} \sim N^{1/(\gamma-1)}$, then the second

---

[3]Note that the spectral dimension of uncorrelated and many other networks is infinite and so the return probability in these networks decays faster than any power of t. See Samukhin, Dorogovtsev, and Mendes (2008) for the corresponding rapidly decaying asymptotics in diffusion on uncorrelated networks.

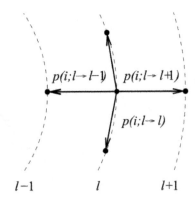

Fig. 10.2 The vertices in a network are additionally labelled according to their shortest-path distance ℓ from a target (vertex-attractor). In the biased random walk, the probability of a move from a vertex depends on the direction of this move. Namely, $p(i; \ell \to \ell - 1) > p(i; \ell \to \ell) > p(i; \ell \to \ell + 1)$.

regime for this hub disappears. If, in addition, the exponent γ of the degree distribution of this network is close to 2, then the probability $P_i(t)$ for the hub decays with t particularly slowly.

Using Eq. (10.17), one can obtain the average first passage time T_i of a target vertex i for random walks started from all other vertices in such a network (Hwang, Lee, and Kahng, 2012b):

$$
T_i \sim \begin{cases} N^{2/D_s} & \text{for } D_s < 2, \\ q_i^{-(1-2/D_s)(\gamma-1)} N & \text{for } 2 < D_s < 2(\gamma-1)/(\gamma-2), \\ q_i^{-1} N & \text{for } D_s > 2(\gamma-1)/(\gamma-2). \end{cases} \tag{10.18}
$$

For a hub of degree of the order of $N^{1/(\gamma-1)}$, this gives

$$
T_{\text{hub}} \sim \begin{cases} N^{2/D_s} & \text{for } D_s < 2(\gamma-1)/(\gamma-2), \\ N^{(\gamma-2)/(\gamma-1)} & \text{for } D_s > 2(\gamma-1)/(\gamma-2). \end{cases} \tag{10.19}
$$

The sublinear size dependence $T_{\text{hub}}(N)$, which is realized in the case of $D_s > 2$, means that a random walker usually reaches the hub without visiting all vertices in the network.

10.1.4 Biased random walks

In the unbiased simple random walks on undirected networks discussed, the probability p_i that a walker moves to a nearest neighbour of vertex i with degree q_i is the same for all the neighbours, $p_i = 1/q_i$. In general, a bias assumes that a set of probabilities for moves from neighbouring vertices, say the probability p_{ij} of the move from i to j, deviate from $1/q_i$. This bias can markedly change the random walk (Fronczak and Fronczak, 2009). Let us touch upon one of the versions of biased random walks in which the bias is towards a target vertex. The presence of this bias means that the probability of a move from a vertex in the direction of the target exceeds the probability of a move from this node in the opposite direction, (see Figure 10.2 explaining the notations for the probabilities).[4] Sood and Grassberger (2007) explored the interesting case of the exponential bias, that is the ratio of the probabilities was fixed:

$$\frac{p(i; \ell \to \ell - 1)}{p(i; \ell \to \ell)} = \sqrt{g} = \frac{p(i; \ell \to \ell)}{p(i; \ell \to \ell + 1)}, \tag{10.20}$$

where $g > 1$. This equality assumes that

$$\frac{p(i; \ell \to \ell - 1)}{p(i; \ell \to \ell + 1)} = g. \tag{10.21}$$

The idea was to start a random walk from a uniformly randomly chosen target vertex j, to obtain the mean first-return time for this vertex and then to average it over all positions of the target, namely, $T \equiv \sum_j \langle t_{jj} \rangle / N$. In particular, for an unbiased random walk, Eq. (10.2) provides the following expression for the average first-return time

$$T = \langle q \rangle \left\langle \frac{1}{q} \right\rangle N. \tag{10.22}$$

It turns out that for the biased random walk, the dependence $T(N)$ sharply changes at a certain, critical value of the bias parameter, g_c. This critical value exactly coincides with the mean branching in the network, $g_c = \langle b \rangle$. Assuming that a network is uncorrelated and $\langle b \rangle < \infty$, Sood and Grassberger (2007) and Bénichou and Voituriez (2007) obtained three distinct regimes:

$$T \sim \begin{cases} N^{\ln(\langle b \rangle / g)/\ln\langle b \rangle} & \text{for } 1 \le g < \langle b \rangle, \\ \ln N & \text{for } g = \langle b \rangle, \\ \xrightarrow{N \to \infty} \text{const} & \text{for } g > \langle b \rangle. \end{cases} \tag{10.23}$$

The regime below g_c can be called 'delocalization'. In this regime, in an infinite network, the walks are transient, and a walker escapes the bias. At

---

[4]We discuss only the important case of an attracting target.

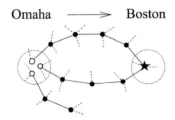

Fig. 10.3 How Stanley Milgram scanned a net of acquaintances in the United States. Notice that some chains of acquaintances were broken off.

the critical value g_c, a 'localization transition' occurs, and above this point, a walker is trapped by the bias. In this regime, the walks are recurrent, that is, most walks approach the target in a finite time. The localization transition disappears if $\langle b \rangle \sim \langle q^2 \rangle$ diverges, that is, hubs hinder this localization.

10.2   Greedy routing

In 1967 Stanley Milgram performed a seminal experiment for measuring distances in a network of acquaintances in the United States (Milgram, 1967; Travers and Milgram, 1969) (Figure 10.3). Milgram queried how many intermediate social links separated two randomly selected (and geographically separated) individuals. Milgram chose two locations: Omaha, Nebraska and Boston, Massachusetts.[5] A target person was chosen at random in Boston. A large enough number of randomly selected residents of Omaha received a letter with the following instructions:

  (i) If you know the target person 'on a personal basis' (his/her name and address were enclosed), send the letter directly to him/her.
 (ii) Otherwise mail a copy of this instruction to your 'personal' acquaintance (someone you know on a first name basis) who is more likely than you to know the target person.

An essential fraction of letters approached the target, after passing through only, on average, 5.5 social links. This is what is known as the 'six degrees of separation'.[6] Note that these six degrees are not the length of the shortest path between two persons in the network of acquaintances but rather a

---

[5]In fact, Milgram made two attempts. The first one, with starting points in Wichita, Kansas and a target person in Sharon, Massachusetts, resulted in only three finished chains, but the second attempt turned out to be more successful.

[6]Thirty-six years after Milgram his experiment was repeated on a greater scale by using the modern opportunity of email (Dodds, Muhamad, and Watts, 2003). Volunteers started 24,163 chains aimed at reaching 18 target persons in 13 countries. Only 384 (!) of

rough estimate from above. Indeed, what was the essence of Milgram's idea if we ignore less important details? The participants in the experiment were asked to forward a letter to those of their acquaintances who were closer to the target person. The targets address was known, and the 'closer', in this idealization, simply means 'geographically closer'. Each participant knew the addresses of his or her acquaintances, and so it was easy to select a proper recipient. In this search process (searching for the shortest route to the target), all participants used very reduced, local information, namely the addresses (geographic coordinates) of their acquaintances, and, of course, the target's address. The participants had no idea about the full structure of their network.

Milgram's algorithm is actually the standard one in computer science, belonging to the class of decentralized search algorithms. A number of routing algorithms exploit geographic information about vertices of communication networks (Karp and Kung, 2000). The simplest geographic routing implements the greedy routing algorithm assuming that:

(i) each vertex in a network has its geographic coordinate, and

(ii) a vertex forwards messages (packets) to that its nearest neighbour in the network, which is geographically closest to the destination.

In the greedy routing, the vertices use only local information, which simplifies the algorithm and makes it ultimately efficient and quick. Speaking more precisely, each vertex knows only: (i) its own geographic coordinates, (ii) the coordinates of its nearest neighbours in the network, and (iii) the coordinates of the destination. With each step, we are getting geographically closer to the destination. On the other hand, 'geographically closer' does not mean 'closer in the network'. So moving in this manner, we easily miss the shortest path to the target. We conclude that Milgram's six degrees of separation are simply the average delivery time (average hop-length) provided by the greedy algorithm for his specific problem. The average greedy routing time $\langle \tau \rangle$ for all source–destination pairs characterizes a network and its geographic embedding (unit time is one step of the algorithm). The question is: by how much does this average delivery time $\langle \tau \rangle$ exceed the shortest path length of the network, $\langle \ell \rangle$?

The greedy routing can be based, with a varying degree of efficacy, on embedding a network in various metric spaces, including Euclidean and

the chains were completed. On the other hand, the successful chains turned out to be an average of about four links, that is, even fewer than 'six degrees'.

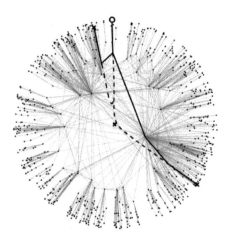

Fig. 10.4 A modelled network embedded in the hyperbolic plane, and greedy routing in it. The source is the circled vertex and the destinations are the crosses. The greedy paths are shown by the solid lines. The dashed curves are the geodesics (hyperbolically straight lines) between the source and the destinations. Adapted from Papadopoulos, Krioukov, Boguñá, and Vahdat (2010).

hyperbolic spaces. Such metric spaces behind a network may be hidden, and special embedding algorithms—computing each's vertex virtual coordinate in a metric space—were developed. See Kleinberg (2007), Boguná, Papadopoulos, and Krioukov (2010), and Serrano, Krioukov, and Boguñá (2008) for embedding a network in the hyperbolic plane which has an advantage of well fitting the structures of numerous scale-free and highly clustered real networks and their models. Although the greedy routing doesn't use the shortest paths between a source and a destination neither in the network nor in the embedding metric space (Figure 10.4) it provides a low-cost solution for traffic of packets without huge routing tables.

In addition to the average greedy routing time, two numbers, the success ratio and the average stretch of successful paths, are used to characterize the efficacy of the greedy routing. The point is that greedy routing may get stuck when there is a vertex closer to the destination than all its nearest neighbours. The *success ratio* p_s is the proportion of paths reaching their destinations. The *average stretch* $\langle \tau/\ell \rangle$ is the average ratio of the hop-length of a successful greedy routing path and the shortest path length between its source and destination.

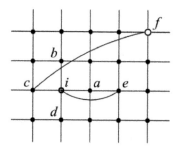

Fig. 10.5 How the greedy algorithm works in Kleinberg's network. Let the goal be to reach the destination (node f), starting from the source vertex i, in the shortest time. The greedy algorithm results in a four-step path passing through vertex e, while the shortest path through vertex c is of only two steps.

10.3 Navigability

The dependence of the average greedy routing time $\langle \tau \rangle$ on the size of a generalized small-world network was explored by Kleinberg (2000b, 2000a, 2006). The questions were how quickly we could find a target in a network by using the greedy algorithm? Or, equivalently, how easily could we navigate through a network? The version of a small-world network used by Kleinberg provided a range of network architectures controlled by a model parameter. The network was a D-dimensional lattice of $N = L \times L \times \ldots \times L$ vertices with added specifically distributed shortcuts (Figure 10.5). In this geometry, all greedy routing paths complete successfully, and the success ratio $p_s = 1$. For a lattice without shortcuts, clearly, $\langle \tau \rangle \sim L$, and it takes a lot of time to reach the target. Apparently, added shortcuts diminish delivery times, but by how much?

Originally, Kleinberg's network was based on a two-dimensional lattice substrate, but here we assume it to be D-dimensional. Each vertex of the lattice has a shortcut to a vertex at the Euclidean distance ℓ drawn from a power-law probability distribution, $p(\ell) \sim \ell^{-\alpha}$.[7] If exponent α equals zero, then shortcuts connect uniformly randomly chosen vertices, and hence we arrive at the standard version of a small-world network. If α is large (short-range shortcuts), then the network, in effect, approaches a D-dimensional lattice, and $\langle \tau \rangle \sim L$. Long-range shortcuts surely diminish the delivery time, but the degree of this decrease strongly depends on α. In particular,

---

[7]The number of shortcuts per vertex can be taken to be equal to an arbitrary finite number, without loss of generality.

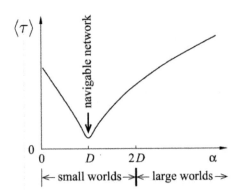

Fig. 10.6 Schematic plot of the average delivery time provided by the greedy algorithm versus exponent α according to Kleinberg. The size of the network is fixed. When $\alpha \leq 2D$, where D is the dimensionality of the underlying lattice, the network is a small world.

uniformly distributed shortcuts, $\alpha = 0$, give a small chance of getting closer to the target and so they they do not substantially improve navigation compared to a pure lattice. Kleinberg studied the size dependence $\langle \tau \rangle (L)$ in the entire range of exponent α values, from zero to infinity, and found dramatically different dependences. Figure 10.6 shows the resulting average delivery time vs. exponent α for a network of a given size. The main finding was that the delivery time of the greedy algorithm has a deep minimum at $\alpha = D$. Kleinberg proved that at this unique point, the size-dependence of the delivery time is very slow—polylogarithmic, $\langle \tau \rangle \sim (\ln L)^2$, while the delivery time increases much faster, as a power of L, $\langle \tau \rangle \sim L^x$, at all other values of α. Exponent x approaches zero as α tends to D. This sharp difference enabled Kleinberg to introduce the notion of *navigability*. In terms of Kleinberg, a network is navigable if the greedy algorithm provides rapid navigation, that is, if the function $\langle \tau \rangle (L)$ increases slower than any power law of L.

Notably, this network is a small world, that is infinite-dimensional, in a wider range of α than one would expect knowing that it is navigable at a single point. This takes place for $0 \leq \alpha \leq 2D$. In this region, except when $\alpha = D$, $\langle \tau \rangle (L)$ is much greater than the average shortest-path length if the network size is large. Compare the analytical expressions for asymptotic dependencies $\langle \tau \rangle (L)$ at various α (Carmi, Carter, Sun, and Ben-Avraham, 2009; Cartozo and De Los Rios, 2009):

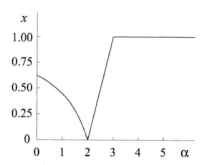

Fig. 10.7 Delivery time exponent x of the power-law dependence $\langle\tau\rangle \sim L^x$ versus exponent α for the two-dimensional Kleinberg model according to Carmi, Carter, Sun, and Ben-Avraham (2009) and Cartozo and De Los Rios (2009). The network is navigable at $\alpha = 2$.

$$
\tau \sim \begin{cases}
L^{(D-\alpha)/(D+1-\alpha)} & \text{for } 0 \le \alpha < D, \\
(\ln L)^2 & \text{for } \alpha = D, \\
L^{\alpha-D} & \text{for } D < \alpha < D+1, \\
L & \text{for } \alpha > D+1.
\end{cases} \tag{10.24}
$$

Figure 10.7 illustrates these formulae showing the dependence $x(\alpha)$ of the exponent in the power law $\langle\tau\rangle \sim L^x$.

In more realistic networks than Kleinberg's model, a fraction of greedy paths fail to reach their destinations, and the success ratio p_s is smaller than 1. Clearly, navigation in large network of N vertices is impossible if $p_s(N{\to}\infty) \to 0$. Consequently, navigability, in general, should assume not only that (i) greedy routs quickly reach their destinations, but also (ii) the condition $p_s(N{\to}\infty) = \text{const} > 0$. Loosely speaking, the 'quick greedy routing' means that the average greedy routing time $\langle\tau\rangle$ does not exceeds dramatically the average short-path length $\langle\ell\rangle$, like $\langle\tau\rangle \sim (\ln N)^2$ vs. $\langle\ell\rangle \sim \ln N$ in the navigability regime of Kleinberg's network. One can suggest that a network is navigable when an imbedding metric space well fits the network's structure, in other words, is 'congruent' with it. Notably, the region of navigability can occur wide, unlike a single point of navigability for Kleinberg's network. Boguñá, Krioukov, and claffy (2009) and Boguná, Papadopoulos, and Krioukov (2010) explored greedy routing on networks with a hidden hyperbolic space (hyperbolic plane) behind them and observed navigability within a broad range of network parameters. They inspected network models on the hyperbolic plane (Section 4.15) and real networks, for which embed-

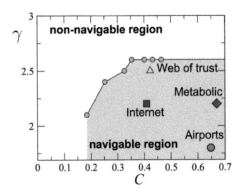

Fig. 10.8 Area (grey) on the plane 'clustering coefficient C—degree distribution exponent γ' in which success ratio remains finite in the infinite network size limit. The plot obtained for a model scale-free clustered network embedded in a hyperbolic plane. The points within the grey area indicate measured values of C and γ for a few real networks. Adapted from Boguñá, Krioukov, and claffy (2009).

ding in a hidden metric space was uncovered algorithmically, and in both cases observed quick greedy routing, $\langle \tau \rangle \sim (\ln N)^{\nu}$, where $\nu > 1$, and a large success ratio in a broad region (Figure 10.8). Thus the hyperbolic embedding well fits the structures of the explored networks. Furthermore, Boguná and Krioukov (2009) studied the greedy routing in ultra-small worlds—scale-free networks with exponent $\gamma < 3$ having an average shortest-path length $\langle \ell \rangle \sim \ln \ln N$—embedded in the hyperbolic plane, and obtained a remarkably short average delivery time $\langle \tau \rangle \sim \ln \ln N \sim \langle \ell \rangle$.

In social networks, navigability can be also treated as the possibility to quickly find a target—*searchability*. Watts, Dodds, and Newman (2002) proposed a model for a hierarchically organized social network demonstrating this property. Figure 10.9 explains their network and social distances between individuals. Acquaintances between individuals i and j are set with probability $p_{ij} \propto e^{-\alpha x_{ij}}$, where x_{ij} is the social distance between i and j and $\alpha \geq 0$ is a parameter of the model.[8] The complete model contains H such hierarchies, $h = 1, 2, \ldots, H$, and so an individual is labelled by a set of $H(l-1)$ coordinates. Two individuals are separated by the social distances $x_{ij}^1, x_{ij}^2, \ldots, x_{ij}^H$ within each of these hierarchies, and acquaintances emerge within hierarchies independently. The average degree of the resulting net-

---

[8]In essence, this model provides embedding of a network of acquaintances in a discrete metric space formed by a hierarchical tree.

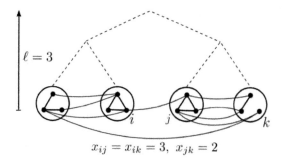

$$x_{ij} = x_{ik} = 3, \; x_{jk} = 2$$

Fig. 10.9 Hierarchically organized social network according to Watts, Dodds, and Newman (2002). Groups of $g = 3$ individuals belong to groups of $b = 2$ groups and so on, forming an ($l = 3$)-level hierarchy. Each individual i is described by a set v_i of $l-1$ its coordinates in this hierarchy. The social distance between two individuals in the same group is set to 1. A social distance x_{ij} between individuals i and j from different groups equals the height of their lowest common ancestor in the hierarchy. Acquaintances are shown by arcs. The probability that two individuals know each other is a function of the social distance between them.

work of acquaintances is set to $\langle q \rangle$. The social distance between individuals i and j accounting for all hierarchies is defined as $y_{ij} \equiv \max_{h} x_{ij}^{h}$.

This metric structure enables one to organize the greedy forwarding of messages to a target using only local information. In this model, message chains break due to a non-zero probability p to lose a message at each step. According to Watts, Dodds, and Newman, a network is searchable if the success ratio p_s exceeds a given threshold r. Simultaneously, this condition, effectively cutting the lengths of message chains, assumes a quick search. Indeed, for a message chain of length τ, the condition $p_s > r$ gives $p_s = \langle (1-p)^{\tau} \rangle > r$, which provides approximately $\langle \tau \rangle < \ln r^{-1} / \ln(1-p)^{-1}$. Thus, in a searchable network, if r and p are fixed, then the average length of a message chain is independent of the population size N. Figure 10.10 shows a large region on the $H-\alpha$ plane in which a network is searchable.

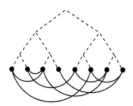

The groups of g individuals can be treated as the bottom branches with g leaves in this tree. See a similar model in Section 13.2.

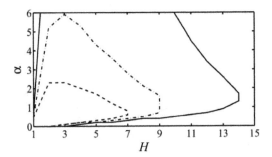

Fig. 10.10 Areas on $H-\alpha$ plane where model networks of different size are searchable. Adapted from Watts, Dodds, and Newman (2002). The solid, dot-dashed, and dashed lines are the boundaries of the searchable network region for population sizes $N = 102{,}400$, $204{,}800$, and $409{,}600$, respectively. The network parameters are $g = 100$ (size of each group), $\langle q \rangle = 99$ (average number of acquaintances of an individual), and $b = 2$ (branching of each hierarchy tree). At each step, a message is lost with probability $p = 0.25$. The seachability threshold is $r = 0.05$.

Notice that even a network based on a single hierarchy, $H = 1$, appears to be searchable for sufficiently small α.

10.4 Google PageRank

Let us touch upon a historically important part of Google's technology,[9] namely the ranking of results of search queries—the *Google PageRank* (Brin and Page, 1998; Page, Brin, Motwani, and Winograd, 1999). A search query is a set of words and numbers. The result of a query is a ranked list of the addresses of pages containing this combination. Without ranking, a huge array of information would be virtually useless for a user. The PageRank algorithm estimates the relative importance of a web page based on its popularity in the WWW. In this way PageRank ranks all pages in the WWW. After computation, this global ranking is used for the ranking of entries in the list of results of each search query independently on the contents of this query. How does the Google PageRank work? The key idea is that the popularity of a web page is proportional to the number of times a crazy web-surfer visits this page when randomly surfing the WWW over an infinite time period. So the problem is essentially reduced to a random walk

---

[9]See a more detailed discussion 'How Google works' on http://www.googleguide.com/google_works.html. At the present time, the Google PageRank is actually only one of about two hundred ranking criteria used together for attaining the final ranking by modern search engines.

on a directed network. In one respect, this random walk differs from the one on an undirected network. In directed networks, there may be clusters connected to the remaining part of a network only by incoming connections. If a walker moves only following directed links they will be trapped finally in one of these clusters. To avoid this, we have to allow our imaginary surfer to restart the process, say, from a random vertex. It is easy to show that the vector of final stationary probabilities of finding the surfer on vertex i, $i = 1, \ldots, N$, at infinite time is proportional to the vector $\tilde{\mathbf{r}}$ with elements satisfying the following set of equations:

$$\tilde{r}_i = \frac{1-\alpha}{N} + \alpha \sum_{j \in \partial i} \frac{\tilde{r}_j}{q_{\text{out},j}} = \frac{1-\alpha}{N} + \alpha \sum_{j:\, q_{\text{out},j} \neq 0} \frac{(A)_{ji}}{q_{\text{out},j}} \tilde{r}_j, \qquad (10.25)$$

Here N is the size of a network, $q_{\text{out},j}$ is the out-degree of vertex j, the sum $\sum_{j \in \partial i}$ is over all incoming connections of vertex i, and finally $1 - \alpha$ is the probability that, instead of moving to one of its nearest neighbours, the walker jumps to a node chosen uniformly at random. The complementary probability $\alpha < 1$, which is the only parameter of this algorithm, was claimed to be about 0.85. It is clear why only out-degrees are present in the sum on the right-hand side of this equation. Indeed, a walker escapes from a vertex along each of the outgoing edges with equal probability. Equations (10.25) are solved numerically by iteration starting from arbitrary initial conditions. This is a simple task even for large networks since these iterations rapidly converge. The sum of the elements of the resulting vector $\tilde{\mathbf{r}}$ may occur smaller than 1 in some networks, so in general $\sum_i \tilde{r}_i \leq 1$, and, strictly speaking, $\tilde{r}_i$ obtained from Eq. (10.25) is not a probability but only proportional to it.

If we wish to describe the temporal evolution of the real probability $r_i(t)$ to find this random walker on vertex i we should write the following equations for this Markov process

$$r_i(t+1) = (1-\alpha)\frac{1}{N} + \alpha \sum_{j:\, q_{\text{out},j} \neq 0} \frac{(A)_{ji}}{q_{\text{out},j}} r_j(t) + \frac{\alpha}{N} \sum_{j:\, q_{\text{out},j}=0} r_j(t) \qquad (10.26)$$

supplied by the initial condition $\sum_i r_i(0) = 1$. Notice the third term on the right-hand side that accounts for the walkers hopping from vertices of zero out-degree to random vertices. Check that $\sum_i r_i(t) = \text{const}$ in this process as is should be. The final steady state $\mathbf{r}(t \to \infty)$ is proportional to $\tilde{\mathbf{r}}$ obtained from Eq. (10.25).

Instead of solving Eq. (10.26), equivalently, one can find the eigenvector $\mathbf{r}_1$ corresponding to the principal eigenvalue λ_1 of the matrix

$$\mathcal{G}_{ij} \equiv (1-\alpha)\frac{1}{N} + \alpha[1-\delta(q_j,0)]\frac{(A^T)_{ij}}{q_j} + \alpha\delta(q_j,0)\frac{1}{N}, \tag{10.27}$$

which is called the *Google matrix*, for detail see Ermann, Frahm, and Shepelyansky (2015),

$$\mathcal{G}\mathbf{r} = \lambda\mathbf{r}. \tag{10.28}$$

One can see that the spectral radius $\lambda_1 = 1$, and for $\alpha < 1$ this eigenvalue is not degenerate.[10] The corresponding eigenvector $\mathbf{r}_1$, irrespectively of normalisation, is called the *PageRank centrality*.

In directed uncorrelated networks, each vertex is completely characterized by its in- and out-degrees. So for such networks, the PageRank of a vertex is described by the probability distribution $P(r; q_{in}, q_{out})$, where q_{in}, q_{out} are the in- and out-degrees of this vertex. Fortunato, Boguñá, Flammini, and Menczer (2006) obtained the first and second moments of this distribution. The first moment, that is the average PageRank of a vertex with q_{in}, q_{out} equals

$$\langle r \rangle(q_{in}, q_{out}) = \frac{1-\alpha}{N} + \frac{\alpha}{N}\frac{q_{in}}{\langle q_{in} \rangle}, \tag{10.29}$$

which properly satisfies the equality $N \sum_{q_{in},q_{out}} P(q_{in}, q_{out})\langle r \rangle(q_{in}, q_{out}) = 1$, while the standard deviation is

$$(\langle r^2 \rangle - \langle r \rangle^2)(q_{in}, q_{out}) = \frac{\alpha^2}{N^2\langle q_{in} \rangle^2}\frac{\frac{1}{\langle q_{in} \rangle}\left\langle \frac{[(1-\alpha)\langle q_{in} \rangle + \alpha q_{in}]^2}{q_{out}} \right\rangle - 1}{1 - \frac{\alpha^2}{\langle q_{in} \rangle}\left\langle \frac{q_{in}}{q_{out}} \right\rangle} q_{in}. \tag{10.30}$$

Notice that $\langle r \rangle$ is determined only by the in-degree of a vertex depending on it linearly, and moreover, the standard deviation is proportional to q_{in}.

---

[10]For positive matrices, the inequalities $\min_i \sum_j (X)_{ij} \leq \lambda_1 \leq \max_i \sum_j (X)_{ij}$ hold, which gives the spectral radius $\lambda_1 = 1$ for the Google matrix having $\sum_j (G)_{ij} = 1$.

11

Temporal Networks

11.1 The concept of a temporal network

When a process takes place on an evolving network or this network serves as an evolving substrate of a dynamical system, two time scales naturally emerge: (i) the shortest time of structural changes in a local neighbourhood of each vertex, and (ii) the shortest time (time step) of a process. The notion of a *temporal network* assumes that local structural changes in an evolving network occur faster than the time step of a process or that these two time scales are comparable.[1] The simplest example of such structural changes is sufficiently frequent emergence and disappearance of edges in a network. A standard example of a process on a network is a random walk, whose shortest time scale is the minimal time a walker stays on a vertex between two moves. Loosely speaking, a temporal network changes locally faster than a process on it or with equal speed. Still, this state of a network can be steady.

[1]The term a 'time-varying network' is also in common use.

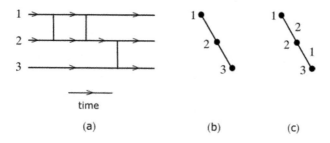

time

(a)   (b)   (c)

Fig. 11.1 (a) Event representation of a small temporal network of three vertices, 1, 2, and 3. Events have zero duration. The arrows on the time trajectories for vertices indicate causality. There is a temporal walk from vertex 1 to vertex 3 while there is no temporal walk from vertex 3 to vertex 1. (b) Unweighted and (c) weighted time-aggregated counterparts of this temporal network.

Using the common language of this research field, temporal edges (acts of interaction between two vertices) are treated as *events* having specified durations and separated in time, where time is continuous (Holme and Saramäki, 2012; Holme, 2015; Masuda and Lambiotte, 2016). In this representation, a temporal network is defined by a set of N vertices, $i = 1, 2, \ldots, N$, and the time-ordered list of events $(e_1, e_2, \ldots, e_N)$, where each event e_a, $a = 1, 2, \ldots, N$, is described by four numbers: $e_a = (i_a, j_a, t_a, \tau_a)$, indicating two vertices, i_a and j_a, between which an edge exists during the time interval $[t_a, t_a + \tau_a)$. Note that one pair of vertices can appear in multiple events. When the durations of all events tend to 0, then each event e_a can be described by three numbers, $e_a = (i_a, j_a, t_a)$, and the network is often called an *instant-event temporal network*. Figure 11.1 graphically demonstrates an instant-event temporal network and explains two versions—unweighted and weighted—of the corresponding 'static' *aggregated network*. For a temporal network with events having finite durations, one can introduce an aggregated network in which an edge weight is a fraction of time taken by the full sequence of events between two vertices. If events for each involved pair of vertices happen with diverging frequency, then this version of an aggregated network can be substituted for a temporal network as an effective substrate of processes and dynamical models.

Furthermore, Figure 11.1 demonstrates that walks in temporal networks are also temporal ('causal') and they may appear to be effectively directed even in undirected temporal networks. In this example, there is a temporal walk from vertex 1 to vertex 3 while there is no such a walk from vertex 3 to vertex 1. Based on this directedness, one can introduce temporal directed connected components (Nicosia, Tang, Musolesi, Russo, Mascolo, and Latora, 2012, 2013) resembling those in 'static' networks (Section 6.6). It is natural to treat the time trajectories of vertices in the event representation of a temporal network (Figure 11.1a) as unidirected bonds in an extra space dimension. This enabled Parshani, Dickison, Cohen, Stanley, and Havlin (2010) to reduce the problem to well-developed classical directed percolation of condensed matter theory, with specific percolation transition and critical phenomena.

In another representation, a temporal network is a sequence of periodically taken instantaneous snapshots—graphs (Krings, Karsai, Bernhardsson, Blondel, and Saramäki, 2012). This picture provides a natural mapping of a temporal network to a multilayer one.

In the null model of an uncorrelated temporal network, events connecting different pairs of vertices happen independently, and a sequence of events for each individual pair is a random process, for example, a Poisson having an exponential distribution of inter-event times. For each two vertices, i and j, one can introduce the activity rate a_{ij} showing the frequency of interactions between these two vertices, and then the temporal network is defined by the full set of activity rates for all $N(N-1)/2$ pairs of vertices. Furthermore, one can fix a given aggregated network and generate independent random sequences of events only for those pairs of vertices that are adjacent in the aggregated network. Masuda and Lambiotte (2016) call this model the *stochastic temporal network model*. The reader can find a more advanced statistical ensemble view of temporal networks in Gauvin, Génois, Karsai, Kivelä, Takaguchi, Valdano, and Vestergaard (2018).

An alternative, *vertex activity driven model* of temporal networks was proposed by Perra, Gonçalves, Pastor-Satorras, and Vespignani (2012), but see also Liu, Perra, Karsai, and Vespignani (2014). In this model, each vertex from time to time becomes active in line with a given pattern of activity and generates a number of temporal links to other vertices. The activity patterns of vertices can significantly differ from each other, generating diverse network structures.

A simple version of this model, particularly convenient for analytical treatment, is described by the following rules. Each vertex i of N vertices in the network has its activity rate a_i drawn from a given probability distribution $\mathcal{F}(a)$, so that the average number of active vertices per unit time in the network equals $\langle a \rangle N$. The network undergoes a sequence of instant changes separated by intervals of length Δ (time steps). This parameter controls the speed of structural changes. The configurations of the network within individual intervals are its snapshots. The evolution starts with N isolated vertices:

(i) At each time step, each vertex i with probability $a_i \Delta$ becomes active and creates m edges between itself and m other uniformly randomly chosen vertices.

(ii) At the next time step, all edges are removed, and the process repeats from step (i).

The average degree of a vertex in a snapshot equals $2m\langle a \rangle \Delta$. If $\langle a \rangle$ and Δ are sufficiently small, then the snapshots are sets of stars, rarely overlapping with each other. If we aggregate the network during a long time $T \gg \Delta$, then the vertices will accumulate the average degree $2m\langle a \rangle T$. Moreover, one can show that the degree distribution $P(q; T)$ of this aggregated network is determined by the probability distribution $\mathcal{F}(a)$,

$$P(q;T) \sim \frac{1}{Tm} \mathcal{F}\left(\frac{q}{Tm}\right) \qquad (11.1)$$

for $q \ll N$. In the next sections, we consider spreading phenomena in networks provided by these models.

The structures of aggregated networks keep a significant piece of information about the full temporal networks. Masuda, Klemm, and Eguíluz (2013) compared the Laplacian spectra of temporal networks obtained in the framework of the snapshot representation with their aggregated counterparts.[2] They observed that these spectra—their eigenvectors and eigenvalues—are surprisingly closely related. Eigenvalues of the spectrum of a temporal network are smaller than the corresponding eigenvalues for the aggregated network. As one could expect, this slows down diffusion in temporal networks compared to aggregated ones, slows down relaxation in dynamical systems, and hinders synchronization. On the other hand, Li, Cornelius, Liu, Wang, and Barabási (2017) showed, rather counterintuitively, that linear dynamical systems on temporal networks can be controlled more easily and rapidly than on the corresponding aggregated networks. The key reason for all these effects is that a temporal network is less connected than its aggregated counterpart. Indeed, any two vertices of a temporal network cannot be better interconnected than the corresponding vertices in the aggregated net due to the directedness of temporal (causal) walks and paths and their finite duration.

11.2 Random walks on temporal networks

Let us consider a version of a continuous random walk on a temporal network. Figure 11.2 explains this random walk—the *passive random walk* (Speidel, Lambiotte, Aihara, and Masuda, 2015). A walker moves from a vertex to a vertex only using events connecting them. Between such events, it stays at a vertex waiting for the next event. The resulting walk is completely determined by the history of a temporal network, in other words, a given realization of a temporal network, and by a starting vertex.

We assume first that a temporal network is provided by the stochastic temporal network model in which events of zero duration are generated by the Poisson processes with the same event rate $\mu = 1/\langle \tau \rangle$ for each of the

---

[2]See Masuda, Klemm, and Eguíluz (2013) for details on the effective Laplacian matrix of a temporal network. See also Section 11.3.

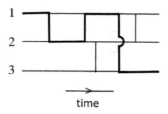

Fig. 11.2 'Passive' random walk from vertex 1 to vertex 3 on an instant-event temporal network. A walker uses relevant events to move from a vertex to a vertex. Between such events, it stays at a vertex waiting for the next event.

edges of the aggregated network, where $\langle \tau \rangle$ is a mean inter-event time for each pair of the adjacent vertices in the aggregated net. Hoffmann, Porter, and Lambiotte (2013) and Speidel, Lambiotte, Aihara, and Masuda (2015) showed that the random walk on this network is particularly simple, and it can be exactly described by the following master equation for the probability $P_i(t)$ that a walker is at vertex i at time t:

$$\frac{dP_i(t)}{dt} = \sum_{j \in \partial i} \left[\frac{1}{\langle \tau \rangle} P_j(t) - \frac{1}{\langle \tau \rangle} P_i(t) \right], \tag{11.2}$$

where the notation $j \in \partial i$ indicates that vertex j is the nearest neighbour of vertex i in the aggregated network, and $1/\langle \tau \rangle$ plays the role of the transition rate for a single walker to move from i to j (or from j to i). If the aggregated network is connected, that is, consists of a single connected component, then the stationary solution of Eq. (11.2) is uniform,

$$P_i(\infty) = \frac{1}{N}, \tag{11.3}$$

where N is the number of vertices in a network. Compare this uniform steady-state distribution $P_i(\infty) = 1/N$ with the corresponding non-uniform steady state of a simple random walk on a 'static' network, Eq. (9.24). The uniform steady-state distribution, Eq. (11.3), results in the following mean time $\mathcal{T}_i$ for which a walker stays at vertex i waiting for the next event,

$$\mathcal{T}_i = \frac{\langle \tau \rangle}{q_i}, \tag{11.4}$$

where q_i is the degree of vertex i in the aggregated network. The derivation of the mean recurrence time $\langle t_{ii} \rangle$ of this walk for vertex i is more demanding (Speidel, Lambiotte, Aihara, and Masuda, 2015). The expression for the mean recurrence time, resembling Eq. (10.2) for a simple random walk, is given by

$$\langle t_{ii} \rangle = \frac{N \langle \tau \rangle}{q_i}. \tag{11.5}$$

In this sense, the passive random walk on this stochastic temporal network exhibits a very similar dynamic to the simple random walk on the corresponding aggregated network.

A situation appears to be more complicated for stochastic temporal networks generated by non-Poisson processes. Speidel, *et al.* showed that in this case the expressions for the steady state probabilities $P_i(\infty)$ and the mean recurrence times, Eqs. (11.3) and (11.5), are fulfilled only approximately. It is worthwhile to explain the reason for this difficulty. For a general stochastic process with a distribution $\mathcal{P}(\tau)$ of inter-event times, the average waiting time $\langle \tau_{\rm w} \rangle$ from a uniformly randomly selected instant till the closest event is greater than one would expect,

$$\langle \tau_{\rm w} \rangle = \frac{\langle \tau^2 \rangle}{2 \langle \tau \rangle}, \tag{11.6}$$

namely, $\langle \tau_{\rm w} \rangle > \langle \tau \rangle / 2$, which is the famous waiting-time paradox (inspection paradox, bus paradox) of queueing theory.[3] The theory of the passive random walk on a temporal network uses the average waiting time of a walker at a vertex and the distribution of its waiting times. The problem is that when events are generated by non-Poisson stochastic processes, then different trajectories (histories) of a random walker lead to different average waiting times that it spends at a vertex. If a walker moves, say, by the trajectory $1 \to 2 \to 1$ (first and second moves in Figure 11.2), then its average waiting time at vertex 2 between these moves is $\langle \tau \rangle$, that is, the average time

---

[3]The distribution $\mathcal{P}_{\rm w}(\tau)$ of waiting time and the average waiting time, Eq. (11.6), can be derived in the following way (Masuda and Hiraoka, 2020). Let us select at random an instant t^*. It falls in an interval of length τ with the probability density

$$p^*(\tau) = \frac{\tau \mathcal{P}_{\rm w}(\tau)}{\langle \tau \rangle}.$$

If t^* falls in an interval of length τ, then the uniform probability density that the waiting time is t equals $1/\tau$ for $0 \leq t \leq \tau$. This gives

$$\mathcal{P}_{\rm w}(\tau) = \int_t^\infty d\tau \frac{1}{\tau} p^*(\tau) = \frac{1}{\langle \tau \rangle} \int_t^\infty d\tau \mathcal{P}(\tau),$$

which readily leads to Eq. (11.6). The distributions $\mathcal{P}_{\rm w}(\tau)$ and $\mathcal{P}(\tau)$ coincide only for the Poisson process, where $\mathcal{P}_{\rm w}(\tau) = \mathcal{P}(\tau) = \frac{1}{\langle \tau \rangle} e^{-\tau / \langle \tau \rangle}$, and $\langle \tau_{\rm w} \rangle = \langle \tau \rangle$.

between two events. On the other hand, if a walker moves by the trajectory $2 \rightarrow 1 \rightarrow 3$ (second and third moves in Figure 11.2), then its average waiting time at vertex 1 equals $\langle \tau^2 \rangle / (2\langle \tau \rangle) \neq \langle \tau \rangle$ according to Eq. (11.6). These two times (and two corresponding waiting time distributions) coincide only if events are generated by a Poisson processes. Otherwise this random walk is non-Markovian and its analytical treatment is involved. A particularly strong difference between $\langle \tau^2 \rangle / (2\langle \tau \rangle)$ and $\langle \tau \rangle$ occurs when the inter-event time distribution $\mathcal{P}(\tau)$ is fat-tailed, for example, power law, that is, when event sequences are 'bursty'. In this situation, $\langle \tau^2 \rangle / (2\langle \tau \rangle) \gg \langle \tau \rangle$, and the walker strongly tends to return to the previous vertex from which it came, as in the trajectory $1 \rightarrow 2 \rightarrow 1$ in Figure 11.2. This tendency effectively slows down the random walk process.

One of the difficulties that researchers face while studying processes on real temporal networks concerns the finite duration of empirical datasets with contact (event) sequences. This duration is often too short to allow a process to reach its asymptotic regime. Starnini, Baronchelli, Barrat, and Pastor-Satorras (2012) performed a set of synthetic extensions of empirical contact sequences to overpass this limit (sequence replication, sequence randomization, etc.). Some of the features of random walks on temporal networks generated in this way appeared to be noticeably different from theoretical predictions. In particular, the mean first passage time $\langle t_{ii} \rangle$ decayed slower with a vertex strength s_i than in Eq. (11.5).[4]

11.3   Epidemic spreading

This section discusses how the rapidly varying connections of temporal networks affect the epidemic threshold of the SIS model. As a starting point, let us apply the heterogeneous mean-field approximation to the SIS spreading process on a network generated by the vertex activity driven model (Section 11.1). This problem was considered by Perra, Gonçalves, Pastor-Satorras, and Vespignani (2012), and we follow their reasoning. Introducing the number $N_a(t)$ of vertices having activity rate a, where $\int da N_a(t) = N$, and the number $I_a(t)$ of infective vertices with activity rate a, one can write the following evolution equation of the heterogeneous mean-field theory:

---

[4]Note that Starnini, *et al.* used a weighted version of an aggregated network with edge weights proportional to a total time of interactions between their end vertices.

$$I_a(t + \Delta) - I_a(t) = -\mu \Delta I_a(t) + \beta m [N_a(t) - I_a(t)] a \Delta \int db \, \frac{I_b(t)}{N}$$

$$+ \beta m [N_a(t) - I_a(t)] \int db \, \frac{I_b(t)}{N} b \Delta, \quad (11.7)$$

where β and μ are infection and recovery rates in the SIS model, respectively. The second term on the right-hand side accounts for the probability that an active susceptible vertex receives a connection from an arbitrary infective vertex. The third term takes into account the probability that an arbitrary susceptible vertex (active or inactive), receives a connection from an infected active vertex. Let us introduce the fraction of infective vertices in the network at time t, $\rho(t) \equiv \int da I_a(t)/N$, and the quantity $w(t) \equiv \int da I_a(t) a/N$. First apply $\int da/N$ to both sides of Eq. (11.7). Neglect small terms, including the terms of the order of $1/N$, keeping in mind that we only need a linear equation to get the epidemic threshold. Then apply $\int da \, a/N$ to both sides of Eq. (11.7), and perform the same approximations. The result are two equations for $\rho(t)$ and $w(t)$,

$$\rho(t + \Delta) - \rho(t) = (-\mu + \beta m \langle a \rangle) \rho(t) + \beta m w(t)$$

$$w(t + \Delta) - w(t) = \beta m \langle a^2 \rangle \rho(t) + (-\mu + \beta m \langle a \rangle) w(t). \quad (11.8)$$

Assuming that the lifetime T of the temporal network is large, the epidemic threshold is determined by the condition that the largest eigenvalue of the Jacobian matrix of these equations equals 0. This results in the epidemic threshold

$$\frac{\beta_c}{\mu} = \frac{1}{m(\langle a \rangle + \sqrt{\langle a^2 \rangle})}. \quad (11.9)$$

As is natural, this expression does not contain the time interval Δ. Indeed, the heterogeneous mean-field theory effectively replaces a network by its annealed counterpart. For a temporal network, this suggests that the heterogeneous mean-field approximation is valid only if structural changes occur quite rapidly compared to the epidemic process, which corresponds to the limit $\Delta \to 0$. One can show that the inverse number, $(\beta_c/\mu)^{-1} = m(\langle a \rangle + \sqrt{\langle a^2 \rangle})$, approximately coincides with the largest eigenvalue of the annealed counterpart of the temporal network—the weighted aggregated network (Speidel, Klemm, Eguíluz, and Masuda, 2016). This weighted network is obtained in the following way. The edges between each two vertices are aggregated during a long period T, and the result is divided by the total number T/Δ of the snapshots forming the temporal network.[5]

To advance beyond the annealed network regime for epidemic spreading, one has to apply a more fine quenched mean-field approximation (Section 7.4). This was done by Valdano, Ferreri, Poletto, and Colizza (2015) for the discrete-time SIS model and by Speidel, Klemm, Eguíluz, and Masuda (2016, 2017) for the continuous-time SIS model. These two approaches are similar; we follow Speidel, *et al.*

In the snapshot representation, a temporal network is a sequence of snapshots. Let the adjacency matrix of the k-th snapshot be A_k, where $k = 1, 2, \ldots, n = T/\Delta$. It describes the 'static' configuration of a temporal network within the time window $(k-1)\Delta \le t < k\Delta$. Defining n matrices S_k for the snapshots as

$$S_k \equiv -\mu I + \beta A_k, \quad k = 1, 2, \ldots, n, \tag{11.10}$$

where I is the identity matrix, one can represent the linearized evolution equation of the quenched mean-field theory, Eq. (7.23), in matrix form within each individual time window:[6]

$$\frac{d}{dt}\mathbf{x}(t) = S_k\mathbf{x}(t), \quad \text{for } (k-1)\Delta \le t < k\Delta, \tag{11.11}$$

where $\mathbf{x}(t)$ is an N-vector each of whose components is the probability $x_i(t)$ that vertex i is infected at time t. The epidemic threshold $\beta_c^{(k)}$ of the SIS model on an individual k-th snapshot graph is provided by the following condition: the largest eigenvalue of the matrix S_k must be 0, that is $-\mu + \beta_c^{(k)}\lambda_1^{(k)} = 0$, where $\lambda_1^{(k)}$ is the largest eigenvalue of the adjacency matrix A_k. Using the solution of this linear equation, see Eq. (9.17), one can obtain the vector $\mathbf{x}(k\Delta)$, knowing $\mathbf{x}((k-1)\Delta)$, namely

$$\mathbf{x}(k\Delta) = e^{S_k\Delta}\mathbf{x}((k-1)\Delta). \tag{11.12}$$

For the lifetime $T = n\Delta$, this yields

$$\mathbf{x}(T) = \mathbf{x}(n\Delta) = W\mathbf{x}(0), \tag{11.13}$$

[5]Note that this annealed counterpart of a temporal network should be distinguished from the standard annealed network whose edge weights are determined by the degrees of their end vertices (Section 4.5). This explains why the expression for the epidemic threshold, Eq. (11.9), markedly differs from the corresponding result of the heterogeneous mean-field theory for 'static' networks, Eq. (7.17).

[6]We consider the linearized equation, since we focus on the epidemic threshold.

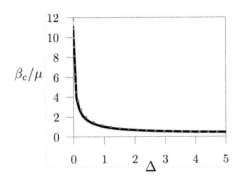

Fig. 11.3 Epidemic threshold for the vertex activity driven model versus the length Δ of the time intervals within which a temporal network is frozen. The solid line shows the result of the numerical solution of Eq. (11.15). The dotted line is calculated using the expression in Eq. (11.16). The parameters of the model are $N = 2000$, $m = 15$, the activity distribution for each vertex is $F(a) \propto a^{-3}$ with the coefficient and the range adjusted to give $\langle a \rangle = 0.0025$. Adapted from Speidel, Klemm, Eguíluz, and Masuda (2016).

where the matrix W is defined as

$$W(\Delta, \beta, \mu) \equiv e^{S_n \Delta} e^{S_{n-1} \Delta} \dots e^{S_1 \Delta}. \tag{11.14}$$

Equation (11.13) provides the threshold condition

$$\lambda_1^{(W)} = 1, \tag{11.15}$$

where $\lambda_1^{(W)}$ is the largest eigenvalue of the matrix W (Valdano, Ferreri, Poletto, and Colizza, 2015). One can check that this condition allows one to recover both limits, namely, of a 'static' network (single snapshot) and of an annealed one (small Δ and small rates β and μ providing a slow spreading process compared to the network dynamics). Importantly, Speidel, *et al.* showed that in this approximation, the epidemic threshold of a temporal network cannot exceed the epidemic threshold of its weighted aggregated counterpart with the adjacency matrix $A^{(a)} = \sum_{k=1}^{n} A_k / n$.[7,8]

[7] In the limit $n \to \infty$, this weighted network is the annealed counterpart of a temporal network. Thus in this limit, the epidemic threshold obtained by using the quenched mean-field approximation cannot exceed the epidemic threshold indicated by the heterogeneous mean-field theory.

[8] Essentially the same chain of reasoning, Eq. (11.11) → Eq. (11.12) → Eq. (11.13) → Eq. (11.14), with the Laplacian matrices L_k of the snapshots substituted for $-S_k$ enabled Masuda, Klemm, and Eguíluz (2013) to conclude that the spectral gap in the Laplacian spectrum of a temporal network is smaller than for the weighted aggregated network $A^{(a)}$.

The interested reader should refer to the papers of Valdano, *et al.* and Speidel, *et al.* for the numerical analysis and consequences of the condition in Eq. (11.15). In the important case of the vertex activity driven model, Eq. (11.15) can be solved analytically, yielding the epidemic threshold

$$\frac{\beta_c}{\mu} \approx \frac{1}{\Delta\sqrt{m}} \ln\left[1 + \frac{e^{\Delta} - 1}{\sqrt{m}(\langle a \rangle + \sqrt{\langle a^2 \rangle})}\right] \qquad (11.16)$$

(Speidel, Klemm, Eguíluz, and Masuda, 2016), where $n \to \infty$ and the temporal network is assumed to be so sparse that stars generated by active vertices rarely touch each other. For small Δ, Eq. (11.16) reduces to the heterogeneous mean-field result, Eq. (11.9). For large Δ, Eq. (11.16) leads to the epidemic threshold $\beta_c/\mu = 1/\sqrt{m}$, whose form becomes clear if we recall that the largest eigenvalue of the adjacency matrix of a single m-star equals $\sqrt{m}$ (Section 2.9). Between these two limits, namely of the annealed network and of the 'static' one, the epidemic threshold decays with increasing Δ (Figure 11.3).

12

Cooperative Systems on Networks

Many of the problems considered in this book are equivalent to models of interacting spins or agents. For instance, the bond percolation problem is equivalent to the 1-state Potts model (Kasteleyn and Fortuin, 1969; Kasteleyn and Fortuin, 1969), the k-core problem is equivalent to the ferromagnetic Ising model in a heterogeneous magnetic field (Section 6.9), and so on. This chapter touches upon a few basic cooperative models demonstrating various behaviours.

12.1   The Ising model

The Hamiltonian of the Ising model with pairwise interaction between the neighbouring spins has the form:

$$\mathcal{H} = -\sum_{i<j} J_{ij} S_i S_j - \sum_i H_i S_i, \tag{12.1}$$

where each spin can be in two states, $S_i = \pm 1$, and H_i is a local magnetic field acting on spin S_i, $i = 1, 2, \ldots, N$. We consider the ferromagnetic Ising model that has positive couplings J_{ij} between spins S_i and S_j. When these spins are placed on the vertices of a simple undirected graph with an adjacency matrix A, the coupling matrix $\mathcal{J}$ with the entries $A_{ij} J_{ij}$ represents a weighted graph. In the simplest situation, all couplings J_{ij} are equal, $J_{ij} = J > 0$.

Since we mainly focus on infinite-dimensional networks with few cycles, let us apply the belief propagation techniques (algorithm) as appropriate in such situations (Mezard and Montanari, 2009). This techniques is similar to message passing for percolation considered in Section 6.4. Consider

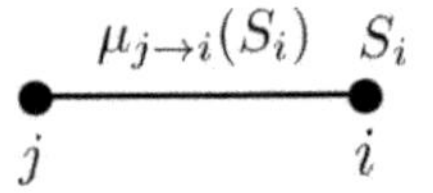

$$\mu_{j \to i}(S_i) \quad S_i$$

Fig. 12.1 Message from vertex j to vertex i.

an undirected simple graph (single realization) with the Ising spins on its vertices. For each edge (i, j), let $\mu_{j \to i}(S_i)$ and $\mu_{i \to j}(S_j)$ be two messages from vertex j to vertex i and from i to j, respectively, having the following meaning. For example, in the message $\mu_{j \to i}(S_i)$, vertex j 'writes' to vertex i (Figure 12.1):

> Dear friend, I don't know either your current state or the local field applied to you, but based on the messages I got from my other friends,
> I believe that the probability that you are in state S_i equals $\mu_{j \to i}(S_i)$.[1]

Clearly,

$$\sum_{S_i = \pm 1} \mu_{j \to i}(S_i) = 1. \tag{12.2}$$

If the messages are known, and they are independent, which is the case in trees and locally tree-like networks and which is an approximation in graphs with cycles, one can easily find the probability $p_i(S_i)$ that vertex i is in state S_i for each vertex. The full set of messages can be obtained from update equations derived by using the same assumption. These expression for the probability $p_i(S_i)$ in terms of messages and the equation for the messages can be written as

$$p_i(S_i) = C_1 e^{\beta H_i S_i} \prod_{j \in \partial i} \mu_{j \to i}(S_i), \quad \langle S_i \rangle = \sum_{S_i = \pm 1} S_i p_i(S_i), \tag{12.3}$$

$$\mu'_{j \to i}(S_i) = C_2 \sum_{S_j = \pm 1} e^{\beta(H_j S_j + J_{ji} S_j S_i)} \prod_{k \in \partial j \setminus i} \mu_{k \to j}(S_j). \tag{12.4}$$

To grasp the form of Eqs. (12.3) and (12.4), examine Figure 12.2 and, also, compare these two equations with Eqs. (6.48) and (6.49) of message passing for percolation. The probabilistic factor $e^{\beta H_i S_i}$ on the right-hand side of Eq. (12.3) is due to a local field H_i at vertex i, where $\beta \equiv 1/T$ is the inverse temperature, if we set the Boltzmann constant $k_B = 1$. Similarly, the factors $e^{\beta H_j S_j} e^{\beta J_{ji} S_j S_i}$ on the right-hand side of Eq. (12.4) are due to a local field H_j at vertex j and due to the interaction $-J_{ji} S_j S_i$ between spins on j and i, respectively.[2]

Due to normalization, Eq. (12.1), the messages $\mu_{j \to i}(S_i)$ can be written in the following form

---

[1]In fact, $\mu_{j \to i}(S_i)$ is the probability to find vertex i in state S_i if (i) the connections of vertex i with all its neighbours except vertex j are removed and (ii) $H_i = 0$.

[2]We present Eq. (12.4) in a self-consistent form. One can write $\mu'_{j \to i}(S_i)$ on the left-hand side, to get it in the update equation form.

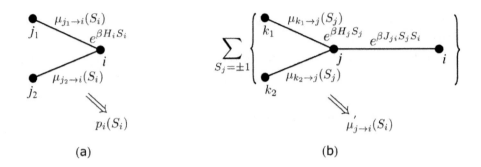

Fig. 12.2 (a) Expression of the probability $p_i(S_i)$ in terms of messages, Eq. (12.3), and (b) equation for messages, Eq. (12.4), in a graphical form.

$$\mu_{j\to i}(S_i) = \frac{e^{\beta h_{j\to i} S_i}}{2\cosh(\beta h_{j\to i})},$$ (12.5)

where the parameter $h_{j\to i}$ has the meaning of an additional field, which spin S_j creates at vertex i. Substituting Eq. (12.5) into Eqs. (12.3) and (12.4), one gets the expression for the average spins in terms of the fields $h_{j\to i}$ and the self-consistency equation for these fields in the form usual for the Bethe–Peierls approximation and the cavity method,

$$\langle S_i \rangle = \tanh\left[\beta\left(H_i + \sum_{j\in\partial i} \beta h_{j\to i}\right)\right],$$ (12.6)

$$\tanh(\beta h_{j\to i}) = \tanh(\beta J_{ji})\tanh\left[\beta\left(H_j + \sum_{k\in\partial j\setminus i} h_{k\to j}\right)\right].$$ (12.7)

For the q-regular Bethe lattice with a uniform coupling J, where $\langle S_i \rangle = \langle S \rangle$ and $h_{j\to i} = h$, Eqs. (12.6) and (12.7) reduce to

$$\langle S \rangle = \tanh[\beta(H + qh)],$$ (12.8)

$$\tanh(\beta h) = \tanh(\beta J)\tanh[\beta(H + qh)]$$ (12.9)

(Baxter, 2016), indicating a continuous (second-order) ferromagnetic phase transition, and for the critical point, we have

$$\tanh(\beta_c J) = \frac{1}{q-1},$$ (12.10)

where $q - 1$ is branching.[3]

---

[3]Note that if we consider the ferromagnetic Ising model on an infinite Cayley tree, we would get $T_c = 0$ as for any proper tree.

Linearizing Eq. (12.7) with a uniform coupling $J_{ji} = J$ and zero local fields, $H_i = 0$, we arrive at the equation

$$h_{j \to i} = \tanh(\beta J) \sum_{k \in \partial j \backslash i} h_{k \to j}. \qquad (12.11)$$

For the $2E$-vector $\mathbf{h}$ with the components $h_{j \to i}$, where E is the number of edges of a graph, this can be written in the matrix form,

$$\mathbf{h} = \tanh(\beta J) B \mathbf{h}, \qquad (12.12)$$

where B is the non-backtracking matrix, Eq. (6.57). Equations (12.11) and (12.12) coincide with the corresponding equations for percolation, Eqs. (6.56) and (6.58), with $\tanh(\beta J)$ substituted for p. Then for the critical point, we have

$$\tanh(\beta_c J) = \frac{1}{\lambda_1}, \qquad (12.13)$$

where λ_1 is the largest eigenvalue of the non-backtracking matrix B, coinciding with the branching b of the non-backtracking expansion (Section 6.3).[4] We can repeat here comments on the message-passing approximation for percolation. In particular, note that the belief-propagation approximation, Eq. (12.13), wrongly predicts a finite critical temperature $T_c = 1/\beta_c$ for finite graphs, if they are not trees.[5] For uncorrelated networks, $\lambda_1 = \langle q(q-1) \rangle / \langle q \rangle$ (Appendix F), and hence

$$\tanh(\beta_c J) = \frac{\langle q \rangle}{\langle q(q-1) \rangle}. \qquad (12.14)$$

In these networks, the divergent second moment of a degree distribution, $\langle q^2 \rangle \to \infty$, ensures the presence of long-range order at any temperature.

The Ising model is a particular case ($s = 2$) of the s-state Potts model, where the interaction between the Potts spins $\sigma_i = 1, 2, \ldots, s$ is described by the Hamiltonian

$$\mathcal{H} = -\sum_{i<j} J_{ij} \delta_{\sigma_i, \sigma_j} - \sum_i H_i \delta_{\sigma_i, 1}, \qquad (12.15)$$

where $\delta_{\sigma_i, \sigma_j}$ is the Kronecker symbol. For $s \geq 3$, the Potts model has a first-order phase transition. Nonetheless, its treatment in the framework of

---

[4]The relation $\tanh(\beta_c J) = 1/b$ for the Ising model on infinite locally tree-like graphs was first proven by Lyons (1989).

[5]Standard belief propagation approximately substitutes a locally tree-like network for an original graph. For networks with motifs, one can make one step further and assume approximately that finite cycles occur only within motifs, treating motifs exactly and the connections between then as tree-like (Yoon, Goltsev, Dorogovtsev, and Mendes, 2011).

belief propagation is similar to the Ising model (Dorogovtsev, Goltsev, and Mendes, 2004). For a more detailed though brief introduction into the belief-propagation techniques for cooperative models on networks, including the treatment of a wide range of thermodynamical quantities and correlations, see Dorogovtsev, Goltsev, and Mendes (2008a).

12.2 Critical phenomena

The first study of the Ising model on complex networks was due to Aleksiejuk, Hołyst, and Stauffer (2002) who focused on the Barabási–Albert networks. To treat the Ising model on random networks, using believe propagation, one must average the believe propagation equations and expressions for thermodynamic quantities over a statistical ensemble. Proceeding in this or a similar way, Dorogovtsev, Goltsev, and Mendes (2002a) and Leone, Vázquez, Vespignani, and Zecchina (2002) precisely solved the Ising model on uncorrelated networks and described the critical singularities. Here we consider a far more simple, though approximate, approach based on the annealed network approximation (Bianconi, 2002).[6] This approximation, which is actually equivalent to the heterogeneous mean-field theory discussed in Chapter 7, provides correct critical singularities for this problem.

Recall that the annealed network mimicking a uniformly random graph with a given degree sequence $(q_1, q_2, \ldots, q_N)$ is a fully connected graph with its edge weights $w_{ij}(q_i, q_j) = q_i q_j/(N\langle q \rangle)$, Eq. (4.20). Exchanging $J_{ij} \to J_{ij} w_{ij}$ in the ferromagnetic Ising model Hamiltonian, Eq. (12.1), we write for the annealed network

$$\mathcal{H} = -\sum_{i<j} J_{ij} \frac{q_i q_j}{N\langle q \rangle} S_i S_j - \sum_i H_i S_i. \tag{12.16}$$

In the limit $N \to \infty$, the resulting problem has the following exact solution:

$$\langle S_i \rangle = \tanh \left[\beta \left(H_i + \frac{q_i}{N\langle q \rangle} \sum_{j=1}^{N} J_{ij} q_j \langle S_j \rangle \right) \right]. \tag{12.17}$$

If the couplings are uniform, $J_{ij} = J > 0$ and $H = 0$, then for the weighted magnetic moment

$$M_w \equiv \frac{\sum_j q_j \langle S_j \rangle}{N\langle q \rangle}, \tag{12.18}$$

---

[6]See Krasnytska, Berche, Holovatch, and Kenna (2016) for more detail.

Table 12.1 Critical singularities of the magnetization $M = \sum_i \langle S_i \rangle$, the spin contribution to the specific heat δC, and the susceptibility χ in the ferromagnetic Ising
model on uncorrelated networks with a degree distribution $P(q) \sim q^{-\gamma}$ for various
values of exponent γ (Dorogovtsev, Goltsev, and Mendes, 2002a). $\tau \equiv |1 - T/T_c|$.
In the normal phase $(T > T_c)$, $\delta C = 0$ as is common for the mean-field theories.

	M	$\delta C(T < T_c)$	χ
$\gamma > 5$	$\tau^{1/2}$	jump at T_c	τ^{-1}
$\gamma = 5$	$\tau^{1/2}/(\ln \tau^{-1})^{1/2}$	$1/\ln \tau^{-1}$	τ^{-1}
$3 < \gamma < 5$	$\tau^{1/(\gamma-3)}$	$\tau^{(5-\gamma)/(\gamma-3)}$	τ^{-1}
$\gamma = 3$	$e^{-2T/\langle q \rangle}$	$T^2 e^{-4T/\langle q \rangle}$	T^{-1}
$2 < \gamma < 3$	$T^{-1/(3-\gamma)}$	$T^{-(\gamma-1)/(3-\gamma)}$	T^{-1}

Eq. (12.17) readily gives the equation

$$M_w = \sum_q \frac{qP(q)}{\langle q \rangle} \tanh(\beta J q M_w). \quad (12.19)$$

The average spin $\langle S_i \rangle$ is expressed in terms of M_w,

$$\langle S_i \rangle = \tanh(\beta J q_i M_w). \quad (12.20)$$

Similarly one can derive the free energy for this Hamiltonian, Eq. (12.16),
and obtain other thermodynamic quantities.

Equation (12.19) provides the critical temperature $T_c = \langle q(q-1) \rangle \langle q \rangle$
deviating from the exact one, Eq. (12.14). Still, in this approximation, T_c
diverges with $\langle q^2 \rangle$, and for a power-law degree distribution $P(q) \sim q^{-\gamma}$, the
critical singularities coincide with the correct ones in Table 12.1 for any
value of exponent γ.[7] Notice that the usual mean-field theory predicting

[7] All these critical singularities can also be obtained in the framework of a phenomenological approach based on the Landau theory of phase transitions (Goltsev, Dorogovtsev,
and Mendes, 2003). Taking into account (i) the symmetry of the order parameter m, (ii)
the analytical properties of the thermodynamic potential $\Phi(m, h, A)$, where h is an external field and A is the control parameter measuring a deviation from the critical point,
and (iii) a network heterogeneity, one can obtain the leading terms of $\Phi(m, h, A)$. For a
ferromagnetic transition on a scale-free network, we get

$$\Phi(m, h, A) = \begin{cases} Am^2 + Bm^4 - hm & \text{if } \gamma > 5, \\ Am^2 + Bm^{\gamma-1} - hm & \text{if } 3 < \gamma < 5, \end{cases}$$

providing the critical singularities.

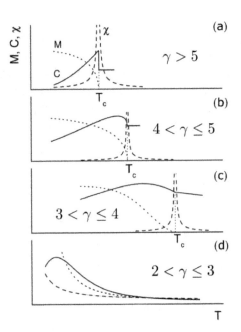

Fig. 12.3 Schematic view of the critical behaviour of the magnetization M (dotted lines), the magnetic susceptibility χ (dashed lines), and the specific heat C (solid lines) for the ferromagnetic Ising model on uncorrelated random networks with a degree distribution $P(q) \sim q^{-\gamma}$. (a) $\gamma > 5$, the standard mean-field critical behaviour. A jump of C disappears as $\gamma \to 5$. (b) $4 < \gamma \le 5$, the ferromagnetic phase transition is of second order. (c) $3 < \gamma \le 4$, the transition becomes of higher order. (d) $2 < \gamma \le 3$, the transition is of infinite order, and $T_c \to \infty$ as $N \to \infty$. Adapted from Dorogovtsev, Goltsev, and Mendes (2008a).

the square-root singularity for the magnetization, $M \propto \sqrt{T_c - T}$, is valid for $\gamma > 5$, while for the percolation problems, the standard mean-field theory works when $\gamma > 4$; see Eqs. (6.26) and (6.35) for bond and site percolation. Figure 12.3 shows the full set of these critical behaviours for different γ.[8] Finally, at the critical point, where the susceptibility diverges, the dependence of the magnetization on magnetic field is nonlinear,

---

[8]The problem of the Ising and Potts models on growing networks is more demanding. The idea is to first grow a random network up to an infinite size and then consider these models on the resulting statistical ensemble. In this setup, the Ising and Potts models exhibit an infinite order phase transition with the BKT singularity (Bauer, Coulomb, and Dorogovtsev, 2005; Khajeh, Dorogovtsev, and Mendes, 2007).

$$M \propto \begin{cases} H^{1/3} & \text{if } \gamma > 5, \\ H^{1/(\gamma-2)} & \text{if } 3 < \gamma < 5. \end{cases} \tag{12.21}$$

Note that, for the sake of brevity, this chapter focuses on cooperative systems on infinite networks. For finite-size scaling for cooperative models on complex networks, see Hong, Ha, and Park (2007) and Dorogovtsev, Goltsev, and Mendes (2008a).

12.3   Games on networks

We pass now from interacting spins on networks to specifically interacting individuals, namely, to players. Let us model interactions within a pair of individuals by one of so-called one-shot *symmetric two-person games*. The goal of a player in these games is to receive the higher payoff than the second player. The players independently choose one of two strategies, for example, to cooperate, C, or to defect, D, and the individual payoffs are awarded according to the combination of the strategies accepted by the players. The awarded payoffs are given by the *payoff matrix*

$$\begin{array}{cc} & \begin{array}{cc} C & D \end{array} \\ \begin{array}{c} C \\ D \end{array} & \begin{pmatrix} R & S \\ T & P \end{pmatrix}, \end{array} \tag{12.22}$$

which defines the game. The entries $M_{\alpha\beta}$ of the matrix are the payoffs (real numbers) awarded to the row player; $\alpha, \beta = 1, 2$, where 1 and 2 are for the strategies C and D, respectively, α is for the first ('row') player, β is for the second ('column') player. The entries of the transpose of the matrix provide the payoffs awarded to the column player. The entries—parameters of a game—are typically denoted by R, S, T, P as in Eq. (12.22). The full set of binary relations between the given values of the parameters indicate which game we play.[9] The relation $T > R > P > S$ corresponds to the most known game, the Prisoner's Dilemma, which we choose for our players. Each of two players ('prisoners') independently decides for himself or herself which is better: to cooperate (remain silent) or to defect (betray)? This decision should be just based on the set of payoffs, Eq. (12.22) for the players.

---

[9]After we fix the relation $R > P$, the possible combinations of the values of the 4 parameters result in $12 = 4!/2$ different games: the Prisoner's Dilemma, $T > R > P > S$, the Snowdrift game, $T > R > S > P$, the Stag Hunt game, $R > T > P > S$, the Harmony game, $R > T > S > P$, etc. (Hauert, 2001; Shutters, 2013). Some of them even have no names. The standard convenient gauge is $R = 1$, $P = 0$.

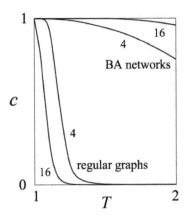

Fig. 12.4 Schematic plot of a final concentration c of cooperators vs. parameter T of the spatial Prisoner's Dilemma on two contrasting networks according to Santos and Pacheco (2005). The other parameters are $R = 1$, $P = 0$, S is negative close to zero. The numbers labelling the curves indicate the mean degrees of the networks.

What is better in the Prisoner's Dilemma: to cooperate or to defect? What is more profitable in terms of payoff? If both players decide to cooperate or both decide to defect, then they will receive equal payoffs, R or P, respectively. If the first player defects, and the second—cooperates, then the first will get the bigger payoff T, while the second will receive the smaller payoff T. So the best action for a player in the Prisoner's Dilemma, irrespective of the strategy of the second player, is to defect. The game becomes much more interesting when players can adopt the strategies of their opponents—the *evolutionary games*, and when there are many players (Nowak, 2006; Szabó and Fath, 2007). For a network of players, at a given moment, each vertex (player) i can be in one of two states, $\sigma_i = C$ or $\sigma_i = D$, depending on the strategy that the player uses at this moment, to cooperate or to defect. Each pair of nearest neighbours independently play the game and receive their payoffs determined by the states of these pairs according to the payoff matrix, Eq. (12.22). To make this system evolve we must allow individual players to change their strategies from time to time, $\sigma_i = C \longleftrightarrow \sigma_i = D$. Suppose that the players have information only about the results of their opponents—neighbours. Then a natural idea for an adaptive player is to adopt the most successful strategy in his close environment (the player himself and his nearest neighbours).

This idea was realized in Nowak and May's (1992) *spatial Prisoner's*

Dilemma. Without going into detail, in this mathematical model, originally formulated for a regular graph, the evolution of the players' strategies appears as follows. Initially, there is a random configuration of defectors and cooperators. For example, the initial concentration of cooperators c may be set to $1/2$. Then all of the pairs of nearest neighbours play the game independently, and after this round, each player i accumulates his payoffs as P_i. After that, for each player i, choose at random one of its immediate neighbours, j, and compare the scores P_i and P_j.[10] If $P_i > P_j$, then leave the state (strategy) of i unchanged. Otherwise, let player i accept the strategy of j with some probability. After this update, pass to the next round, recalculate all the scores, and so on. As a result of this evolution, the concentration c of cooperators approaches a stationary value depending on the parameters of the Prisoner's Dilemma. This evolutionary model was generalized to other, non-regular networks and simulated (Santos and Pacheco, 2005). Figure 12.4 shows the resulting final concentrations of cooperators in contrasting networks (random regular graphs and Barabási–Albert networks). The reader can see that scale-free architectures stimulate and support cooperation. It turns out that players occupying the hubs tend to be cooperators, which makes cooperation dominant in these networks. These observations may, at least partially, explain why cooperation is widespread in heterogeneous populations. One can go even further and couple two evolutionary processes in these systems, namely, the evolutions of players and of a network (Pacheco, Traulsen, and Nowak, 2006). In this generalization, not only does the network influence the course of the game, but also the dynamics of the game changes the structure of the network.

12.4   Phase synchronization

The *Kuramoto model* is the classical paradigm for phase synchronization (Kuramoto, 1975, 1984; Strogatz, 2000; Acebrón, Bonilla, Pérez Vicente, Ritort, and Spigler, 2005). The model describes the dynamics of N coupled phase oscillators with phases $\theta_i(t)$, $i = 1, 2, \ldots, N$ running at natural frequencies ω_i, randomly distributed according to a probability density $g(\omega)$,

$$\frac{d}{dt}\theta_i = \omega_i + \sum_{j=1}^{N} J_{ij} A_{ij} \sin(\theta_j - \theta_i) - h_i \sin \theta_i. \qquad (12.23)$$

----

[10]The original version of the model is deterministic: each player adopts the best strategy in his close environment (Nowak and May, 1992). Here we describe a stochastic version.

Here $J_{ij} > 0$ is coupling between oscillators i and j, A_{ij} are the entries of the adjacency matrix of the network of oscillators, and h is the external field. In the original model, solved by Kuramoto in 1975, this network was an infinite fully connected graph, $A_{ij} = 1$ for all $i \neq j$, $N \to \infty$, all couplings were equal, $J_{ij} = J/N > 0$, and $h_i = h = 0$.

12.4.1 The uniform Kuramoto model

For the sake of clarity, we first briefly discuss the original Kuramoto model. The global state in this model is characterized by the following average:

$$z(t) = r(t)e^{i\psi(t)} = \frac{1}{N} \sum_{j=1}^{N} e^{i\theta_j}, \tag{12.24}$$

which is the complex order parameter, introduced by Kuramoto, $r(t)$ is its absolute value, measuring the phase coherence, and $\psi(t)$ is the average phase. A zero value, $r = 0$, means incoherence. The real part of the equality

$$re^{i(\psi-\theta_i)} = \frac{1}{N} \sum_{j=1}^{N} e^{i(\theta_j-\theta_i)} \tag{12.25}$$

is

$$r \sin(\psi - \theta_i) = \frac{1}{N} \sum_{j=1}^{N} \sin(\theta_j - \theta_i). \tag{12.26}$$

Comparing with Eq. (12.23), one gets N equations,

$$\frac{d}{dt}\theta_i = \omega_i + Jr \sin(\psi - \theta_i). \tag{12.27}$$

The solution for a symmetric natural frequency distribution, $g(\omega) = g(-\omega)$, is more simple. In this case, one can set $\psi = 0$, and the equations are completely uncoupled. For non-symmetric distributions of natural frequencies, phase $\psi(t)$ rotates with frequency Ω, $\psi(t) = \Omega t + \text{const}$, which must be found, slightly complicating the problem (Strogatz and Mirollo, 1991; Basnarkov and Urumov, 2008). For a symmetric $g(\omega)$,

$$\frac{d}{dt}\theta_i = \omega_i - Jr \sin \theta_i. \tag{12.28}$$

In the limit $t \to \infty$, the order parameter approaches a constant value r, and for the oscillators with natural frequencies around $\omega = 0$, satisfying

$|\omega_i| \leq Jr$, the phases θ_i are locked, since Eq. (12.28) has a stationary solution for each of these oscillators. The presence of a finite fraction of oscillators with the locked (in other words, rigidly coupled) phases for $r > 0$
corresponds to spontaneous phase synchronization. Other oscillators, hav-
ing natural frequencies $|\omega_i| > Jr$ are 'drifting' in the sense that their phases
non-uniformly change in time. So the system in the phase with $r > 0$ is
'partially synchronized'. For a symmetric distribution of natural frequen-
cies, the averages over drifting oscillators are zero, $\langle \ldots \rangle_{\text{drift}} = 0$, and only
$\langle \ldots \rangle_{\text{lock}}$ matters. This leads to the following self-consistency equation for
the order parameter:

$$r = \langle e^{i\theta} \rangle_{\text{lock}} = \langle \cos \theta \rangle_{\text{lock}} = \int_{-Jr}^{Jr} d\omega g(\omega) \cos \theta$$

$$= \int_{-Jr}^{Jr} d\omega g(\omega) \sqrt{1 - \left(\frac{\omega}{Jr}\right)^2}. \tag{12.29}$$

For a unimodal distribution,

$$g(\omega) \cong g(0) - A\omega^2, \tag{12.30}$$

where A is a constant, Eq. (12.29) provides the continuous (second-order)
phase transition (Figure 12.5a), at the critical point

$$J_c = \frac{2}{\pi g(0)}, \tag{12.31}$$

with the square-root critical singularity

$$r \propto \sqrt{J - J_c}, \tag{12.32}$$

that is, the critical exponent β of the order parameter equals $1/2$, like, for
example, in the Ising model above the upper critical dimension. The critical
singularity of the order parameter is determined by the behaviour of $g(\omega)$
near $\omega = 0$. More generally, for $g(\omega) - g(0) \propto \omega^\epsilon$, Eq. (12.29) leads to the
critical singularity

$$r \propto (J - J_c)^{1/\epsilon}. \tag{12.33}$$

If $\epsilon \to \infty$, then the critical exponent $\beta \to 0$, indicating a discontinuity in
this limit, which has to be treated differently. The Kuramoto model with
the flat distribution of natural frequencies

$$g(\omega) = \frac{1}{2\omega_0} \theta(\omega + \omega_0)\theta(\omega_0 - \omega), \tag{12.34}$$

where $\theta(x)$ is the step function, corresponding to such situation, was first
considered by Strogatz, Mirollo, and Matthews (1992) and then, in more

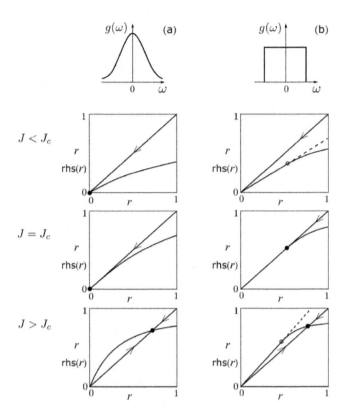

Fig. 12.5 Graphical solution of Eq. (12.29) for two distinct unimodal distributions, resulting in (a) continuous and (b) hybrid phase transitions.

detail, by Pazó (2005). Equation (12.29) for this distribution of natural frequencies (Figure 12.5b), near the critical point $J_c = 4\omega_0/\pi$ has the form

$$r = \frac{J}{J_c}r - Br(r - r_c)^{3/2},\qquad(12.35)$$

where $r_c = \omega_0/J_c = \pi/4$ and B is a constant. This leads to the discontinuity $r(J \le J_c) = 0$, $r(J = J_c+0) = r_c$ accompanied by the singularity of the order parameter

$$r - r_c \propto (J - J_c)^{2/3}.\qquad(12.36)$$

This hybrid transition (Figure 12.5b), differs from that in the k-core problem and in interdependent networks (Sections 6.7 and 8.2, respectively), which was actually the stability limit of a first-order phase transition. In the Kuramoto model with the flat distribution of natural frequencies, the discontinuity is determined by the width of the flat part of the distribution, while the critical singularity of the order parameter depends on a shape of the 'outer' parts of the distribution.[11] The flat distribution of natural frequencies sits between a unimodal and bimodal distributions, for the latter of which (where $g(\omega)$ is concave at $\omega = 0$), the phase transition is first order (Kuramoto, 1984).[12]

For a uniform field h in Eq. (12.1), one can introduce the susceptibility $\chi = dr(h)/dh|_{h=0}$ (Shinomoto and Kuramoto, 1986). If the distribution of natural frequencies is unimodal, Eq. (12.30), and hence the synchronization phase transition is continuous, then the Curie–Weiss law holds near the critical point, $\chi \propto 1/|J-J_c|$. Studying the relaxation of small perturbations around a stationary state, $r + \delta r(t)$, is greatly simplified by using the Ott–

---

[11]If outside of the flat region of the distribution of natural frequencies, we have

$$g(\omega)/g(0) - 1 \overset{|\omega-\omega_0|\ll1}{\propto} -\theta(|\omega| - \omega_0)|\omega - \omega_0|^{\kappa},$$

then the critical singularity of the order parameter is

$$r - r_c \propto (J - J_c)^{2/(3+2\kappa)}.$$

[12]In fact, the case of a multi-modal distribution of natural frequencies is more difficult than one would expect (Kuramoto, 1984; Montbrió, Kurths, and Blasius, 2004; Martens, Barreto, Strogatz, Ott, So, and Antonsen, 2009). For example, for a symmetric bimodal distribution, synchrony emerges at the peaks, and the phases of two groups of oscillators with frequencies at the peaks first are locked almost independently, forming two weakly coupled 'giant oscillators', and 'for even stronger coupling they will eventually be entrained to each other to form a single giant oscillator' (Kuramoto, 1984). For the distribution

$$g(\omega) \propto [(\omega - \omega_0)^2 + \Delta^2]^{-1} + [(\omega + \omega_0)^2 + \Delta^2]^{-1},$$

which is bimodal when $\omega_0/\Delta > 1/\sqrt{3}$, the phase diagram of the model appears as follows:

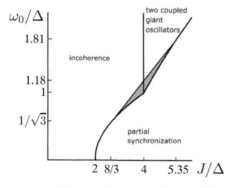

The shaded area is the region of hysteresis. Adapted from Martens, Barreto, Strogatz, Ott, So, and Antonsen (2009). Hysteresis occurs in the shaded area.

Antonsen ansatz (Ott and Antonsen, 2008).[13] Close to the critical point, the relaxation is exponential, $\delta r(t) \sim e^{-t/\tau}$, where the relaxation time $\tau \propto 1/|J - J_c|$, while the critical relaxation is power law, $\delta r(t) \sim t^{-1/2}$. On the other hand, in the case of a flat distribution of natural frequencies, Eq. (12.34), the relaxation is power law in the whole incoherence phase, $\delta r(t) \sim t^{-1}$ (Strogatz, Mirollo, and Matthews, 1992).

12.4.2 The Kuramoto model on complex networks

The simplest way to treat the Kuramoto model on complex networks is by applying the annealed network approximation (Rodrigues, Peron, Ji, and Kurths, 2016); see Eq. (12.16) for the Ising model. Placing the oscillators on a fully connected graph with edge weights $w_{ij}(q_i, q_j) = q_i q_j / (N\langle q \rangle)$ we arrive at the Kuramoto model with a degree-dependent coupling,

$$\frac{d}{dt}\theta_i = \omega_i + \frac{J q_i}{N\langle q \rangle} \sum_{j=1}^{N} \sin(\theta_j - \theta_i). \tag{12.37}$$

Introducing a weighted order parameter,

$$\tilde{r}(t)e^{i\tilde{\psi}(t)} = \frac{1}{N\langle q \rangle} \sum_{j=1}^{N} q_i e^{i\theta_j}, \tag{12.38}$$

we rewrite Eq. (12.37) as

$$\frac{d}{dt}\theta_i = \omega_i + J\tilde{r}q_i \sin(\tilde{\psi} - \theta_i). \tag{12.39}$$

Hence in the coherent state, the oscillators with $|\omega_i| \le J\tilde{r}q_i$ are synchronized. The phases of these oscillators are locked, $\sin\theta_i = \omega_i/(J\tilde{r}q_i)$, where we set $\tilde{\psi} = 0$. Thus oscillators on vertices with large degrees synchronize

---

[13]In terms of the probability density $f(\theta, \omega, t)$, that is, the density of oscillators with natural frequency ω, having phase θ at time t, $\int_0^{2\pi} d\theta f(\theta, \omega, t) = g(\omega)$, the complex order parameter takes the form

$$z(t) = \int_{-\infty}^{\infty} d\omega \int_0^{2\pi} d\theta f(\theta, \omega, t)e^{i\theta}.$$

The Ott–Antonsen ansatz assumes the following form of the density $f(\theta, \omega, t)$:

$$f(\theta, \omega, t) \approx \frac{1}{2\pi}g(\omega)\left\{1 + \sum_{n=1}^{\infty}\left[(\alpha(\omega, t)e^{i\theta})^n + \text{c.c.}\right]\right\}.$$

easier than those on vertices with small degrees. Assuming a symmetric unimodal distribution of natural frequencies, $g(\omega)$, we get the equation for $\tilde{r}$ in
a way similar to the derivation of Eq. (12.29),

$$\tilde{r} = \sum_q \frac{qP(q)}{\langle q \rangle} \int_{-J\tilde{r}q}^{J\tilde{r}q} d\omega g(\omega) \sqrt{1 - \left(\frac{\omega}{J\tilde{r}q}\right)^2}, \qquad (12.40)$$

where $P(q)$ is the degree distribution of the network. For a convex symmetric
$g(\omega)$, this equation provides the continuous phase transition at the critical
coupling

$$J_c = \frac{2\langle q \rangle}{\pi g(0)\langle q^2 \rangle}, \qquad (12.41)$$

predicting (partial) synchronization for any non-zero coupling for the degree
distributions with a diverging second moment, $\langle q^2 \rangle \to \infty$. In this situation,
synchronization cannot be destroyed by uniformly random failures. For a
power-law degree distribution $P(q) \sim q^{-\gamma}$, the critical singularities coincide
with those of the Ising model for any γ (Oh, Lee, Kahng, and Kim, 2007)
(Section 12.2). The square-root critical singularity of the order parameter
occurs when $\langle q^4 \rangle < \infty$, that is, when $\gamma > 5$. When $\langle q^2 \rangle < \infty$, that is, $\gamma > 3$,
the Curie–Weiss law for the susceptibility $\chi \propto 1/|J - J_c|$ and the relaxation
time $\tau \propto 1/|J - J_c|$ holds (Yoon, Sindaci, Goltsev, and Mendes, 2015) as in
the homogeneous Kuramoto model.

   This picture qualitatively changes when the natural frequencies of oscil-
lators correlate with the degrees of their vertices. Simulations by Gómez-
Gardenes, Gómez, Arenas, and Moreno (2011) and Leyva, Sevilla-Escoboza,
Buldú, Sendina-Nadal, Gómez-Gardeñes, Arenas, Moreno, Gómez, Jaimes-
Reátegui, and Boccaletti (2012) revealed a first-order synchronization phase
transition in this system even with a unimodal distribution of natural fre-
quencies (see also Boccaletti, Almendral, Guan, Leyva, Liu, Sendiña-Nadal,
Wang, and Zou, 2016).[14] The theory of this phenomenon was developed by
Coutinho, Goltsev, Dorogovtsev, and Mendes (2013).

   Let the natural frequencies ω_i and degrees q_i be linearly related, $\omega_i =
aq_i + b$, where a and b are constants, and the distribution of natural fre-
quencies $g(\omega)$ be, say, Cauchy–Lorentz or Gaussian. Applying the annealed
network approximation, one can derive equations for the order parameter.
The coupling between natural frequencies and degrees makes this problem

---

[14]A similar phenomenon occurs if the interaction strength depends on natural frequencies, that is, in a frequency-weighted network (Zhang, Hu, Kurths, and Liu, 2013; Song,
Um, Park, and Kahng, 2020).

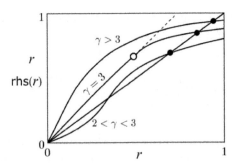

Fig. 12.6 Graphical solution of the equation $r = \text{rhs}(r)$ for the order parameter in the Kuramoto model with linearly coupled natural frequencies and degrees on a scale-free network (compare with Figure 12.5). All three functions $\text{rhs}(r)$ for different γ are shown above a transition, in the partially synchronized phase. Adapted from Coutinho, Goltsev, Dorogovtsev, and Mendes (2013).

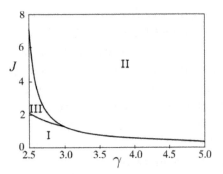

Fig. 12.7 Phase diagram of Kuramoto model with linearly coupled natural frequencies and degrees in the $\gamma-J$ plane. I—incoherence. II—partial synchronization. III—area of hysteresis. The hybrid transition occurs at $\gamma = 3$. Adapted from Coutinho, Goltsev, Dorogovtsev, and Mendes (2013).

non-symmetric in contrast to the situations discussed earlier, which demands more cumbersome derivations. Loosely speaking, in this mean-field theory, the coupling between ω_i and q_i effectively reshapes the distribution of natural frequencies. This produces phase transitions similar to those occurring in the original, homogeneous Kuramoto model with a bimodal and flat distributions. The equations for the order parameter show that a phase transition in this system occurs when the first moment of the degree distribution is finite, $\langle q \rangle < \infty$, that is for $\gamma > 2$. The graphical solutions of these equations are demonstrated in Figure 12.6 for different γ. The figure explains that for $\gamma > 3$, the transition is continuous (the critical singularity of the order parameter is square root) and for $2 < \gamma < 3$, the transition is first order. Finally, for $\gamma = 3$, the transition appears to be similar to that in the uniform Kuramoto model with a flat distribution of natural frequencies. This hybrid transition combines a discontinuity and the singularity of the order parameter, $r - r_c \propto (J - J_c)^{2/3}$, Eq. (12.36). Figure 12.7 shows the resulting phase diagram of this system in the $\gamma - J$ plane.

13

Inference and Reconstruction

13.1 Finding the root of a growing tree

Inference is generally about drawing conclusions about the whole on the basis of a sample. Statistical inference is 'the process of deducing properties of an underlying distribution by analysis of data' (Zdeborová and Krzakala, 2016). More specifically, in statistical mechanics inference usually implies concluding characteristics of a statistical ensemble (or its model, which is practically the same) on the basis of a sample (Clauset, Moore, and Newman, 2006). Here we touch upon a more restricted problem. Consider a branching process taking place on a given graph, which started from some unknown initial vertex, a root. At some instant, an observer makes a snapshot of this process and records its result—a tree subgraph of the substrate graph. The questions are: is it possible to guess the root from this observation; and, when it is possible, what is the best root-finding algorithm? The answers to these questions depend on the branching process and on the substrate graph. Remarkably, root finding is possible for a wide range of branching processes and substrate graphs.

Shah and Zaman (2011) proposed the maximum likelihood estimate of the source for what they called rumour spreading on tree or locally tree-like networks. In fact, by rumour spreading they meant the SI model process, where only the order in which vertices become infected turns out to be significant. When the degrees of vertices of a regular locally tree-like substrate are sufficiently large, this process can be naturally substituted with a recursive growing tree without any substrate, and the problem is reformulated as finding the root of a recursive tree, generated by some model, for example, the random recursive tree, a preferential attachment recursive tree, etc. For the problems of this sort, Shah and Zaman showed that their source estimator, the *rumour centrality* of a vertex in a resulting tree is effective in a wide range of situations, allowing us to find the source with finite probability even in infinite trees. The rumour centrality of vertex i in a tree $\mathcal{T}$ of N vertices is defined in the following way

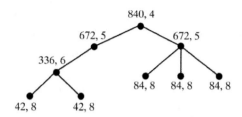

Fig. 13.1 Rumour centralities R_i and the sizes M_i of maximum subtrees among those rooted at the neighbouring vertices of i for a sample tree of 9 vertices, Eqs. (13.1) and (13.2), respectively. Numbers at vertices show the pairs R_i, M_i. In particular, the rumour centrality of the top vertex, which is the most probable root, is $R = \dfrac{9!}{9 \times 4 \times 4 \times 3 \times 1 \times 1 \times 1 \times 1 \times 1} = 840$. Notice that the vertex with the largest rumour centrality R has the smallest M.

$$R_i \equiv \frac{N!}{\prod_{j \in T} N_j(i)}, \tag{13.1}$$

where $N_j(i)$ is the size of the subtree of the tree T, rooted at j and pointing away from i. In particular, $N_i(i) = N$ (Figure 13.1). For a given labelled recursive tree T with a root at vertex i, R_i equals exactly the total number of possible orders of attaching the vertices. One can represent a recursive tree by a string of labels of vertices according to the order of attachment, where the first entry is i. Each of these strings is a particular *history of a tree* started from vertex i. Then the rumour centrality R_i gives the number of these strings-histories, which explains the meaning of this metric. The vertex with the largest rumour centrality is supposed to be the source. The fraction $p_i = R_i / \sum_{j=1}^{N} R_j$ is the proportion of histories started with vertex i among all histories resulting in the observed tree T.[1]

---

[1]The direct computing of all R_i (or p_i) of an undirected tree T of N vertices takes $O(N^2)$ operations, but Shah and Zaman (2011) proposed the following much faster algorithm of $O(N)$ operations, in essence, based on message passing.

Choose an arbitrary vertex, say, s, in the tree T. To all edges of T, ascribe directions away from s. For each edge $j \to k$ find the number of descendants, $d_{j \to k}$, of vertex k, including k,

$$d_{j \to k} = 1 + \sum_{l \in \partial k \backslash j} d_{k \to l}.$$

Here $\partial k \backslash j$ denotes all the nearest neighbours of vertex k except vertex j. This recursion enables one to compute the numbers $d_{j \to k}$ for all edges by $O(N)$ operations, starting from the edges adjacent to the leaves of the tree T. Then, set $p_s = 1$ and, starting from s, progressively compute

Shah and Zaman (2011, 2016) proved that for most tree and locally tree-like network substrates, except for, for example, a chain, the rumour centrality enables one to find the rumour source with a finite probability by using the rumour-spreading tree recorded at time $t \equiv N \to \infty$. The rumour centrality allows one to roughly estimate the detection probability $p_i^{(d)}$ for the rumour source, although one should apply more refined approaches to get sufficiently precise values. The reader can easily estimate that for a chain substrate, with observation at large time N, the detection probability for the central vertex decays as $N^{-1/2}$, and hence the rumour source cannot be found for $N \to \infty$. For other tree and locally tree-like substrates, allowing the rumour source detection, Shah and Zaman showed that in the limit $N \to \infty$, detection probability for vertices k, ranked in the descending order started from the one with the largest $p(d)$, approaches the exponential decay, $p_k^{(d)}(N \to \infty) \sim e^{-Ck}$.

Let us focus now on recursive trees, growing without any substrate. Section 5.1 explained that in the random recursive trees, for a given size N, each labelled tree (the member of this ensemble) occurs with equal probability. Consequently, any sequence of attached vertices resulting in a given tree has the same probability. In other words, all possible histories of a given tree in this model are equiprobable. The same is true for recursive trees growing with linear preferential attachment, see Eq. (5.12). This important property significantly simplifies the inference of roots and related problems for such recursive trees. Both for random recursive trees and recursive trees growing with the linear preferential attachment, Bubeck, Devroye, and Lugosi (2017) showed that the same rumour centrality metric, Eq. (13.1), or its slight variations, can be used as a root estimator in the root finding problem. Furthermore, it turned out that even a more simple metric can be often used as well. This is the size of the largest subtree among those rooted at the nearest neigbours of vertex i and not including i,

$$M_i \equiv \max_{j \in \partial i} N_j(i) \tag{13.2}$$

(Figure 13.1). For these recursive trees, Bubeck, *et al.* proved that the root can be found with a finite probability even in the limit of an infinite tree.

$$p_k = p_j \left(\frac{d_{j \to k}}{N - d_{j \to k}} \right)$$

for each child k of vertex j in the directed tree until p_k for all vertices will be found. Finally, normalize the resulting p_k to 1.

More refined and powerful message-passing techniques allow one not only to find the root of a given recursive tree, but also to infer other elements of its history (Cantwell, St-Onge, and Young, 2021).[2] These strong methods are applicable to a much wider class of recursive trees than those with equiprobable histories.[3] Notably, Young, St-Onge, Laurence, Murphy, Hébert-Dufresne, and Desrosiers (2019) demonstrated that the set of recursive trees with a recoverable history is surprisingly large. Furthermore, Cantwell, *et al.* found a way, by using message passing, to infer a growth model from a single sample of a recursive tree.

Finding a root or pieces of a history of a network is essentially about comparison between different histories of a tree. Therefore the number $\mathcal{N}$ of possible histories of a recursive tree is one of the key numbers in the outlined theories. The number of histories of a tree can be interpreted as its history degeneracy, and hence it is natural to consider the logarithm $\ln \mathcal{N}$ for a given recursive tree and the average $\langle \ln \mathcal{N} \rangle$ for a statistical ensemble of recursive trees. For the recursive trees with equiprobable histories, this is the entropy of a statistical ensemble. For the long chain of N vertices, where a history is not recoverable, the number of histories is minimal,

$$\ln \mathcal{N}_{\text{chain}}(N) = \ln \left[\sum_{i=1}^{N} \binom{N-1}{i-1} \right] = \ln(2^{N-1}) \cong N \ln 2. \qquad (13.3)$$

For the star graph of N vertices, clearly having the best history recoverability, the number of histories is maximal,

$$\ln \mathcal{N}_{\text{star}}(N) = \ln[2(N-1)!] \cong N \ln N - N. \qquad (13.4)$$

A numerical study of a set of models of random recursive trees with linear and power-law preferential attachment showed that the average $\langle \ln \mathcal{N} \rangle$ has the asymptotics

$$\langle \ln \mathcal{N} \rangle \cong N \ln N - cN, \qquad (13.5)$$

where c is a constant depending on a growth model (Timár, da Costa, Dorogovtsev, and Mendes, 2020).[4] In one limit of the spectrum of these

---

[2]Network archaeology (uncovering pieces of network history from present-day observations) is a rapidly developing research direction with numerous practical applications (Navlakha and Kingsford, 2011).

[3]See Timár, da Costa, Dorogovtsev, and Mendes (2020) for a less-precise but rapid algorithm, also based on message passing, allowing the history inference of a very large trees.

[4]Magner, Grama, Sreedharan, and Szpankowski (2017) proved that for proportional preferential attachment, $\langle \ln \mathcal{N} \rangle = [N + o(1)] \ln N$.

models, which is a star graph, $c = 1$. On the other hand, when a growth model approaches the second limit, namely, a chain, $c \to \infty$.

13.2 Finding missing links

In many real-world situations, measurements do not provide accurate or complete information about all vertices and edges of a network. Employed data sets contain errors and omissions. The fundamental problem is how to estimate network structure from available data, that is, to reconstruct a network (Newman, 2018*b*, 2018*a*). We touch upon here a special case of this problem. Let the available information about a simple graph of N vertices be incomplete, namely only E of its edges are known certainly. The straightforward way to learn the full structure of this network is to perform $N(N-1)/2 - E$ additional measurements—checks—for the remaining pairs of vertices providing all missing edges. One can however apply a far more efficient approach if the available information is sufficient to guess how this network is organized, in other words, to infer a model fitting the measured structure of a network reasonably well. By using this model, one can obtain the probabilities of connection for the $N(N-1)/2 - E$ remaining pairs of vertices and restrict the additional measurements only to vertex pairs for which this probability is high, exceeding a specified threshold. The number of such pairs is typically small, which ensures the efficacy of the approach.

A somewhat related *link-prediction problem* for evolving, in particular social, networks was formulated by Liben-Nowell and Kleinberg (2007): 'Given a snapshot of a social network at time t, we seek to accurately predict the edges that will be added to the network during the interval from time t to a given future time t'.' In this problem, the model of an evolving network should be inferred from its snapshot, allowing one to find the probabilities of the connections in the near future. Sometimes, however, it appears to be sufficient to know that a network belongs to some class, for example, to social networks, and to use empirical observations collected for these networks. Section 5.3 mentioned Newman's (2001*a*) observation that the probability of emergence of an edge between two vertices in a collaboration network increases with the number of their common nearest neighbours. Liben-Nowell and Kleinberg used this number as an edge predictor for ranking the potential future connections and found that it works well for the studied networks of collaborations from the arXiv.

Zhou, Lü, and Zhang (2009) applied a similar local predictor to the problem of missing edges in a set of rather diverse networks, including social networks, power grids, protein–protein interaction networks, etc. This simple predictor indicated potential missing edges with a reasonable accuracy in the majority of these networks.

A far more advanced technique of Clauset, Moore, and Newman (2008) is applicable to a wide class of networks exhibiting hierarchical organization, which means that a network can be divided into blocks, these blocks can be divided into blocks of blocks, and so on. This set of divisions can be represented by a rooted hierarchical tree—a dendrogram, as discussed in Section 9.6 (Figure 9.12). All splits in this tree are into two branches. These splits are the internal vertices r of the dendrogram. The splitting process ends with the full set of the vertices of the original network, which are the leaves of the dendrogram. To describe network structures of this sort Clauset, Moore, and Newman used the following model of a hierarchical random graph. This random graph is defined by a dendrogram to each of whose internal vertices, a given number $0 \leq p_r \leq 1$ is ascribed. The leaves of the dendrogram are the vertices of this random graph. The probability that a pair of these vertices, i and j, is connected by an edge equals p_r, where r is the lowest common ancestor of i and j (compare with Figure 10.9). In this technique, the known network edges are statistically fitted to this hierarchical random graph model. For the sake of brevity, we do not discuss how this fitting can be performed. The result is a dendrogram with the full set of probabilities p_r, enabling one to find the probabilities of connections for all pairs of vertices and so to indicate likely missing edges.

Guimerà and Sales-Pardo (2009) developed another method for indicating missing and spurious edges. In this approach, an observed network is treated as a realization of an underlying stochastic block model. These authors exploited the following idea. The probability t_{ij} that a given edge (ij) 'truly' exists (given the observation of the whole network) can be expressed in terms of a sum over a family of stochastic block models generated by all possible partitions of the observed network. In practice, the direct summation over all partitions, without sampling, is possible only for tiny networks. The Metropolis algorithm enables one to sample the partitions significantly contributing to the sum and to compute the probabilities t_{ij}. For an even more powerful and widely applicable method based on Bayesian inference combined with community detection, see Peixoto (2018).

14
What's Next?

The science of complex networks has been explosively expanding over the last twenty-plus years, resulting in a number of new research directions. Some have already reached maturity, some became less exciting for the research community with time, and none of the directions has disappeared. Figure 14.1 gives an idea of how this field developed in the respect of objects, phenomena, and processes, putting aside applications. One can see the recent trends and the topics attracting particular attention and efforts of nu-

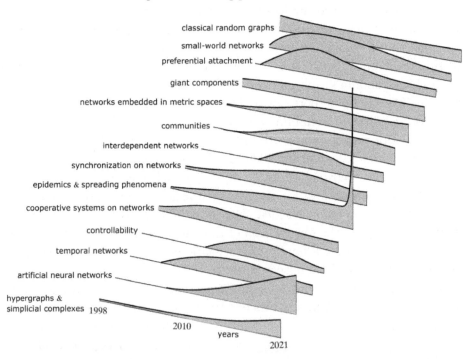

Fig. 14.1 Evolution of the research topics on complex networks (objects, phenomena, and processes) during last 20 years. The curves roughly estimate the flow of works on the topics.

merous academicians, whose studies on complex networks were essentially induced and multiplied by three major issues: the Big data problem, the Artificial Intelligence algorithms, and the COVID-19 pandemic. We suggest that these key problems, in addition to the growing needs of biotechnology and pharmaceutical industries, will largely determine the progress of the science of complex networks in the forthcoming years.

Further reading

Here we list a number of books and reviews on complex networks for further study.

Books about complex networks and random graphs

Barabási, A.-L. (2003). *Linked: The New Science of Networks*. American Association of Physics Teachers, College Park, MD.

Barabási, A.-L. (2016). *Network Science*. Cambridge University Press, Cambridge.

Barrat, A., Barthelemy, M., and Vespignani, A. (2008). *Dynamical Processes on Complex Networks*. Cambridge University Press, Cambridge.

Barthélemy, M. (2014). *Spatial Networks*. Springer, Berlin.

Baxter, G. J., da Costa, R. A., Dorogovtsev, S. N., and Mendes, J. F. F. (2021). *Weak Multiplex Percolation*. Cambridge University Press, Cambridge.

Bianconi, G. (2018). *Multilayer Networks: Structure and Function*. Oxford University Press, Oxford.

Bianconi, G. (2021). *Higher-Order Networks*. Cambridge University Press, Cambridge.

Blanchard, P. and Volchenkov, D. (2008). *Mathematical Analysis of Urban Spatial Networks*. Springer, Berlin.

Blanchard, P. and Volchenkov, D. (2011). *Random Walks and Diffusions on Graphs and Databases: An Introduction*. Springer, Berlin.

Bollobás, B. (1985). *Random Graphs*. Academic Press, New York.

Bollobás, B. (2012). *Graph Theory: An Introductory Course*. Springer, Berlin.

Bondy, J. A. and Murty, U. S. R. (1976). *Graph Theory with Applications*. The Macmillan Press Ltd., London.

Caldarelli, G. (2007). *Scale-Free Networks: Complex Webs in Nature and Technology*. Oxford University Press, Oxford.

Caldarelli, G. and Catanzaro, M. (2012). *Networks: A Very Short Introduction*. Oxford University Press, Oxford.

Caldarelli, G. and Chessa, A. (2016). *Data Science and Complex Networks: Real Case Studies with Python*. Oxford University Press, Oxford.

Chung, F. R. K. (1997). *Spectral Graph Theory*. American Mathematical Society, Providence, RI.

Cimini,G., Mastrandrea R., and Squartini, T. (2021). *Reconstructing Networks*. Cambridge University Press, Cambridge.

Clark, J. and Holton, D. A. (1991). *A First Look at Graph Theory*. World Scientific, Singapore.

Cohen, R. and Havlin, S. (2010). *Complex Networks: Structure, Robustness and Function*. Cambridge University Press, Cambridge.

Cozzo, E., De Arruda, G. F., Rodrigues, F. A., and Moreno, Y. (2018). *Multiplex Networks: Basic Formalism and Structural Properties*. Springer, Berlin.

Cvetković, D. M., Doob, M., and Sachs, H. (1998). *Spectra of Graphs: Theory and Applications*. Wiley, New York.

Dehmer, M. (2010). *Structural Analysis of Complex Networks*. Springer, Berlin.

Dehmer, M. and Emmert-Streib, F. (2009). *Analysis of Complex Networks: From Biology to Linguistics*. John Wiley & Sons, New Jersey.

Dehmer, M. and Emmert-Streib, F. (2014). *Quantitative Graph Theory: Mathematical Foundations and Applications*. CRC Press, New York.

Dorogovtsev, S. N. (2010). *Lectures on Complex Networks*. Oxford University Press, Oxford.

Dorogovtsev, S. N. and Mendes, J. F. F. (2003). *Evolution of Networks: From Biological Nets to the Internet and WWW*. Oxford University Press, Oxford.

Drmota, M. (2009). *Random Trees: An Interplay Between Combinatorics and Probability*. Springer, Berlin.

Durrett, R. (2007). *Random Graph Dynamics*. Cambridge University Press, Cambridge.

Easley, D. and Kleinberg, J. (2010). *Networks, Crowds, and Markets: Reasoning About a Highly Connected World*. Cambridge University Press, Cambridge.

El Gamal, A. and Kim, Y.-H. (2011). *Network Information Theory*. Cambridge University Press, Cambridge.

Estrada, E. (2012). *The Structure of Complex Networks: Theory and Applications*. Oxford University Press, Oxford.

Estrada, E. and Knight, P. A. (2015). *A First Course in Network Theory*. Oxford University Press, Oxford.

Even, S. (2011). *Graph Algorithms*. Cambridge University Press, Cambridge.

Frieze, A. and Karoński, M. (2016). *Introduction to Random Graphs*. Cambridge University Press, Cambridge.

Goldenberg, A., Zheng, A. X., Fienberg, S. E., and Airoldi, E. M. (2010). *A Survey of Statistical Network Models*. Now Publishers Inc, Norwell, MA.

Harris, J. K. (2013). *An Introduction to Exponential Random Graph Modeling*. Sage Publications, Los Angeles, CA.

Heydenreich, M. and Van der Hofstad, R. (2017). *Progress in High-dimensional Percolation and Random Graphs*. Springer, Berlin.

Holme, P. and Saramäki, J. (2013). *Temporal Networks*. Springer, Berlin.

Iványi, A. (2010). *Graph Theory*. `http://compalg.inf.elte.hu/~tony/Oktatas/TDK/FINAL/`.

Kiss, I., Miller, J. C., and Simon, P. L. (2017). *Mathematics of Epidemics on Networks*. Springer, Berlin.

Krapivsky, P. L., Redner, S., and Ben-Naim, E. (2010). *A Kinetic View of Statistical Physics*. Cambridge University Press, Cambridge.

Lambiotte, R. and Schaub, M. T. (2021). *Modularity and Dynamics on Complex Networks*. Cambridge University Press, Cambridge.

Latora, V., Nicosia, V., and Russo, G. (2017). *Complex Networks: Principles, Methods and Applications*. Cambridge University Press, Cambridge.

Lovász, L. (2012). *Large Networks and Graph Limits*. American Mathematical Society, Providence, RI.

Lyons, R. and Peres, Y. (2017). *Probability on Trees and Networks*. Cambridge University Press, Cambridge.

Masuda, N. and Lambiotte, R. (2016). *A Guide to Temporal Networks*. World Scientific, Singapore.

Menczer, F., Fortunato, S., and Davis, C. A. (2020). *A First Course in Network Science*. Cambridge University Press, Cambridge.

Newman, M. E. J. (2010). *Networks: An Introduction*. Oxford University Press, Oxford.

Pastor-Satorras, R. and Vespignani, A. (2007). *Evolution and Structure of the Internet: A Statistical Physics Approach*. Cambridge University Press, Cambridge.

Penrose, M. (2003). *Random Geometric Graphs*. Oxford University Press, Oxford.

Porter, M. A. and Gleeson, J. P. (2016). *Dynamical Systems on Networks: A Tutorial*. Springer, Cham.

Serrano, M. A. and Boguñá, M. (2021). *The Shortest Path to Network Geometry*. Cambridge University Press, Cambridge.

Van Der Hofstad, R. (2016). *Random Graphs and Complex Networks*. Volume 1. Cambridge University Press, Cambridge.

Van Dooren, P. (2009). *Graph Theory and Applications*. http://www.hamilton.ie/ollie/Downloads/Graph.pdf.

Van Mieghem, P. (2010). *Graph Spectra for Complex Networks*. Cambridge University Press, Cambridge.

Watts, D. J. (2004). *Six Degrees: The Science of a Connected Age*. W. W. Norton & Company, New York.

West, D. B. (1996). *Introduction to Graph Theory*. Prentice Hall, Upper Saddle River, NJ.

Whittle, P. (2007). *Networks: Optimisation and Evolution*. Cambridge University Press, Cambridge.

Wu, C. W. (2007). *Synchronization in Complex Networks of Nonlinear Dynamical Systems*. World Scientific, Singapore.

Reviews on complex networks

Aggarwal, C. and Subbian, K. (2014). Evolutionary network analysis: A survey. *ACM Computing Surveys (CSUR)*, **47**, 1.

Albert, R. and Barabási, A.-L. (2002). Statistical mechanics of complex networks. *Rev. Mod. Phys.*, **74**, 47.

Arenas, A., Díaz-Guilera, A., Kurths, J., Moreno, Y., and Zhou, C. (2008). Synchronization in complex networks. *Phys. Rep.*, **469**, 93.

Barabási, A.-L., Gulbahce, N., and Loscalzo, J. (2011). Network medicine: A network-based approach to human disease. *Nat. Rev. Genet.*, **12**, 56.

Barthélemy, M. (2011). Spatial networks. *Phys. Rep.*, **499**, 1.

Battiston, F., Cencetti, G., Iacopini, I., Latora, V., Lucas, M., Patania, A., Young, J.-G., and Petri, G. (2020). Networks beyond pairwise interactions: Structure and dynamics. *Phys. Rep.*, **874**, 1.

Ben-Naim, E., Krapivsky, P. L., and Redner, S. (2004). Extremal properties of random structures. In *Complex Networks* (ed. E. Ben-Naim, H. Frauenfelder, and

Z. Toroczkai), pp. 211–233. Springer, Berlin.

Boccaletti, S., Almendral, J. A., Guan, S., Leyva, I., Liu, Z., Sendiña-Nadal, I., Wang, Z., and Zou, Y. (2016). Explosive transitions in complex networks structure and dynamics: Percolation and synchronization. *Phys. Rep.*, **660**, 1.

Boccaletti, S., Bianconi, G., Criado, R., Del Genio, C. I., Gómez-Gardeñes, J., Romance, M., Sendiña-Nadal, I., Wang, Z., and Zanin, M. (2014). The structure and dynamics of multilayer networks. *Phys. Rep.*, **544**, 1.

Boguñá, M., Bonamassa, I., De Domenico, M., Havlin, S., Krioukov, D., and Serrano, M. Á. (2021). Network geometry. *Nature Rev. Phys.*, **3**, 114.

Braunstein, L. A., Wu, Z., Chent, Y., Buldyrev, S. V., Kalisky, T., Sreenivasan, S., Cohen, R., López, E., Havlin, S., and Stanley, H. E. (2007). Optimal path and minimal spanning trees in random weighted networks. *Int. J. Bifurc. Chaos Appl. Sci. Eng.*, **17**, 2215.

Burioni, R., Cassi, D., and Vezzani, A. (2008). Random walks and physical models on infinite graphs: An introduction. In *Random Walks and Geometry*, pp. 35–72. De Gruyter.

Castellano, C., Fortunato, S., and Loreto, V. (2009). Statistical physics of social dynamics. *Rev. Mod. Phys.*, **81**, 591.

Costa, L. da F., Rodrigues, F. A., Travieso, G., and Villas Boas, P. R. (2007). Characterization of complex networks: A survey of measurements. *Adv. Phys.*, **56**, 167.

de Arruda, G. F., Rodrigues, F. A., and Moreno, Y. (2018). Fundamentals of spreading processes in single and multilayer complex networks. *Phys. Rep.*, **756**, 1.

Dorogovtsev, S. N., Goltsev, A. V., and Mendes, J. F. F. (2008). Critical phenomena in complex networks. *Rev. Mod. Phys.*, **80**, 1275.

Dorogovtsev, S. N. and Mendes, J. F. F. (2002). Evolution of networks. *Adv. Phys.*, **51**, 1079.

Drmota, M. and Gittenberger, B. (2009). Analytic combinatorics on random graphs. In *Analysis of Complex Networks: From Biology to Linguistics* (ed. M. Dehmer and F. Emmert-Streib), pp. 425–450. John Wiley & Sons, New Jersey.

D'Souza, R. M., Gómez-Gardeñes, J., Nagler, J., and Arenas, A. (2019). Explosive phenomena in complex networks. *Adv. Phys.*, **68**, 123.

Ermann, L., Frahm, K. M., and Shepelyansky, D. L. (2015). Google matrix analysis of directed networks. *Rev. Mod. Phys.*, **87**, 1261.

Fortunato, S. (2010). Community detection in graphs. *Phys. Rep.*, **486**, 75.

Fortunato, S. and Hric, D. (2016). Community detection in networks: A user guide. *Phys. Rep.*, **659**, 1.

Holme, P. (2015). Modern temporal network theory: A colloquium. *Eur. Phys. J. B*, **88**, 1.

Holme, P. and Saramäki, J. (2012). Temporal networks. *Phys. Rep.*, **519**, 97.

Huang, X., Chen, D., Ren, T., and Wang, D. (2021). A survey of community detection methods in multilayer networks. *Data Min. Knowl. Discov.*, **35**, 1.

Jalili, M. and Perc, M. (2017). Information cascades in complex networks. *J. Complex Netw.*, **5**, 665.

Kivelä, M., Arenas, A., Barthelemy, M., Gleeson, J. P., Moreno, Y., and Porter,

M. A. (2014). Multilayer networks. *J. Complex Netw.*, **2**, 203.

Lee, D., Kahng, B., Cho, Y. S., Goh, K.-I., and Lee, D.-S. (2018). Recent advances of percolation theory in complex networks. *J. Korean Phys. Soc.*, **73**, 152.

Lee, C. and Wilkinson, D. J. (2019). A review of stochastic block models and extensions for graph clustering. *Appl. Netw. Sci.*, **4**, 1.

Li, M., Liu, R.-R., Lü, L., Hu, M.-B., Xu, S., and Zhang, Y.-C. (2021). Percolation on complex networks: Theory and application. *Phys. Rep.*, **907**, 1.

Liu, X., Li, D., Ma, M., Szymanski, B. K., Stanley, H. E., and Gao, J. (2020). Network resilience. *arXiv:2007. 14464.*

Lü, L., Medo, M., Yeung, C. H., Zhang, Y.-C., Zhang, Z.-K., and Zhou, T. (2012). Recommender systems. *Phys. Rep.*, **519**, 1.

Lü, L., Chen, D., Ren, X.-L., Zhang, Q.-M., Zhang, Y.-C., and Zhou, T. (2016). Vital nodes identification in complex networks. *Phys. Rep.*, **650**, 1.

Mariani, M. S., Ren, Z.-M., Bascompte, J., and Tessone, C. J. (2019). Nestedness in complex networks: observation, emergence, and implications. *Phys. Rep.*, **813**, 1.

Masuda, N., Porter, M. A., and Lambiotte, R. (2017). Random walks and diffusion on networks. *Phys. Rep.*, **716**, 1.

Mata, A. S. (2020). Complex networks: A mini-review. *Brazilian J. Phys.*, **50**, 658.

Newman, M. E. J. (2003). The structure and function of complex networks. *SIAM Review*, **45**, 167.

Pastor-Satorras, R., Castellano, C., Van Mieghem, P., and Vespignani, A. (2015). Epidemic processes in complex networks. *Rev. Mod. Phys.*, **87**, 925.

Perc, M., Jordan, J. J., Rand, D. G., Wang, Z., Boccaletti, S., and Szolnoki, A. (2017). Statistical physics of human cooperation. *Phys. Rep.*, **687**, 1.

Rodrigues, F. A., Peron, T. K. DM., Ji, P., and Kurths, J. (2016). The Kuramoto model in complex networks. *Phys. Rep.*, **610**, 1.

Sarkar, C. and Jalan, S. (2018). Spectral properties of complex networks. *Chaos*, **28**, 102101.

Schaeffer, S. E. (2007). Graph clustering. *Computer Science Review*, **1**, 27.

Strauss, D. (1986). On a general class of models for interaction. *SIAM Review*, **28**, 513.

Strogatz, S. H. (2001). Exploring complex networks. *Nature*, **410**, 268.

Susca, V. A. R., Vivo, P., and Kühn, R. (2021). Cavity and replica methods for the spectral density of sparse symmetric random matrices. *SciPost Phys. Lect. Notes*, **33**, 1.

Valdez, L. D., Shekhtman, L., La Rocca, C. E., Zhang, X., Buldyrev, S. V., Trunfio, P. A., Braunstein, L. A., and Havlin, S. (2020). Cascading failures in complex networks. *J. Complex Netw.*, **8**, cnaa013.

Wang, Z., Bauch, C. T., Bhattacharyya, S., d'Onofrio, A., Manfredi, P., Perc, M., Perra, N., Salathé, M., and Zhao, D. (2016). Statistical physics of vaccination. *Phys. Rep.*, **664**, 1.

Zdeborová, L. and Krzakala, F. (2016). Statistical physics of inference: Thresholds and algorithms. *Adv. Phys.*, **65**, 453.

Zhang, Z.-K., Liu, C., Zhan, X.-X., Lu, X., Zhang, C.-X., and Zhang, Y.-C. (2016).

Dynamics of information diffusion and its applications on complex networks. *Phys. Rep.*, **651**, 1.

Appendix A
Adjacency Matrix for Hypergraphs

One can introduce an adjacency matrix for a uniform hypergraph in a quite straight-forward way, similar to ordinary graphs. Clearly, for an m-uniform hypergraph with N vertices, this is an m-order N-dimensional hypermatrix (tensor) whose non-zero elements indicate edges (we later indicate the values of these elements). For uniform multi-hypergraphs, the adjacency matrix construction is only a little bit more complicated than for simple uniform hypergraphs, and it is also straightforward (Pearson and Zhang, 2015). As for non-uninform hypergraphs, which have a collection of hyperedges of various cardinalities, the situation is not that clear since one has to decide how to treat edges of cardinalities lower than maximum. For these 'difficult' hypergraphs, the adjacency matrix was defined by Banerjee, Char, and Mondal (2017), but see also Ouvrard, Goff, and Marchand-Maillet (2017), in the following way.

For the sake of simplicity, we focus on simple non-uniform hypergraphs. For such a hypergraph of N vertices, having edges of the maximum cardinality m, introduce a hypermatrix $\hat{A}$ of order m, $\overbrace{N \times N \times \ldots \times N}^{m}$, with the elements

$$\hat{A} = (a_{i_1 i_2 \ldots i_m}), \quad 1 \le i_1, i_2, \ldots, i_m \le N. \tag{A.1}$$

For each edge of cardinality $l \le m$, connecting vertices $r_1, r_2, \ldots, r_l$, set all the elements

$$a_{k_1 k_2 \ldots k_m} = \frac{l}{c(l,m)} \quad \text{with} \quad c(l,m) = \sum_{\substack{j_1, j_2, \ldots, j_l \ge 1 \\ \sum_a^l j_a = m}} \frac{m!}{j_1! j_2! \ldots j_l!}, \tag{A.2}$$

where the multisets $k_1, k_2, \ldots, k_m$ are chosen in all possible ways from the set $r_1, r_2, \ldots, r_l$ with the constraint that each of the elements of $r_1, r_2, \ldots, r_l$ must be present in each of the multisets $k_1, k_2, \ldots, k_m$ at least once. Set the remaining elements of the hypermatrix to zero. The denominator in Eq. (A.2) $c(l,m)$ is the number of such possible combinations of l different

388

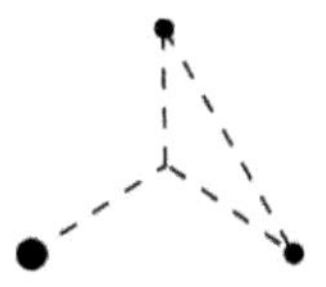

Fig. A.1 A simple non-uniform hypergraph with three edges of cardinalities 1, 2, and 3. The 1-edge adjacent to vertex 1 is shown as a big vertex.

elements placed in the sequences of m members. The summation in this denominator runs over all possible multiplicities $j_1, j_2, \ldots, j_l \geq 1$, $\sum_a^l j_a = m$, of the members of the set $r_1, r_2, \ldots, r_l$ in the multiset $k_1, k_2, \ldots, k_m$. Note that with this definition for simple hypergraphs, each entry of the adjacency hypermatrix can correspond, at most, to a single edge, as is should be.

In particular, for an m-uniform simple hypergraph, this definition, Eq. (A.2), is reduced to

$$a_{k_1 k_2 \ldots k_m} = \frac{1}{(m-1)!} \tag{A.3}$$

for all $m!$ permutations $k_1, k_2, \ldots, k_m$. The same is true for the edges of maximum cardinality in a non-uniform hypergraph. If such an edge is incident to vertex i, then, according to Eq. (A.2), the corresponding hypermatrix element is $a_{\underbrace{ii\ldots i}_{m}} = 1$.

This definition leads to the proper expression for the degree of a vertex (the number of edges attached to it) in term of the adjacency hypermatrix elements:

$$q_i = \sum_{i_2, i_3, \ldots, i_m}^{N} a_{i i_2 \ldots i_m} \tag{A.4}$$

and to the equality for the number of edges:

$$\sum_l l E_l = \sum_i q_i = \sum_{i_1, i_2, \ldots, i_m}^{N} a_{i_1 i_2 \ldots i_m}. \tag{A.5}$$

Here E_l in the number of edges of cardinality l in the graph.

For example, for the hypergraph shown in Figure A.1, we have $a_{111} = 1$ (1-edge), $a_{223} = a_{232} = a_{322} = a_{332} = a_{323} = a_{233} = 1/3$ (2-edge), $a_{123} = a_{132} = a_{213} = a_{231} = a_{312} = a_{321} = 1/2$ (3-edge), and the degrees of the vertices $q_1 = q_2 = q_3 = 2$.

Appendix B
Spectra of symmetric normalized Laplacians of sample graphs

Here we demonstrate spectra of symmetric normalized Laplacian matrices $\mathcal{L}$ for few sample graphs shown in Figure 2.15.[1] One can compare them with spectra of Laplacian matrices L of the same graphs outlined in Section 2.9. For the fully connected graph of N vertices (Figure 2.15a), we have

$$\lambda_1 = 0, \quad \lambda_{2 \leq \alpha \leq N} = \frac{N}{N-1},$$
$$\mathbf{v}_1 = \mathbf{1},$$
$$\mathbf{v}_\alpha = (1, 0, 0, \dots, -1, 0, \dots)^T, \tag{B.1}$$

where $\mathbf{1}$ is the column vector with all entries equal 1.

For the q-star with the central vertex 1 and q leaves (Figure 2.15b), the spectrum is

$$\lambda = 0, \overbrace{1, 1, \dots, 1}^{q-1}, 2,$$
$$\mathbf{v}_1 = (\sqrt{q}, 1, 1, \dots, 1)^T,$$
$$\mathbf{v}_{2 \leq \alpha \leq q} = (0, 1, 0, 0, \dots, -1, 0, 0, \dots)^T,$$
$$\mathbf{v}_{q+1} = (\sqrt{q}, -1, -1, \dots, -1)^T. \tag{B.2}$$

Here in $\mathbf{v}_{2 \leq \alpha \leq q}$, the second entry equals 1 and the $\alpha + 1$-th one equals -1.

For the ring of 8 vertices (Figure 2.15c),

$$\lambda = 0, 1 - \sqrt{\frac{1}{2}}, 1 - \sqrt{\frac{1}{2}}, 1, 1, 1 + \sqrt{\frac{1}{2}}, 1 + \sqrt{\frac{1}{2}}, 2,$$
$$\mathbf{v}_1 = \mathbf{1},$$
$$\mathbf{v}_2 = (1, \sqrt{2}, 1, 0, -1, -\sqrt{2}, -1, 0)^T, \quad \dots . \tag{B.3}$$

---

[1]The set of eigenvalues of the normalized Laplacian matrix $\tilde{\mathcal{L}}$ coincides with that of $\mathcal{L}$, while eigenvectors are different.

For the chain of 8 vertices (Figure 2.15d),

$$\lambda \approx 0, 0.099, 0.377, 0.777, 1.223, 1.623, 1.901, 2,$$
$$\mathbf{v}_1 = (1, \sqrt{2}, \sqrt{2}, \sqrt{2}, \sqrt{2}, \sqrt{2}, \sqrt{2}, 1)^T,$$
$$\mathbf{v}_2 \approx (1, 1.274, 0.882, 0.315, -0.315, -0.882, -1.274, -1)^T, \ldots \quad (B.4)$$

For the ring of 8 vertices with opposite vertices linked (Figure 2.15e),

$$\lambda = 0, \frac{2}{3}, \frac{2}{3}, \frac{4-\sqrt{2}}{3}, \frac{4-\sqrt{2}}{3}, \frac{4}{3}, \frac{4+\sqrt{2}}{3}, \frac{4+\sqrt{2}}{3},$$
$$\mathbf{v}_1 = 1,$$
$$\mathbf{v}_2 = (1, 0, -1, 0, 1, 0, -1, 0)^T, \ldots \quad (B.5)$$

Finally, for the Cayley tree of 10 vertices of degree 3 (Figure 2.15f),

$$\lambda = 0, 1 - \sqrt{\frac{2}{3}}, 1 - \sqrt{\frac{2}{3}}, 1, 1, 1, 1, 1 + \sqrt{\frac{2}{3}}, 1 + \sqrt{\frac{2}{3}}, 2,$$
$$\mathbf{v}_1 = (\sqrt{3}, \sqrt{3}, \sqrt{3}, \sqrt{3}, 1, 1, 1, 1, 1, 1)^T,$$
$$\mathbf{v}_2 = (0, \sqrt{2}, -\sqrt{2}, 0, 1, 1, -1, -1, 0, 0)^T, \ldots \quad (B.6)$$

Figure 2.16 shows these spectra together with the spectra of the adjacency matrix A and the Laplacian matrix L.

Appendix C
Generating Functions

In random graph theory, the generating functions technique is ideally suited for locally tree-like uncorrelated networks. As well, it is the main approach for the solution of discrete difference equations. Here we list the basic formulae for generating functions. See Wilf (2013) for useful details.

The definition of the generating function of a distribution $P(q)$ is [1]

$$G(z) \equiv \sum_{q=0}^{\infty} P(q)z^q. \tag{C.1}$$

The inverse of $G(z)$, that is, the coefficients of the Taylor series for $G(z)$, is

$$P(q) = \frac{1}{q!} \frac{d^q G(z)}{dz^q}\bigg|_{z=0} = \frac{1}{2\pi i} \oint_C dz \frac{G(z)}{z^{q+1}}. \tag{C.2}$$

Here C is a contour around 0, which does not enclose the singularities of $G(z)$.

The limiting theorems are

$$G(0) = P(0) \tag{C.3}$$

and, if the finite limit $P(q \to \infty)$ exists,

$$\lim_{z \to 1} (1 - z)G(z) = P(q \to \infty). \tag{C.4}$$

The generating function $G_\Sigma(z)$ of the distribution

$$Q(s) = \sum_{q_1,\dots,q_n} \prod_{i=1}^{n} P_i(q_i)\delta(s, \sum_{i=1}^{n} q_i) \tag{C.5}$$

of the sum of n random variables, $s = \sum_{i=1}^{n} q_i$, distributed with distributions $P_i(q_i)$, $i = 1, \dots, n$, equals the product

---

[1] This coincides with the Z-transform with the changed sign of the power in the sum.

$$G_\Sigma(z) = \prod_{i=1}^{n} G_i(z), \tag{C.6}$$

where $G_i(z) \equiv \sum_{q=0} P_i(q)z^q$. In particular, the generating function of the convolution is the product of two generating functions:

$$\sum_{v=0}^{q} P_a(v)P_b(q-v) \rightarrow G_a(z)G_b(z), \text{ where } P_a(q) \rightarrow G_a(z) \text{ and } P_b(q) \rightarrow G_b(z). \tag{C.7}$$

The formula for moments is

$$\langle q^n \rangle \equiv \sum_{q=0} q^n P(q) = \left[\left(z\frac{d}{dz} \right)^n G(z) \right]_{z=1}, \tag{C.8}$$

so $\langle q \rangle = G'(1)$, $\langle q^2 \rangle = G'(1) + G''(1)$, and so on, while

$$G(1) = 1, \ G'(1) = \langle q \rangle, \ G''(1) = \langle q(q-1) \rangle, \ G'''(1) = \langle q(q-1)(q-2) \rangle, \text{ etc.} \tag{C.9}$$

Here are some of useful relations:

$$\sum_{q=0} P(q-1)z^q = zG(z),$$

$$\sum_{q=0} P(q+1)z^q = \frac{1}{z}[G(z) - P(0)],$$

$$\sum_{q=0} qP(q)z^q = zG'(z), \tag{C.10}$$

$$\sum_{q=0} (q-1)P(q-1)z^q = z^2 G'(z),$$

$$\sum_{q=0} (q+1)P(q+1)z^q = G'(z),$$

$$\frac{1}{n!} \sum_{q=0} \binom{q}{n} P(q)z^q = z^n \frac{d^n}{dz^n} G(z).$$

In particular, for a typical sum in the k-core problems, we have

$$\sum_{q} P(q) \sum_{s=k}^{q} \binom{q}{s} x^s (1-x)^{q-s} = 1 - \sum_{s=0}^{k-1} x^s \frac{G^{(s)}(1-x)}{s!}, \tag{C.11}$$

where $G^{(s)}(x)$ denotes the s-th derivative of $G(x)$ over x.

C GENERATING FUNCTIONS

Let $P(q = 0) = 0$. Then

$$\sum_{q=1}^{\infty} \frac{P(q)}{q} = \sum_{q=1}^{\infty} P(q) \int_0^{\infty} dz\, z^{q-1} = \int_0^{\infty} dz \frac{G(z)}{z}. \tag{C.12}$$

The generating function of the Poisson distribution is

$$P(q) = e^{-\langle q \rangle} \langle q \rangle^q / q! \;\rightarrow\; G(z) = e^{\langle q \rangle (z-1)}, \tag{C.13}$$

its derivative equals

$$G_1(z) \equiv \frac{G'(z)}{\langle q \rangle} = e^{\langle q \rangle (z-1)}; \tag{C.14}$$

and so, for the Poisson distribution, it holds that $\langle q^2 \rangle = \langle q \rangle^2 + \langle q \rangle$. The average number of the second-nearest neighbours of a vertex in infinite uncorrelated graphs is $z_2 = \langle q^2 \rangle - \langle q \rangle$. Hence in such graphs with the Poisson degree distribution, $z_2 = \langle q \rangle^2 \equiv z_1^2$ and the average branching $\langle b \rangle = (\langle q^2 \rangle - \langle q \rangle) / \langle q \rangle = \langle q \rangle$.

The generating function of the exponential function is

$$e^{-q/r} \;\rightarrow\; G(z) = \frac{1}{1 - e^{-1/r} z}, \tag{C.15}$$

that is, it has a pole at $z = e^{1/r} > 1$.

For a power-law degree distribution $P(q) \cong A q^{-\gamma}$, with a minimum degree q_0, we can write (we set $z = 1 - x$, assuming $x \ll 1$):

$$G(1 - x) = \sum_q A q^{-\gamma} (1 - x)^q \cong \int_{q_0}^{\infty} dq\, A q^{-\gamma} e^{-qx}. \tag{C.16}$$

Let $y = qx$, then

$$G(1 - x) \cong x^{\gamma - 1} \int_{q_0 x}^{\infty} dy\, A y^{-\gamma} e^{-y}. \tag{C.17}$$

Integrating by parts twice gives, for a noninteger γ,

$$G(1 - x) \cong 1 - \langle q \rangle x + A\Gamma(1 - \gamma) x^{\gamma - 1} + O(x^2). \tag{C.18}$$

The term with $x^{\gamma - 1}$ is either the leading, the second, or the third, etc., term depending on the value of exponent γ. Keeping only the leading terms in small x, we find, for a non-integer γ, that for $3 < \gamma < 4$, $G(1-x) \cong 1 - \langle q \rangle x + \frac{1}{2} \langle q(q-1) \rangle x^2 + A\Gamma(1-\gamma)x^{\gamma-1}$, (C.19)

for $2 < \gamma < 3$, $G(1-x) \cong 1 - \langle q \rangle x + A\Gamma(1-\gamma)x^{\gamma-1}$, (C.20)

for $1 < \gamma < 2$, $G(1-x) \cong 1 + A\Gamma(1-\gamma)x^{\gamma-1} + Bx$, (C.21)

where the coefficient B of the linear term for $1 < \gamma < 2$ is no longer equal to the average degree $\langle q \rangle$, which diverges, but instead depends on a specific form of the distribution,

$$B = -\left\{ \sum_q q[P(q) - Aq^{-\gamma}] - \zeta(\gamma - 1) \right\}.$$ (C.22)

Note that $\Gamma(z) < 0$ for $z \in (-1, 0)$ while $\Gamma(z) > 0$ for $z \in (-2, -1)$.

For an integer γ in $P(q) \cong Aq^{-\gamma}$, using Eq. (C.16), one can get the following singularities:

for $\gamma = 4$, $G(1-x) \cong 1 + A\frac{1}{6}(1-x)^3 \ln(1-x)$

$$- G'(1)(1-x) + \frac{1}{2}G''(1)(1-x)^2 - \frac{1}{6}G'''(1)(1-x)^3 + \ldots,$$

for $\gamma = 3$, $G(1-x) \cong 1 - A\frac{1}{2}(1-x)^2 \ln(1-x)$

$$- G'(1)(1-x) + \frac{1}{2}G''(1)(1-x)^2 + \ldots,$$

for $\gamma = 2$, $G(1-x) \cong 1 + A(1-z) \ln(1-x) - G'(1)(1-x) + \ldots.$ (C.23)

In a similar way, we can easily obtain the singularities of the generating functions of more complex fat-tailed distributions, for example, of $P(q) \cong Aq^{-\gamma} \ln^\delta q$.

For the distribution

$$P(q) = \left[\frac{\langle q \rangle (\gamma - 2)}{2(\gamma - 1)} \right]^{\gamma-1} \frac{\Gamma(q - \gamma + 1, \frac{\langle q \rangle (\gamma-2)}{2(\gamma-1)})}{\Gamma(q+1)}$$ (C.24)

having the power-law asymptotics $P(q) \sim q^{-\gamma}$ for $\gamma > 2$, where $\Gamma(s, x)$ is the upper incomplete gamma function, the generating function can be written explicitly,

$$G(z) = (\gamma - 1)E_n \left[(1 - z)\frac{\langle q \rangle (\gamma - 2)}{2(\gamma - 1)} \right],$$ (C.25)

where $E_n(x) = \int_1^\infty dy\, y^{-n} e^{-xy}$ is the exponential integral.

Finally, the generating function of a power law with an exponential cut-off is

$$[1 - \delta(q,0)]q^{-\gamma}e^{-q/r} \to G(z) = \text{Li}_\gamma(e^{-1/r}z), \qquad \text{(C.26)}$$

where $\text{Li}_\gamma(x) = \sum_{k=1}^{\infty} k^{-\gamma}x^k$ is the polylogarithm function. Near zero, $\text{Li}_\gamma(x) \cong x$, while near the special point $x = 1$,

$$\text{Li}_\gamma(e^{-\mu}) \cong \zeta(\gamma) + \Gamma(1-\gamma)\mu^{\gamma-1} + \sum_{k=1}^{\infty} \frac{\zeta(\gamma-k)}{k!}(-\mu)^k, \qquad \text{(C.27)}$$

where $\ln(1/x) = \mu \ll 1$. Here $\zeta(\gamma)$ is the Riemann zeta function. Consequently, the point of singularity of $G(z)$ in this case is $z^* = e^{1/r}$, and

$$r = 1/\ln z^*. \qquad \text{(C.28)}$$

In a number of interesting situations, for example, Eq. (3.29), a generating function is obtained implicitly as the solution of the functional equation

$$G(z) = zF(G(z)), \qquad \text{(C.29)}$$

where the function $F(x)$ satisfies the following conditions: (i) $F(0) \neq 0$ and (ii) $\left[\dfrac{d^n F(x)}{dx^n}\right]_{x=0} < \infty$ for $n \geq 0$. In this case the useful *Lagrange inversion formula* gives the solution of Eq. (C.29) in the form of the series expansion:

$$G(z) = \sum_{q=1}^{\infty} \frac{1}{q!}\left[\frac{d^{q-1}}{dG^{q-1}}(F(G))^q\right]_{G=0} z^q, \qquad \text{(C.30)}$$

whose coefficients provide the original of the generating function.

Appendix D
Hyperscaling Relations for Percolation

Section 3.5 introduced some of the critical exponents associated with the birth of a giant connected component, which is a version of a general percolation problem. The full set of critical exponents for percolation are discussed in Stauffer and Aharony (1991). Here we outline relations between exponents for pairwise correlation functions, ν and η, only for the sake of brevity focusing on percolation in lattices, although the relations are far more general. The set of scaling forms describing size dependencies contains these exponents. These scaling forms represent the *hyperscaling* (Privman, 1990). The set of relations between these and other exponents, containing space dimension, are the *hyperscaling relations*.

Let the problem be ordinary (site or bond, either will do) percolation on a D-dimensional lattice, and the deviation from the percolation threshold be $\delta = |p - p_c|$, where p is the fraction of retained sites (or bonds). Near the threshold, the pairwise correlation function $c(r) = \langle s(0)s(r) \rangle$ for the spin variable $s = 0, 1$ decays with distance r (measured over the lattice) between two vertices as

$$c(r) \sim e^{r/\xi(\delta)}, \tag{D.1}$$

where the correlation length has the following singularity at the percolation threshold:

$$\xi \sim \delta^{-\nu}, \tag{D.2}$$

which defines the critical exponent ν. At the threshold, the decay is power-law,

$$c(r) \sim r^{-(D-2+\eta)}, \tag{D.3}$$

which defines the critical exponent η.

For $D \leq D_{uc}$, where D_{uc} is the upper critical dimension of a given problem with a continuous phase transition, the hyperscaling relations between the critical exponents ν, η, σ, introduced in Eq. (3.43), and β, where the relative size of the percolation cluster (order parameter) $S \sim \delta^\beta$, have the form

$$1/D_\mathrm{f} = \sigma\nu, \tag{D.4}$$

$$D_\mathrm{f} = D - \beta/\nu, \tag{D.5}$$

$$D - 2 + \eta = 2\beta/\nu. \tag{D.6}$$

Here D_f is the *fractal dimension* of the percolation cluster at criticality, which shows how the number of vertices $M(L)$ of the percolation cluster occurring within a D-dimensional hypercube $L \times L \times \ldots \times L$ in the lattice diverges at $p \downarrow p_c$ as $L \to \infty$: $M(L) \sim L^{D_\mathrm{f}}$. The distance L is measured over the lattice but not over the percolation cluster itself, in contrast to how the dimension D_H was introduced in Section 1.3. So the two dimensions for the percolation cluster at criticality, D_f and D_H, mentioned in Section 3.4 should not be confused.

Let us briefly explain how these relations were derived (Stauffer and Aharony, 1991).

(i) According to the scaling form of the distribution $\mathcal{P}(s,t)$, Eq. (3.43), the critical features are determined by typical cluster sizes $s \sim \delta^{-1/\sigma}$. In the critical region, the clusters are fractals,

$$\delta^{-1/\sigma} \sim s \sim \xi^{D_\mathrm{f}} \sim \delta^{-D_\mathrm{f}\nu}, \tag{D.7}$$

so we readily arrive at Eq. (D.4).

(ii) Consider a hypercube of L^D vertices and estimate the number of vertices $M(L)$ of the percolation cluster falling inside this hypercube above p_c. For $L \ll \xi$, this number is $M \sim L^{D_\mathrm{f}}$, while for $L \gg \xi$ it is $M \sim SL^D$. So for $L \sim \xi$,

$$\xi^{D_\mathrm{f}} \sim \delta^\beta \xi^D, \tag{D.8}$$

which gives Eq. (D.5).

(iii) Finally, in general, the spin–spin correlation function near a continuous phase transition decays as $r^{-(D-2+\eta)}$ until the spin separation r approaches the correlation radius ξ. So we can estimate

$$\xi^{-(D-2+\eta)} \sim S^2 \sim \delta^{2\beta}, \tag{D.9}$$

which gives Eq. (D.5).

Furthermore, Eq. (D.2) directly leads to the following finite size effect:

$$p_c(\infty) - p_c(N) \propto N^{-1/(D\nu)}, \tag{D.10}$$

where $p_c(\infty)$ is the critical point value in the infinite system in which the transition is well defined and $p_c(N)$ is, in particular, the position of the maximum of the susceptibility for the system of N vertices (or another observable diverging at the critical point in the corresponding infinite system).

What is the form of these relations, Eqs. (D.1)–(D.3) and (D.10) when $D \geq D_{uc}$, that is, in the mean-field theory regime? It is well known how to answer this question for general cooperative models with a continuous phase transition. Above the upper critical dimension, in relations of this sort, one should substitute D_{uc} for D and set the standard mean-field theory values for ν and η (Privman, 1990). One should note that, in contrast to other critical exponents, the exponents γ, ν, and η are common for general cooperative models with a continuous phase transition in the mean-field theory regime,

$$\gamma = 1, \quad \nu = \frac{1}{2}, \quad \eta = 0. \tag{D.11}$$

For the ordinary percolation problem in this regime (and, as we saw, for classical random graphs), the critical exponents $\beta = 1$, $\tau = \frac{5}{2}$, and $\sigma = \frac{1}{2}$. In general complex networks, even in infinite-dimensional ones, these three exponents can differ from ordinary percolation. So we do not set them equal to these three values here. Then, recalling Eqs. (3.46) and (3.48), we get[1]

$$D_{uc} = 2 + 4\beta, \tag{D.12}$$

$$D_f = 2 + 2\beta \tag{D.13}$$

and

$$p_c(\infty) - p_c(N) \propto N^{-2/D_{uc}}. \tag{D.14}$$

For ordinary percolation (including the case of the classical random graphs), Eqs. (D.12) and (D.13) give $D_{uc} = 6$ and $D_f = 4$, respectively. Combining the scaling form of the probability distribution $\mathcal{P}(s,p)$ that a uniformly randomly chosen vertex belongs to a cluster of s vertices, $\mathcal{P}(s,p) = s^{1-\tau}f(s\delta^{1/\sigma})$, where $\delta = |p - p_c|$, Eq. (3.43), and Eqs. (D.11) or (D.14), we arrive at the following distribution $\mathcal{P}_c(s,N)$ within the scaling window:

[1]From Eqs. (D.4)–(D.6), we obtain three relations for the dimensions D_{uc} and D_f, but only two of them are independent.

- Below the upper critical dimension $D_u = 6$, within the scaling window $|p - p_c| \lesssim N^{-1/(D\nu)}$, the power-law distribution $\mathcal{P}_c(s, N) \sim s^{1-\tau}$ has the size-dependent cut-off $s_{cut} \sim N^{1/(D\nu\sigma)}$.
- Above the upper critical dimension, within the scaling window $|p - p_c| \lesssim N^{1/3}$, the power-law distribution $\mathcal{P}_c(s, N) \sim s^{-3/2}$ has the cut-off $s_{cut} \sim N^{2/3}$.

Above D_{uc}, the fractal dimension D_f of the percolation cluster at criticality and the Hausdorff dimension of the percolation cluster, D_H, are related in the following simple way:

$$D_H = D_f/2 = 1 + \beta \tag{D.15}$$

(Cohen, Havlin, and Ben-Avraham, 2003b). The reason is that above D_{uc}, the paths within the percolation cluster at criticality can be seen as random walks on the embedding lattice. So distances between vertices of the percolation cluster measured over the lattices scale as the square root of the corresponding distances measured over the percolation cluster, which results in the relation: $D_H = D_f/2$. For ordinary percolation and for the classical random graphs, this gives $D_H = 2$, as mentioned in Section 3.4.

Let us rank the sizes of clusters in descending order, $s^{(1)}, s^{(2)}, \ldots$. Then, for $D \leq D_{uc} = 2 + 4\beta$, within the scaling window $|p - p_c| \lesssim N^{-1/(D\nu)}$, the cluster sizes are asymptotically related to N in the following way,

$$s^{(1 \leq a < \infty)}(N) \sim N^{D_f/D}. \tag{D.16}$$

For $D \geq D_{uc} = 2 + 4\beta$, within the scaling window $|p - p_c| \lesssim N^{-2/D_{uc}}$,

$$s^{(1 \leq a < \infty)}(N) \sim N^{D_f/D_{uc}}, \tag{D.17}$$

where $D_f = 2 + 2\beta$.

Appendix E

Degree Distribution of a Damaged Network

Let p be the fraction of undeleted edges, so for the original, undamaged network, $p = 1$, and the degree distribution $P_1(q) \equiv P(q)$. In the bond problem (removal of edges), the resulting distribution $P_p(q)$ is

$$P_p(q) = \sum_{k \geq q} P(k) \binom{k}{q} p^q (1-p)^{q-k}, \qquad (E.1)$$

so

$$\langle q \rangle_p = p \langle q \rangle \qquad (E.2)$$

and

$$\langle q^2 \rangle_p = p^2 \langle q^2 \rangle + p(1-p)\langle q \rangle. \qquad (E.3)$$

On a scale-free distribution, $P(q) = Aq^{-\gamma}$, the damage acts in the following way. The low-degree part of the distribution increases. In particular, (additional) vertices of degree 0 and especially importantly ones of degree 1 emerge. The high-degree asymptotics stays $\propto q^{-\gamma}$, but its amplitude A_p decreases. This amplitude can be directly obtained from Eq. (E.1) but let us use a simple reasoning possible for $p \ll 1$. Under edge removal, for large q, the number of (surviving) vertices with degrees $k > q$ in the damaged network should be equal to the the number of vertices with degrees $k > q/p$ in the original network. After integration, this gives

$$A_p q^{\gamma-1} \cong A p^{\gamma-1} q^{\gamma-1}. \qquad (E.4)$$

So

$$A_p \cong A p^{\gamma-1}. \qquad (E.5)$$

Appendix F
Non-backtracking Matrix

The non-backtracking matrix B has the entries

$$B_{i \leftarrow j, k \leftarrow l} = \delta_{j,k}(1 - \delta_{i,l}). \tag{F.1}$$

B is a non-symmetric non-negative matrix. According to the Perron-Frobenius theorem, its principal eigenvalue λ_1 is real and positive, and all elements of the corresponding eigenvector have the same sign, if B is irreducible.

Let us focus on undirected simple graphs and denote the leading eigenvalues of the adjacency and non-backtracking matrices as $\lambda_1^{(A)}$ and $\lambda_1^{(B)}$, respectively. Then

$$\lambda_1^{(A)} \geq \lambda_1^{(B)} \tag{F.2}$$

(Karrer, Newman, and Zdeborová, 2014). For example, for a random q-regular graph, the adjacency matrix has q non-zero entries in each row and column, while the non-backtracking matrix has $q - 1$ non-zero entries in each row and column. Consequently, for this network,

$$\lambda_1^{(A)} = q, \quad \lambda_1^{(B)} = q - 1. \tag{F.3}$$

For a tree,

$$\mathrm{Tr}B^n = \sum_i (\lambda_i^{(B)})^n = 0 \tag{F.4}$$

for any power $n \geq 1$, since closed non-backtracking walks are impossible in it. Hence all eigenvalues of the non-backtracking matrix are zero in a tree.

For uncorrelated networks, in the annealed network approximation, $A_{ij} \approx q_i q_j / (N\langle q \rangle)$, one can evaluate the elements of the principal eigenvector of the non-backtracking matrix as

$$v_{i \leftarrow j} \sim q_j - 1, \tag{F.5}$$

and the largest eigenvalue as

$$\lambda_1^{(B)} = \frac{\langle q(q-1) \rangle}{\langle q \rangle} < \lambda_1^{(A)} = \frac{\langle q^2 \rangle}{\langle q \rangle} \tag{F.6}$$

(Martin, Zhang, and Newman, 2014; Pastor-Satorras and Castellano, 2020).

The spectrum of the non-backtracking matrix contains $2E$ eigenvalues, of which at least $2(E-N)$ eigenvalues are equal to ± 1. Each tree connected component of a graph, or a tree branch dangling off it add zero eigenvalues to the spectrum of the non-backtracking matrix. Each connected component with a single cycle add -1, 0, or 1 eigenvalues (Krzakala, Moore, Mossel, Neeman, Sly, Zdeborová, and Zhang, 2013). The $2N \times 2N$ matrix

$$M = \begin{pmatrix} A & I-D \\ I & \mathbf{0} \end{pmatrix},$$ (F.7)

where A, I, and D are the adjacency, identity, and degree matrices, respectively, has the spectrum consisting just of those $2N$ eigenvalues of B which differ from ± 1 (Krzakala, Moore, Mossel, Neeman, Sly, Zdeborová, and Zhang, 2013; Saade, Krzakala, and Zdeborová, 2014). So these $2N$ eigenvalues of the matrix B are the roots of the polynomial

$$\det[(\lambda^2 - 1)I - \lambda A + D] = 0,$$ (F.8)

where the matrix $H(\lambda) = (\lambda^2 - 1)I - \lambda A + D$ is a symmetric real matrix, called the Bethe Hessian operator or the deformed Laplacian (Bass, 1992). The complexity of finding the full spectrum of the matrix M is determined by the number of its non-zero entries, $2(E+N)$, which is of the same order as the number of non-zero entries of the adjacency matrix, $2E$. The principal eigenvector $\mathbf{v}^{(M)}$ of the matrix M has the following $2N$ components:

$$v_i^{(M)} = x_i^{(\mathrm{in})} \equiv \sum_{j \in \partial i} v_{i \leftarrow j}, \quad v_{N+i}^{(M)} = x_i^{(\mathrm{out})} \equiv \sum_{j \in \partial i} v_{j \leftarrow i},$$ (F.9)

$i = 1, 2, \ldots, N$, where $v_{i \leftarrow j}$ are the components of the principal eigenvector of the non-backtracking matrix. The first N components of $\mathbf{v}^{(M)}$ are the non-backtracking centralities $x_i^{(\mathrm{in})} \equiv x_i$ of the vertices in a network. For uncorrelated networks, the non-backtracking centralities are estimated as

$$x_i^{(\mathrm{in})} \sim q_i$$ (F.10)

or, in more rigorous terms, introducing the average non-backtracking centrality of a vertex of degree q, as $\langle x^{(\mathrm{in})}(q) \rangle \approx q/(N\langle q \rangle)$. Clearly, the principal eigenvalues of the matrices M and B coincide, $\lambda_1^{(M)} = \lambda_1^{(B)}$, and hence

$$\lambda_1^{(B)} x_i^{(\mathrm{out})} = (q_i - 1)x_i^{(\mathrm{in})}.$$ (F.11)

Construct the non-backtracking expansion of a finite graph G in the way explained in Figure 6.6b, but this time begin unfolding not from a vertex but from edge $j \leftarrow i$, following the opposite direction to the arrow. Consider the resulting infinite tree branch and classify its edges (numerous replicas of the edges of the graph G, which we keep with the same labels as in G) according to their distance ℓ from the initial edge $j \leftarrow i$. Let $n(\ell, j \leftarrow i)$ be the number of edges of the branch at distance ℓ from edge $j \leftarrow i$ and $n_{l \leftarrow k}(\ell, j \leftarrow i)$ be the number of the copies of edge $l \leftarrow k$ at distance ℓ from $j \leftarrow i$. Then $n_{l \leftarrow k}(\ell, j \leftarrow i)$ satisfies the equation:

$$n_{l \leftarrow k}(\ell + 1, j \leftarrow i) = \sum_{l' \leftarrow k'} n_{l' \leftarrow k'}(\ell, j \leftarrow i) B_{l \leftarrow k, l' \leftarrow k'}. \tag{F.12}$$

Introduce the ratio

$$f_{l \leftarrow k}(\ell, j \leftarrow i) = \frac{n_{l \leftarrow k}(\ell, j \leftarrow i)}{n(\ell, j \leftarrow i)} \tag{F.13}$$

and rewrite Eq. (F.12) as

$$\frac{n(\ell+1, j \leftarrow i)}{n(\ell, j \leftarrow i)} f_{l \leftarrow k}(\ell+1, j \leftarrow i) = n(\ell, j \leftarrow i) \sum_{l' \leftarrow k'} f_{l' \leftarrow k'}(\ell, j \leftarrow i) B_{l \leftarrow k, l' \leftarrow k'}, \tag{F.14}$$

The limit

$$b = \lim_{\ell \to \infty} \frac{n(\ell + 1, j \leftarrow i)}{n(\ell, j \leftarrow i)} \tag{F.15}$$

is independent of the starting edge $j \leftarrow i$. This is the branching b of the non-backtracking expansion ('global branching'). Introducing the limiting vector $\mathbf{f}^{(\infty)}$ with the components $f_{l \leftarrow k}^{(\infty)} = f_{l \leftarrow k}(\ell \to \infty, j \leftarrow i)$, also independent of the starting edge $j \leftarrow i$, enables one to represent Eq. (F.14) in the matrix form:

$$b\mathbf{f}^{(\infty)} = B\mathbf{f}^{(\infty)}. \tag{F.16}$$

So the largest eigenvalue of the non-backtracking matrix of graph G equals the branching of the non-backtracking expansion of this graph,

$$\lambda_1^{(B)} = b. \tag{F.17}$$

Furthermore, for infinite locally tree-like networks with degree–degree correlations between the nearest neighbouring vertices, the following equality holds

$$\lambda_1^{(B)} = \lambda_1^{(\text{branching matrix})} = b, \tag{F.18}$$

where $\lambda_1^{(\text{branching matrix})}$ is the largest eigenvalue of the branching matrix introduced in Eq. (4.40). For uncorrelated networks, $b = \langle q^2 \rangle / \langle q \rangle - 1$ (Timár, da Costa, Dorogovtsev, and Mendes, 2021).

The non-backtracking expansion also allows one to interpret the components of the principal eigenvector of the matrix M (and so the non-backtracking centralities of the vertices). Let $n_k(\ell, j \leftarrow i)$ be the number of the copies of the graph's vertex k at distance ℓ from the starting edge $j \leftarrow i$ of the non-backtracking expansion. Furthermore, for each vertex at distance ℓ from the starting edge $j \leftarrow i$ let us find the label of its parent vertex in the shell $\ell - 1$ and denote the total number of the parents with label k for all vertices of the shell ℓ, and denote the result by $\tilde{n}_k(\ell, j \leftarrow i)$. Consider the ratio

$$g_k(\ell, j \leftarrow i) = \frac{n_k(\ell, j \leftarrow i)}{n(\ell, j \leftarrow i)}, \quad g_{N+k}(\ell, j \leftarrow i) = \frac{\tilde{n}_k(\ell, j \leftarrow i)}{n(\ell, j \leftarrow i)}, \quad \text{(F.19)}$$

$k = 1, 2, \ldots, N$. Introducing the limiting vector $\mathbf{g}^{(\infty)}$ with the components $g_k^{(\infty)} = g_k(\ell \to \infty, j \leftarrow i)$, $k = 1, 2, \ldots, 2N$, independent of the starting edge $j \leftarrow i$ of the non-backtracking expansion, one can get the equation:

$$b\mathbf{g}^{(\infty)} = M\mathbf{g}^{(\infty)}, \quad \text{(F.20)}$$

demonstrating that $\mathbf{g}^{(\infty)}$ is the principal eigenvector of the matrix M. Its first N components, $k = 1, 2, \ldots, N$, that is, the non-backtracking centralities $x_k^{(in)}$, are interpreted as the fraction of vertices of label k (copies of vertex k in the original graph) on the boundary at infinity of the non-backtracking expansion. The other N components, $x_{N+k}^{(out)}$, $k = 1, 2, \ldots, N$, are interpreted as the probability that the parent of a randomly chosen vertex on the boundary at infinity of the non-backtracking expansion has label k.

Appendix G
Treating General Interdependent Networks

Let us derive equations for a giant mutually connected component in more general uncorrelated interdependent networks than multiplexes in Section 8.2, with an arbitrary distribution of interlinks connecting vertices in different layers (Figure 8.3). We consider a generalization of the configuration model for a two-layer network completely described by two joint probability distributions $P_a(q_a, k_a)$ and $P_b(q_b, k_b)$ for vertices in layers a and b, respectively. For example, $P_a(q_a, k_a)$ is the probability that a vertex in layer a has q_a adjacent edges within this layer and k_a interlinks to vertices in layer b. We tend the total number of vertices in two layers, $N = N_a + N_b$, to infinity.

The distributions $P_a(q_a, k_a)$ and $P_b(q_b, k_b)$ determine the relation between N_a and N_b. Using the condition $N_a \langle k_a \rangle = N_b \langle k_b \rangle$, we get the ratio

$$\frac{N_a}{N_b} = \frac{\sum_{q_b, k_b} k_b P_b(q_b, k_b)}{\sum_{q_a, k_a} k_a P_a(q_a, k_a)}. \tag{G.1}$$

This network is locally tree-like similarly to the standard configuration model for one-layer networks, which enables us to introduce four probabilities X_a, Y_a, X_b, and Y_b, explained in Figure G.1. The self-consistency equations for these probabilities, shown in graphical form in Figure G.1b, can be written out as

$$X_a = \sum_{q_a, k_a} \frac{q_a P_a(q_a, k_a)}{\langle q_a \rangle} [1 - (1 - X_a)^{q_a - 1}][1 - (1 - Y_a)^{k_a}],$$

$$Y_a = \sum_{q_b, k_b} \frac{k_b P_b(q_b, k_b)}{\langle k_b \rangle} [1 - (1 - X_b)^{q_b}],$$

$$X_b = \sum_{q_b, k_b} \frac{q_b P_b(q_b, k_b)}{\langle q_b \rangle} [1 - (1 - X_b)^{q_b - 1}][1 - (1 - Y_b)^{k_b}],$$

$$Y_b = \sum_{q_a, k_a} \frac{k_a P_a(q_a, k_a)}{\langle k_a \rangle} [1 - (1 - X_a)^{q_a}]. \tag{G.2}$$

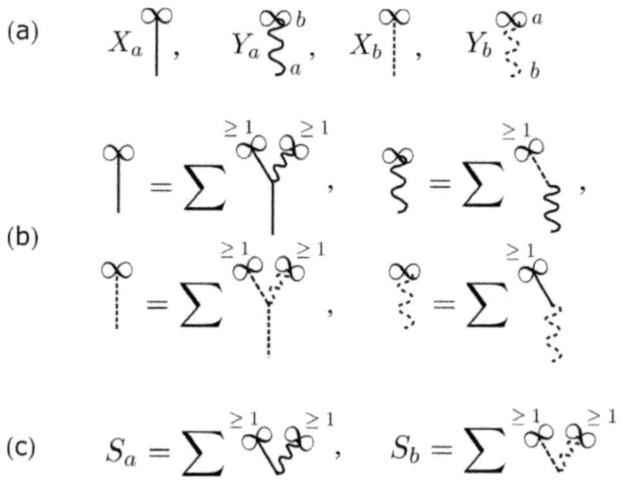

G TREATING GENERAL INTERDEPENDENT NETWORKS

(a)

(b)

(c) $S_a = \sum$, $S_b = \sum$

Fig. G.1 (a) Diagrammatic notations for the probabilities X_a, Y_a, X_b, and Y_b. Here X_a and X_b are the probabilities that if we choose uniformly at random an edge within layer a or b, respectively, and follow it in one of two directions, then we occur in an infinite tree defined by (b). Y_a is the probability that if we choose uniformly at random one of edges interconnecting the layers and follow it from layer a to layer b, then we occur in an infinite tree defined by (b). Y_b is the probability that if we choose uniformly at random one of edges interconnecting the layers and follow it from layer b to layer a, then we occur in an infinite tree defined by (b). (b) Graphical representations of the self-consistency equations for these probabilities, Eq. (G.2), which also define these probabilities. (c) Graphical representation of the expressions of the relative numbers of vertices S_a and S_b within layers a and b, belonging to the giant mutually connected component, in terms of these four probabilities, Eq. (G.3).

The expressions of the relative numbers of vertices S_a and S_b within layers a and b, respectively, in terms of the probabilities X_a, Y_a, X_b, and Y_b (Figure G.1c) have the following form:

$$S_a = \sum_{q_a, k_a} P_a(q_a, k_a)[1 - (1 - X_a)^{q_a}][1 - (1 - Y_a)^{k_a}],$$

$$S_b = \sum_{q_b, k_b} P_b(q_b, k_b)[1 - (1 - X_b)^{q_b}][1 - (1 - Y_b)^{k_b}]. \qquad (G.3)$$

In the particular case of independent degrees q_a and k_a, q_b and k_b, the joint degree-degree distributions are factored into the products $P_a(q_a, k_a) =$

G TREATING GENERAL INTERDEPENDENT NETWORKS

$P_a(q_a)Q_a(k_a)$ and $P_b(q_b, k_b) = P_b(q_b)Q_b(k_b)$. Introducing the generating functions of the distributions $P_a(q)$, $P_b(q)$, $Q_a(k)$, and $Q_b(k)$,

$$G_{a,b}(z) \equiv \sum_q P_{a,b}(q)z^q, \quad R_{a,b}(z) \equiv \sum_k Q_{a,b}(k)z^k,$$

$$G_{1a,b}(z) \equiv \frac{G'_{a,b}(z)}{\langle q_{a,b} \rangle}, \tag{G.4}$$

we rewrite Eqs. (G.2) and (G.3) in a compact form,

$$X_a = [1 - G_{1a}(1 - X_a)][1 - R_a(1 - Y_a)],$$
$$Y_a = 1 - G_b(1 - X_b),$$
$$X_b = [1 - G_{1b}(X_b)][1 - R_b(1 - Y_b)],$$
$$Y_b = 1 - G_a(1 - X_a), \tag{G.5}$$

and

$$S_a = [1 - G_a(1 - X_a)][1 - R_a(1 - Y_a)],$$
$$S_b = [1 - G_b(1 - X_b)][1 - R_a(1 - Y_a)]. \tag{G.6}$$

Furthermore, assuming that the network is symmetric, that is $P_a(q) = P_b(q) = P(q)$ and $Q_a(k) = Q_b(k) = Q(k)$, we have $N_a = N_b = N/2$, $X_a = X_b = X$, $Y_a = Y_b = Y$, and $S_a = S_b = S$. This greatly simplifies these equations, and we finally get

$$X = [1 - G_1(1 - X)][1 - R(1 - Y)], \tag{G.7}$$
$$Y = 1 - G(1 - X) \tag{G.8}$$

and

$$S = [1 - G(1 - X)][1 - R(1 - Y)] = Y[1 - R(1 - Y)]. \tag{G.9}$$

Let the distributions be power law, $P(q) \sim q^{-\gamma}$ and $Q(k) \sim k^{-\tilde{\gamma}}$, which enables us to use convenient asymptotics from Appendix C for the generating functions and to immediately arrive at the same conclusions as for multiplexes if both exponents are greater than 2, $\gamma, \tilde{\gamma} > 2$. The hybrid phase transition is present for these values of the exponents.

To observe a significant difference from multiplex networks, one has to consider the case of a really fat-tailed degree distribution $Q(k)$ with

$1 < \tilde{\gamma} < 2$. The first moment of this distribution diverges, although we can fix it with a cut-off, which we set to a sufficiently large value. Substituting Eq. (G.8) into Eq. (G.7), we get the consistency equation for the probability X in the following form: $X = \mathrm{rhs}(X)$. For small X, we have $\mathrm{rhs}(X) \sim X^{\gamma+\tilde{\gamma}-3}$, where we used the asymptotics for generating functions from Appendix C, assuming $2 < \gamma < 3$ and $1 < \tilde{\gamma} < 2$. Reviewing Figure 6.16 for hybrid transitions, we conclude that a hybrid transition is impossible when $\mathrm{rhs}(X) \sim X^{\alpha}$ with $\alpha < 1$ for small X, since then $\mathrm{rhs}(X) > X$ in the vicinity of the fixed point $X = 0$, and hence this fixed point is unstable. Consequently, for

$$\gamma - 2 < 2 - \tilde{\gamma} > 0, \tag{G.10}$$

the hybrid transition disappears, and the giant mutually connected component exists for all (non-zero) values of a control parameter. Thus the fat-tailed distribution of interlinks between the layers increases the threshold value of γ, below which this network is hyper-resilient against random damage.

References

Acebrón, J. A., Bonilla, L. L., Pérez Vicente, C. J., Ritort, F., and Spigler, R. (2005). The Kuramoto model: A simple paradigm for synchronization phenomena. *Rev. Mod. Phys.*, **77**, 137.

Achlioptas, D., D'Souza, R. M., and Spencer, J. (2009). Explosive percolation in random networks. *Science*, **323**, 1453.

Adamcsek, B., Palla, G., Farkas, I. J., Derényi, I., and Vicsek, T. (2006). CFinder: Locating cliques and overlapping modules in biological networks. *Bioinformatics*, **22**, 1021.

Adler, J. (1991). Bootstrap percolation. *Phys. A: Stat. Mech. Appl.*, **171**, 453.

Adler, J. and Aharony, A. (1988). Diffusion percolation. I. Infinite time limit and bootstrap percolation. *J. Phys. A: Math. Gen.*, **21**, 1387.

Aldous, D. J. (1991). Random walk covering of some special trees. *J. Math. Anal. Appl.*, **157**, 271.

Aleksiejuk, A., Hołyst, J. A., and Stauffer, D. (2002). Ferromagnetic phase transition in Barabási–Albert networks. *Phys. A: Stat. Mech. Appl.*, **310**, 260.

Alexander, S. and Orbach, R. (1982). Density of states on fractals: 'fractons'. *J. Phys. Lett.*, **43**, 625.

Almaas, E., Kovacs, B., Vicsek, T., Oltvai, Z. N., and Barabási, A.-L. (2004). Global organization of metabolic fluxes in the bacterium *Escherichia coli*. *Nature*, **427**, 839.

Alon, N., Avin, C., Koucký, M., Kozma, G., Lotker, Z., and Tuttle, M. R. (2011). Many random walks are faster than one. *Combin. Probab. Comput.*, **20**, 481.

Altarelli, F., Braunstein, A., Dall'Asta, L., Lage-Castellanos, A., and Zecchina, R. (2014a). Bayesian inference of epidemics on networks via belief propagation. *Phys. Rev. Lett.*, **112**, 118701.

Altarelli, F., Braunstein, A., Dall'Asta, L., Wakeling, J. R., and Zecchina, R. (2014b). Containing epidemic outbreaks by message-passing techniques. *Phys. Rev. X*, **4**, 021024.

Alvarez-Hamelin, J. I., Dall'Asta, L., Barrat, A., and Vespignani, A. (2006). Large-scale networks fingerprinting and visualization using the k-core decomposition. In *NIPS'05: Proceedings of the 18th International Conference on Neural Information Processing Systems 18* (ed. Y. Weiss, B. Schölkopf, and J. C. Platt), pp. 41–50.

Ambjørn, J., Durhuus, B., Jonsson, T., and Jonsson, O. (1997). *Quantum Geometry: A Statistical Field Theory Approach*. Cambridge University Press, Cambridge.

Anand, K. and Bianconi, G. (2009). Entropy measures for networks: Toward an information theory of complex topologies. *Phys. Rev. E*, **80**, 045102.

Anand, K. and Bianconi, G. (2010). Gibbs entropy of network ensembles by cavity methods. *Phys. Rev. E*, **82**, 011116.

Andrade Jr, J. S., Herrmann, H. J., Andrade, R. F. S., and Da Silva, L. R. (2005). Apollonian networks: Simultaneously scale-free, small world, Euclidean, space filling, and with matching graphs. *Phys. Rev. Lett.*, **94**, 018702.

Antonioni, A. and Tomassini, M. (2012). Degree correlations in random geometric graphs. *Phys. Rev. E*, **86**, 037101.

Avetisov, V., Gorsky, A., Nechaev, S., and Valba, O. (2020). Localization and non-ergodicity in clustered random networks. *J. Complex Netw.*, **8**, cnz026.

Avetisov, V., Hovhannisyan, M., Gorsky, A., Nechaev, S., Tamm, M., and Valba, O. (2016). Eigenvalue tunneling and decay of quenched random network. *Phys. Rev. E*, **94**, 062313.

Azimi-Tafreshi, N., Dorogovtsev, S. N., and Mendes, J. F. F. (2014*a*). Giant components in directed multiplex networks. *Phys. Rev. E*, **90**, 052809.

Azimi-Tafreshi, N., Gómez-Gardeñes, J., and Dorogovtsev, S. N. (2014*b*). k-core percolation on multiplex networks. *Phys. Rev. E*, **90**, 032816.

Azimi-Tafreshi, N., Osat, S., and Dorogovtsev, S. N. (2019). Generalization of core percolation on complex networks. *Phys. Rev. E*, **99**, 022312.

Banerjee, A., Char, A., and Mondal, B. (2017). Spectra of general hypergraphs. *Linear Algebra Appl.*, **518**, 14.

Banerjee, A. and Char, A. (2017). On the spectrum of directed uniform and non-uniform hypergraphs. *arXiv:1710.06367*.

Barabási, A.-L. and Albert, R. (1999). Emergence of scaling in random networks. *Science*, **286**, 509.

Barahona, M. and Pecora, L. M. (2002). Synchronization in small-world systems. *Phys. Rev. Lett.*, **89**, 054101.

Barlow, M. T., Ding, J., Nachmias, A., and Peres, Y. (2011). The evolution of the cover time. *Combin. Probab. Comput.*, **20**, 331.

Barrat, A., Barthelemy, M., Pastor-Satorras, R., and Vespignani, A. (2004). The architecture of complex weighted networks. *PNAS*, **101**, 3747.

Barthélemy, M., Barrat, A., Pastor-Satorras, R., and Vespignani, A. (2004). Velocity and hierarchical spread of epidemic outbreaks in scale-free networks. *Phys. Rev. Lett.*, **92**, 178701.

Barthélemy, M., Barrat, A., Pastor-Satorras, R., and Vespignani, A. (2005*a*). Characterization and modeling of weighted networks. *Phys. A: Stat. Mech. Appl.*, **346**, 34.

Barthélemy, M., Barrat, A., Pastor-Satorras, R., and Vespignani, A. (2005*b*). Dynamical patterns of epidemic outbreaks in complex heterogeneous networks. *J. Theor. Biol.*, **235**, 275.

Barzel, B. and Barabási, A.-L. (2013). Universality in network dynamics. *Nature Phys.*, **9**, 673.

Basnarkov, L. and Urumov, V. (2008). Kuramoto model with asymmetric distribution of natural frequencies. *Phys. Rev. E*, **78**, 011113.

Bass, H. (1992). The Ihara-Selberg zeta function of a tree lattice. *Int. J. Math.*, **3**, 717.

Bauer, M., Coulomb, S., and Dorogovtsev, S. N. (2005). Phase transition with the Berezinskii–Kosterlitz–Thouless singularity in the Ising model on a growing network. *Phys. Rev. Lett.*, **94**(20), 200602.

Bauer, M. and Golinelli, O. (2001*a*). Core percolation in random graphs: A critical phenomena analysis. *Eur. Phys. J. B*, **24**, 339.

Bauer, M. and Golinelli, O. (2001*b*). Random incidence matrices: Moments of the spectral density. *J. Stat. Phys.*, **103**, 301.

Baxter, G. J., Bianconi, G., da Costa, R. A., Dorogovtsev, S. N., and Mendes, J. F. F. (2016). Correlated edge overlaps in multiplex networks. *Phys. Rev. E*, **94**, 012303.

Baxter, G. J., da Costa, R. A., Dorogovtsev, S. N., and Mendes, J. F. F. (2020). Exotic critical behavior of weak multiplex percolation. *Phys. Rev. E*, **102**, 032301.

Baxter, G. J., Dorogovtsev, S. N., Goltsev, A. V., and Mendes, J. F. F. (2010). Bootstrap percolation on complex networks. *Phys. Rev. E*, **82**, 011103.

Baxter, G. J., Dorogovtsev, S. N., Goltsev, A. V., and Mendes, J. F. F. (2011). Heterogeneous *k*-core versus bootstrap percolation on complex networks. *Phys. Rev. E*, **83**, 051134.

Baxter, G. J., Dorogovtsev, S. N., Goltsev, A. V., and Mendes, J. F. F. (2012). Avalanche collapse of interdependent networks. *Phys. Rev. Lett.*, **109**, 248701.

Baxter, G. J., Dorogovtsev, S. N., Lee, K.-E., Mendes, J. F. F., and Goltsev, A. V. (2015). Critical dynamics of the *k*-core pruning process. *Phys. Rev. X*, **5**, 031017.

Baxter, G. J., Dorogovtsev, S. N., Mendes, J. F. F., and Cellai, D. (2014). Weak percolation on multiplex networks. *Phys. Rev. E*, **89**, 042801.

Baxter, R. J. (2016). *Exactly Solved Models in Statistical Mechanics*. Elsevier, Amsterdam.

Ben-Naim, E. and Krapivsky, P. L. (2005). Kinetic theory of random graphs: From paths to cycles. *Phys. Rev. E*, **71**, 026129.

Bender, E. A. and Canfield, E. R. (1978). The asymptotic number of labeled graphs with given degree sequences. *J. Comb. Theory Ser. A*, **24**, 296.

Bénichou, O. and Voituriez, R. (2007). Comment on 'Localization transition of biased random walks on random networks'. *Phys. Rev. Lett.*, **99**, 209801.

Berestycki, N., Lubetzky, E., Peres, Y., and Sly, A. (2018). Random walks on the random graph. *Ann. Probab.*, **46**, 456.

Berezinskii, V. L. (1971). Destruction of long-range order in one-dimensional and two-dimensional systems having a continuous symmetry group. I. Classical systems. *Sov. Phys. JETP*, **32**, 493.

Berg, J. and Lässig, M. (2002). Correlated random networks. *Phys. Rev. Lett.*, **89**, 228701.

Bhat, U., Krapivsky, P. L., Lambiotte, R., and Redner, S. (2016). Densification and structural transitions in networks that grow by node copying. *Phys. Rev. E*, **94**, 062302.

Bhattacharya, K., Mukherjee, G., Saramäki, J., Kaski, K., and Manna, S. S. (2008). The international trade network: Weighted network analysis and modelling. *J. Stat. Mech.: Theor. Exp.*, **2008**, P02002.

Bialas, P., Burda, Z., Jurkiewicz, J., and Krzywicki, A. (2003). Tree networks with causal structure. *Phys. Rev. E*, **67**, 066106.

Bialas, P. and Oleś, A. K. (2008). Correlations in connected random graphs. *Phys. Rev. E*, **77**, 036124.

Bianconi, G. (2002). Mean field solution of the Ising model on a Barabási–Albert network. *Phys. Lett. A*, **303**, 166.

Bianconi, G. (2005). Emergence of weight-topology correlations in complex scale-free networks. *EPL*, **71**, 1029.

Bianconi, G. (2007*a*). The entropy of randomized network ensembles. *EPL*, **81**, 28005.

Bianconi, G. (2007*b*). A statistical mechanics approach for scale-free networks and finite-scale networks. *Chaos*, **17**, 026114.

Bianconi, G. (2009). Entropy of network ensembles. *Phys. Rev. E*, **79**, 036114.

Bianconi, G. (2017). Fluctuations in percolation of sparse complex networks. *Phys. Rev. E*, **96**, 012302.

Bianconi, G. (2018*a*). *Multilayer Networks: Structure and Function*. Oxford University Press, Oxford.

Bianconi, G. (2018*b*). Rare events and discontinuous percolation transitions. *Phys. Rev. E*, **97**, 022314.

Bianconi, G. (2019). Large deviation theory of percolation on multiplex networks. *J. Stat. Mech.: Theor. Exp.*, **2019**, 023405.

Bianconi, G. and Barabási, A.-L. (2001*a*). Bose–Einstein condensation in complex networks. *Phys. Rev. Lett.*, **86**, 5632.

Bianconi, G. and Barabási, A.-L. (2001*b*). Competition and multiscaling in evolving networks. *EPL*, **54**, 436.

Bianconi, G. and Capocci, A. (2003). Number of loops of size h in growing scale-free networks. *Phys. Rev. Lett.*, **90**, 078701.

Bianconi, G., Darst, R. K., Iacovacci, J., and Fortunato, S. (2014). Triadic closure as a basic generating mechanism of communities in complex networks. *Phys. Rev. E*, **90**, 042806.

Bianconi, G. and Dorogovstev, S. N. (2020). The spectral dimension of simplicial complexes: A renormalization group theory. *J. Stat. Mech.: Theor. Exp.*, **2020**, 014005.

Bianconi, G., Dorogovtsev, S. N., and Mendes, J. F. F. (2015). Mutually connected component of networks of networks with replica nodes. *Phys. Rev. E*, **91**, 012804.

Bianconi, G. and Dorogovtsev, S. N. (2014). Multiple percolation transitions in a configuration model of network of networks. *Phys. Rev. E*, **89**, 062814.

Bianconi, G., Gulbahce, N., and Motter, A. E. (2008). Local structure of directed networks. *Phys. Rev. Lett.*, **100**, 118701.

Bianconi, G. and Gulbahce, N. (2008). Algorithm for counting large directed loops. *J. Phys. A: Math. Theor.*, **41**, 224003.

Bianconi, G. and Marsili, M. (2005). Loops of any size and Hamilton cycles in random scale-free networks. *J. Stat. Mech.: Theor. Exp.*, **2005**, P06005.

Bianconi, G. and Marsili, M. (2006*a*). Emergence of large cliques in random scale-free networks. *EPL*, **74**, 740.

Bianconi, G. and Marsili, M. (2006b). Number of cliques in random scale-free network ensembles. *Physica D*, **224**, 1.

Biroli, G. (2007). Jamming: A new kind of phase transition? *Nature Phys.*, **3**, 222.

Bizhani, G., Paczuski, M., and Grassberger, P. (2012). Discontinuous percolation transitions in epidemic processes, surface depinning in random media, and Hamiltonian random graphs. *Phys. Rev. E*, **86**, 011128.

Blondel, V. D., Guillaume, J.-L., Lambiotte, R., and Lefebvre, E. (2008). Fast unfolding of communities in large networks. *J. Stat. Mech.: Theor. Exp.*, **2008**, P10008.

Boccaletti, S., Almendral, J. A., Guan, S., Leyva, I., Liu, Z., Sendiña-Nadal, I., Wang, Z., and Zou, Y. (2016). Explosive transitions in complex networks structure and dynamics: Percolation and synchronization. *Phys. Rep.*, **660**, 1.

Bogacz, L., Burda, Z., and Wacław, B. (2006). Homogeneous complex networks. *Phys. A: Stat. Mech. Appl.*, **366**, 587.

Boguñá, M., Krioukov, D., and claffy, kc (2009). Navigability of complex networks. *Nature Phys.*, **5**, 74.

Boguñá, M., Pastor-Satorras, R., and Vespignani, A. (2003). Epidemic spreading in complex networks with degree correlations. In *Statistical Mechanics of Complex Networks* (ed. R. Pastor-Satorras, J. M. Rubí, and A. Díaz-Guilera), pp. 127–147. Springer, Berlin.

Boguñá, M., Bonamassa, I., De Domenico, M., Havlin, S., Krioukov, D., and Serrano, M. Á. (2021). Network geometry. *Nature Rev. Physics*, **3**, 114.

Boguná, M. and Krioukov, D. (2009). Navigating ultrasmall worlds in ultrashort time. *Phys. Rev. Lett.*, **102**, 058701.

Boguná, M., Papadopoulos, F., and Krioukov, D. (2010). Sustaining the Internet with hyperbolic mapping. *Nature Commun.*, **1**, 1.

Boguñá, M., Pastor-Satorras, R., and Vespignani, A. (2003). Epidemic spreading in complex networks with degree correlations. *Lect. Note. Phys.*, **625**, 127.

Boguñá, M. and Pastor-Satorras, R. (2002). Epidemic spreading in correlated complex networks. *Phys. Rev. E*, **66**, 047104.

Boguñá, M. and Pastor-Satorras, R. (2003). Class of correlated random networks with hidden variables. *Phys. Rev. E*, **68**, 036112.

Boguñá, M. and Serrano, M. Á. (2005). Generalized percolation in random directed networks. *Phys. Rev. E*, **72**, 016106.

Bollobás, B. (1980). A probabilistic proof of an asymptotic formula for the number of labelled regular graphs. *Eur. J. Comb.*, **1**, 311.

Bollobás, B. (1984). The evolution of sparse graphs. In *Graph Theory and Combinatorics: Proc. Cambridge Combinatorial Conference in Honour of Paul Erdős*, pp. 335–357. Academic Press.

Bollobás, B. (1985). *Random Graphs*. Academic Press, New York.

Bollobás, B. and Riordan, O. (2009). Clique percolation. *Random Struct. Algor.*, **35**, 294.

Bollobás, B. and Riordan, O. M. (2003). Mathematical results on scale-free random graphs. In *Handbook of Graphs and Networks: From the Genome to the Internet* (ed. S. Bornholdt and H. G. Schuster), pp. 1–34. Wiley-VCH Germany, Weinheim.

Borgs, C., Chayes, J. T., Cohn, H., and Holden, N. (2017). Sparse exchangeable graphs and their limits via graphon processes. *J. Mach. Learn. Res.*, **18**, 7740.

Borgs, C., Chayes, J. T., Kesten, H., and Spencer, J. (2001). The birth of the infinite cluster: Finite-size scaling in percolation. *Commun. Math. Phys.*, **224**, 153.

Böttger, H. and Bryksin, V. V. (1985). *Hopping Conduction in Solids*. VCH Weinheim.

Brandes, U. (2001). A faster algorithm for betweenness centrality. *J. Math. Sociol.*, **25**, 163.

Brin, S. and Page, L. (1998). The anatomy of a large-scale hypertextual web search engine. In *Proceedings of the Seventh International World Wide Web Conference*, pp. 107–117. Elsevier, Amsterdam.

Brockmann, D. and Helbing, D. (2013). The hidden geometry of complex, network-driven contagion phenomena. *Science*, **342**, 1337.

Broder, A., Kumar, R., Maghoul, F., Raghavan, P., Rajagopalan, S., Stata, R., Tomkins, A., and Wiener, J. (2000). Graph structure in the web. *Comput. Netw.*, **33**, 309.

Bubeck, S., Devroye, L., and Lugosi, G. (2017). Finding Adam in random growing trees. *Random Struct. Algor.*, **50**, 158.

Buldyrev, S. V., Parshani, R., Paul, G., Stanley, H. E., and Havlin, S. (2010). Catastrophic cascade of failures in interdependent networks. *Nature*, **464**, 1025.

Burda, Z., Correia, J. D., and Krzywicki, A. (2001). Statistical ensemble of scale-free random graphs. *Phys. Rev. E*, **64**, 046118.

Burda, Z., Jurkiewicz, J., and Krzywicki, A. (2004). Network transitivity and matrix models. *Phys. Rev. E*, **69**, 026106.

Butler, S. and Chung, F. (2006). Spectral graph theory. In *Handbook of Linear Algebra*, p. 47. CRC Press, Boca Raton, FL.

Caldarelli, G., Capocci, A., De Los Rios, P., and Munoz, M. A. (2002). Scale-free networks from varying vertex intrinsic fitness. *Phys. Rev. Lett.*, **89**, 258702.

Callaway, D. S., Hopcroft, J. E., Kleinberg, J. M., Newman, M. E. J., and Strogatz, S. H. (2001). Are randomly grown graphs really random? *Phys. Rev. E*, **64**, 041902.

Callaway, D. S., Newman, M. E. J., Strogatz, S. H., and Watts, D. J. (2000). Network robustness and fragility: Percolation on random graphs. *Phys. Rev. Lett.*, **85**, 5468.

Cantwell, G. T. and Newman, M. E. J. (2019). Message passing on networks with loops. *PNAS*, **116**, 23398.

Cantwell, G. T., St-Onge, G., and Young, J.-G. (2021). Inference for growing trees. *Phys. Rev. Lett.*, **126**, 038301.

Capocci, A., Servedio, V. D. P., Caldarelli, G., and Colaiori, F. (2005). Detecting communities in large networks. *Phys. A: Stat. Mech. Appl.*, **352**, 669.

Carmi, S., Carter, S., Sun, J., and Ben-Avraham, D. (2009). Asymptotic behavior of the Kleinberg model. *Phys. Rev. Lett.*, **102**, 238702.

Carmi, S., Havlin, S., Kirkpatrick, S., Shavitt, Y., and Shir, E. (2007). A model of Internet topology using k-shell decomposition. *PNAS*, **104**, 11150.

Cartozo, C. C. and De Los Rios, P. (2009). Extended navigability of small world networks: Exact results and new insights. *Phys. Rev. Lett.*, **102**, 238703.

Castellano, C. and Pastor-Satorras, R. (2006). Non-mean-field behavior of the contact process on scale-free networks. *Phys. Rev. Lett.*, **96**, 038701.

Castellano, C. and Pastor-Satorras, R. (2010). Thresholds for epidemic spreading in networks. *Phys. Rev. Lett.*, **105**, 218701.

Castellano, C. and Pastor-Satorras, R. (2017). Relating topological determinants of complex networks to their spectral properties: Structural and dynamical effects. *Phys. Rev. X*, **7**, 041024.

Catanzaro, M., Boguñá, M., and Pastor-Satorras, R. (2005*a*). Diffusion-annihilation processes in complex networks. *Phys. Rev. E*, **71**, 056104.

Catanzaro, M., Boguñá, M., and Pastor-Satorras, R. (2005*b*). Generation of uncorrelated random scale-free networks. *Phys. Rev. E*, **71**, 027103.

Cayley, A. (1889). A theorem on trees. *J. Math.*, **23**, 376.

Cellai, D., Dorogovtsev, S. N., and Bianconi, G. (2016). Message passing theory for percolation models on multiplex networks with link overlap. *Phys. Rev. E*, **94**, 032301.

Cellai, D., Lawlor, A., Dawson, K. A., and Gleeson, J. P. (2011). Tricritical point in heterogeneous k-core percolation. *Phys. Rev. Lett.*, **107**, 175703.

Cellai, D., Lawlor, A., Dawson, K. A., and Gleeson, J. P. (2013*a*). Critical phenomena in heterogeneous k-core percolation. *Phys. Rev. E*, **87**, 022134.

Cellai, D., López, E., Zhou, J., Gleeson, J. P., and Bianconi, G. (2013*b*). Percolation in multiplex networks with overlap. *Phys. Rev. E*, **88**, 052811.

Chalupa, J., Leath, P. L., and Reich, G. R. (1979). Bootstrap percolation on a Bethe lattice. *J. Phys. C: Solid State Phys.*, **12**, L31.

Chatterjee, S. and Durrett, R. (2009). Contact processes on random graphs with power law degree distributions have critical value 0. *Ann. Probab.*, **37**, 2332.

Chauhan, S., Girvan, M., and Ott, E. (2009). Spectral properties of networks with community structure. *Phys. Rev. E*, **80**, 056114.

Chen, M., Kuzmin, K., and Szymanski, B. K. (2014). Community detection via maximization of modularity and its variants. *IEEE Trans. Comput. Soc. Syst.*, **1**, 46.

Cho, Y. S., Lee, J. S., Herrmann, H. J., and Kahng, B. (2016). Hybrid percolation transition in cluster merging processes: Continuously varying exponents. *Phys. Rev. Lett.*, **116**, 025701.

Chung, F. and Lu, L. (2002). Connected components in random graphs with given expected degree sequences. *Ann. Comb.*, **6**, 125.

Chung, F. and Lu, L. (2004). Coupling online and offline analyses for random power law graphs. *Internet Math.*, **1**, 409.

Chung, F. R. K. (1997). *Spectral Graph Theory*. American Mathematical Society, Providence, RI.

Clauset, A., Moore, C., and Newman, M. E. J. (2006). Structural inference of hierarchies in networks. In *ICML Workshop on Statistical Network Analysis*, pp. 1–13. Springer, Berlin.

Clauset, A., Moore, C., and Newman, M. E. J. (2008). Hierarchical structure and the prediction of missing links in networks. *Nature*, **453**, 98.

Clauset, A., Newman, M. E. J., and Moore, C. (2004). Finding community structure in very large networks. *Phys. Rev. E*, **70**, 066111.

Coghi, F., Radicchi, F., and Bianconi, G. (2018). Controlling the uncertain response of real multiplex networks to random damage. *Phys. Rev. E*, **98**, 062317.

Cohen, R., Ben-Avraham, D., and Havlin, S. (2002). Percolation critical exponents in scale-free networks. *Phys. Rev. E*, **66**, 036113.

Cohen, R., Erez, K., Ben-Avraham, D., and Havlin, S. (2000). Resilience of the Internet to random breakdowns. *Phys. Rev. Lett.*, **85**, 4626.

Cohen, R., Havlin, S., and Ben-Avraham, D. (2003*a*). Efficient immunization strategies for computer networks and populations. *Phys. Rev. Lett.*, **91**, 247901.

Cohen, R., Havlin, S., and Ben-Avraham, D. (2003*b*). Structural properties of scale free networks. In *Handbook of Graphs and Networks: From the Genome to the Internet* (ed. S. Bornholdt and H. G. Schuster), pp. 85–110. Wiley-VCH GmbH & Co., Weinheim.

Cohen, R. and Havlin, S. (2002). Ultra small world in scale-free networks. *Phys. Rev. Lett.*, **90**, 058701.

Colizza, V., Flammini, A., Serrano, M. Á., and Vespignani, A. (2006). Detecting rich-club ordering in complex networks. *Nature Phys.*, **2**, 110.

Colizza, V., Pastor-Satorras, R., and Vespignani, A. (2007). Reaction–diffusion processes and metapopulation models in heterogeneous networks. *Nature Phys.*, **3**, 276.

Colomer-de-Simón, P. and Boguñá, M. (2014). Double percolation phase transition in clustered complex networks. *Phys. Rev. X*, **4**, 041020.

Condamin, S., Bénichou, O., Tejedor, V., Voituriez, R., and Klafter, J. (2007). First-passage times in complex scale-invariant media. *Nature*, **450**, 77.

Cooper, C., Frieze, A., and Vera, J. (2004). Random deletion in a scale-free random graph process. *Internet Math.*, **1**, 463.

Cooper, C. and Frieze, A. (2005). The cover time of random regular graphs. *SIAM J. Discrete Math.*, **18**, 728.

Cooper, C. and Frieze, A. (2007*a*). The cover time of sparse random graphs. *Random Struct. Algor.*, **30**, 1.

Cooper, C. and Frieze, A. (2007*b*). The cover time of the preferential attachment graph. *J. Comb. Theory Ser. B*, **97**, 269.

Costa, L. da F., Rodrigues, F. A., Travieso, G., and Villas Boas, P. R. (2007). Characterization of complex networks: A survey of measurements. *Adv. Phys.*, **56**, 167.

Coutinho, B. C., Goltsev, A. V., Dorogovtsev, S. N., and Mendes, J. F. F. (2013). Kuramoto model with frequency-degree correlations on complex networks. *Phys. Rev. E*, **87**, 032106.

Coutinho, B. C., Wu, A.-K., Zhou, H.-J., and Liu, Y.-Y. (2020). Covering problems and core percolations on hypergraphs. *Phys. Rev. Lett.*, **124**, 248301.

da Costa, R. A., Dorogovtsev, S. N., Goltsev, A. V., and Mendes, J. F. F. (2010). Explosive percolation transition is actually continuous. *Phys. Rev. Lett.*, **105**, 255701.

da Costa, R. A., Dorogovtsev, S. N., Goltsev, A. V., and Mendes, J. F. F. (2014). Solution of the explosive percolation quest: Scaling functions and critical exponents. *Phys. Rev. E*, **90**, 022145.

da Costa, R. A., Dorogovtsev, S. N., Goltsev, A. V., and Mendes, J. F. F. (2015). Solution of the explosive percolation quest. II. Infinite-order transition produced by the initial distributions of clusters. *Phys. Rev. E*, **91**, 032140.

da Silva, D. C., Bianconi, G., da Costa, R. A., Dorogovtsev, S. N., and Mendes, J. F. F. (2018). Complex network view of evolving manifolds. *Phys. Rev. E*, **97**, 032316.

Daley, D. J. and Gani, J. (2001). *Epidemic Modelling: An Introduction*. Cambridge University Press, Cambridge.

Daley, D. J. and Kendall, D. G. (1964). Epidemics and rumours. *Nature*, **204**, 1118.

Daley, D. J. and Kendall, D. G. (1965). Stochastic rumours. *IMA J. Appl. Math.*, **1**, 42.

Dall, J. and Christensen, M. (2002). Random geometric graphs. *Phys. Rev. E*, **66**, 016121.

Danon, L., Diaz-Guilera, A., Duch, J., and Arenas, A. (2005). Comparing community structure identification. *J. Stat. Mech.: Theor. Exp.*, **2005**, P09008.

Danziger, M. M., Bonamassa, I., Boccaletti, S., and Havlin, S. (2019). Dynamic interdependence and competition in multilayer networks. *Nature Phys.*, **15**, 178.

de Souza, D. R. and Tomé, T. (2010). Stochastic lattice gas model describing the dynamics of the SIRS epidemic process. *Phys. A: Stat. Mech. Appl.*, **389**, 1142.

Decelle, A., Hüttel, J., Saade, A., and Moore, C. (2014). Computational complexity, phase transitions, and message-passing for community detection. *arXiv:1409.2290*.

Decelle, A., Krzakala, F., Moore, C., and Zdeborová, L. (2011a). Asymptotic analysis of the stochastic block model for modular networks and its algorithmic applications. *Phys. Rev. E*, **84**, 066106.

Decelle, A., Krzakala, F., Moore, C., and Zdeborová, L. (2011b). Inference and phase transitions in the detection of modules in sparse networks. *Phys. Rev. Lett.*, **107**, 065701.

den Hollander, F. (2000). *Large Deviations*. American Mathematical Society, Providence, RI.

Derényi, I., Palla, G., and Vicsek, T. (2005). Clique percolation in random networks. *Phys. Rev. Lett.*, **94**, 160202.

Deroulers, C. and Monasson, R. (2004). Field-theoretic approach to metastability in the contact process. *Phys. Rev. E*, **69**, 016126.

Destri, C. and Donetti, L. (2002). The spectral dimension of random trees. *J. Phys. A: Math. Gen.*, **35**, 9499.

Dettmann, C. P. and Georgiou, O. (2016). Random geometric graphs with general connection functions. *Phys. Rev. E*, **93**, 032313.

Di Muro, M. A., Valdez, L. D., Rêgo, H. H. A., Buldyrev, S. V., Stanley, H. E., and Braunstein, L. A. (2017). Cascading failures in interdependent networks with multiple supply-demand links and functionality thresholds. *Sci. Rep.*, **7**, 1.

Di Muro, M. A., Valdez, L. D., Stanley, H. E., Buldyrev, S. V., and Braunstein, L. A. (2019). Insights into bootstrap percolation: Its equivalence with k-core percolation and the giant component. *Phys. Rev. E*, **99**, 022311.

Dickman, R. and Vidigal, R. (2002). Quasi-stationary distributions for stochastic processes with an absorbing state. *J. Phys. A: Math. Gen.*, **35**, 1147.

Dodds, Peter S., Muhamad, R., and Watts, D. J. (2003). An experimental study of search in global social networks. *Science*, **301**, 827.

Dodds, P. S. and Watts, D. J. (2004). Universal behavior in a generalized model of contagion. *Phys. Rev. Lett.*, **92**, 218701.

Dodds, P. S. and Watts, D. J. (2005). A generalized model of social and biological contagion. *J. Theor. Biol.*, **232**, 587.

Dommers, S., Van Der Hofstad, R., and Hooghiemstra, G. (2010). Diameters in preferential attachment models. *J. Stat. Phys.*, **139**, 72.

Donetti, L. and Munoz, M. A. (2004). Detecting network communities: A new systematic and efficient algorithm. *J. Stat. Mech.: Theor. Exp.*, **2004**, P10012.

Dorogovtsev, S. N. (2004). Clustering of correlated networks. *Phys. Rev. E*, **69**, 027104.

Dorogovtsev, S. N., Ferreira, A. L., Goltsev, A. V., and Mendes, J. F. F. (2010). Zero Pearson coefficient for strongly correlated growing trees. *Phys. Rev. E*, **81**, 031135.

Dorogovtsev, S. N., Goltsev, A. V., and Mendes, J. F. F. (2002*a*). Ising model on networks with an arbitrary distribution of connections. *Phys. Rev. E*, **66**, 016104.

Dorogovtsev, S. N., Goltsev, A. V., and Mendes, J. F. F. (2002*b*). Pseudofractal scale-free web. *Phys. Rev. E*, **65**, 066122.

Dorogovtsev, S. N., Goltsev, A. V., and Mendes, J. F. F. (2004). Potts model on complex networks. *Eur. Phys. J. B*, **38**, 177.

Dorogovtsev, S. N., Goltsev, A. V., and Mendes, J. F. F. (2006). k-core organization of complex networks. *Phys. Rev. Lett.*, **96**, 040601.

Dorogovtsev, S. N., Goltsev, A. V., and Mendes, J. F. F. (2008*a*). Critical phenomena in complex networks. *Rev. Mod. Phys.*, **80**, 1275.

Dorogovtsev, S. N., Goltsev, A. V., Mendes, J. F. F., and Samukhin, A. N. (2003*a*). Spectra of complex networks. *Phys. Rev. E*, **68**, 046109.

Dorogovtsev, S. N., Mendes, J. F. F., and Samukhin, A. N. (2000). Structure of growing networks with preferential linking. *Phys. Rev. Lett.*, **85**, 4633.

Dorogovtsev, S. N., Mendes, J. F. F., and Samukhin, A. N. (2001*a*). Anomalous percolation properties of growing networks. *Phys. Rev. E*, **64**, 066110.

Dorogovtsev, S. N., Mendes, J. F. F., and Samukhin, A. N. (2001*b*). Giant strongly connected component of directed networks. *Phys. Rev. E*, **64**, 025101.

Dorogovtsev, S. N., Mendes, J. F. F., and Samukhin, A. N. (2001*c*). Size-dependent degree distribution of a scale-free growing network. *Phys. Rev. E*, **63**, 062101.

Dorogovtsev, S. N., Mendes, J. F. F., and Samukhin, A. N. (2003*b*). Metric structure of random networks. *Nucl. Phys. B*, **653**, 307.

Dorogovtsev, S. N., Mendes, J. F. F., and Samukhin, A. N. (2003*c*). Principles of statistical mechanics of uncorrelated random networks. *Nucl. Phys. B*, **666**, 396.

Dorogovtsev, S. N. and Mendes, J. F. F. (2000). Scaling behaviour of developing and decaying networks. *EPL*, **52**, 33.

Dorogovtsev, S. N. and Mendes, J. F. F. (2001*a*). Effect of the accelerating growth of communications networks on their structure. *Phys. Rev. E*, **63**, 025101.

Dorogovtsev, S. N. and Mendes, J. F. F. (2001*b*). Scaling properties of scale-free evolving networks: Continuous approach. *Phys. Rev. E*, **63**, 056125.

Dorogovtsev, S. N. and Mendes, J. F. F. (2003). *Evolution of Networks: From Biological Nets to the Internet and WWW*. Oxford University Press, Oxford.

Dorogovtsev, S. N., Mendes, J. F. F., Samukhin, A. N., and Zyuzin, A. Y. (2008*b*). Organization of modular networks. *Phys. Rev. E*, **78**, 056106.

Drmota, M. (2008). *Recursive trees*. https://dmg.tuwien.ac.at/drmota/recursivetrees.pdf.

Drmota, M. (2009). *Random Trees: An Interplay Between Combinatorics and Probability*. Springer, Berlin.

D'Souza, R. M., Krapivsky, P. L., and Moore, C. (2007). The power of choice in growing trees. *Eur. Phys. J. B*, **59**, 535.

Duan, D., Lv, C., Si, S., Wang, Z., Li, D., Gao, J., Havlin, S., Stanley, H. E., and Boccaletti, S. (2019). Universal behavior of cascading failures in interdependent networks. *PNAS*, **116**, 22452.

Durhuus, B. (2009). Hausdorff and spectral dimension of infinite random graphs. *Acta Phys. Polon.*, **40**, 3509.

Edwards, S. F. and Jones, R. C. (1976). The eigenvalue spectrum of a large symmetric random matrix. *J. Phys. A: Math. Gen.*, **9**, 1595.

Eidsaa, M. and Almaas, E. (2013). *s*-core network decomposition: A generalization of *k*-core analysis to weighted networks. *Phys. Rev. E*, **88**, 062819.

Eidsaa, M. and Almaas, E. (2016). Investigating the relationship between *k*-core and *s*-core network decompositions. *Phys. A: Stat. Mech. Appl.*, **449**, 111.

Eldan, R., Rácz, M. Z., and Schramm, T. (2017). Braess's paradox for the spectral gap in random graphs and delocalization of eigenvectors. *Random Struct. Algor.*, **50**, 584.

Elon, Y. (2008). Eigenvectors of the discrete Laplacian on regular graphs—a statistical approach. *J. Phys. A: Math. Theor.*, **41**, 435203.

Erdős, P. and Gallai, T. (1960). Gráfok előírt fokszámú pontokkal. *Matematikai Lapok*, **11**, 264.

Ergün, G. and Rodgers, G. J. (2002). Growing random networks with fitness. *Phys. A: Stat. Mech. Appl.*, **303**, 261.

Eriksen, K. A., Simonsen, I., Maslov, S., and Sneppen, K. (2003). Modularity and extreme edges of the internet. *Phys. Rev. Lett.*, **90**, 148701.

Ermann, L., Frahm, K. M., and Shepelyansky, D. L. (2015). Google matrix analysis of directed networks. *Rev. Mod. Phys.*, **87**, 1261.

Erzan, A. and Tuncer, A. (2020). Explicit construction of the eigenvectors and eigenvalues of the graph Laplacian on the Cayley tree. *Linear Algebra Appl.*, **586**, 111.

Fan, J. and Chen, X. (2014). General clique percolation in random networks. *EPL*, **107**, 28005.

Faqeeh, A., Melnik, S., and Gleeson, J. P. (2015). Network cloning unfolds the effect of clustering on dynamical processes. *Phys. Rev. E*, **91**, 052807.

Farkas, I. J., Derényi, I., Barabási, A.-L., and Vicsek, T. (2001). Spectra of 'real-world' graphs: Beyond the semicircle law. *Phys. Rev. E*, **64**, 026704.

Feng, Q., Su, C., and Hu, Z. (2005). Branching structure of uniform recursive trees. *Sci. China Ser. A Math.*, **48**, 769.

Fernholz, D. and Ramachandran, V. (2003). The giant k-core of a random graph with a specified degree sequence. Technical report, University of Texas at Austin.

Ferreira, R. S., Da Costa, R. A., Dorogovtsev, S. N., and Mendes, J. F. F. (2016). Metastable localization of diseases in complex networks. *Phys. Rev. E*, **94**, 062305.

Flake, G. W., Lawrence, S., Giles, C. L., and Coetzee, F. M. (2002). Self-organization and identification of web communities. *Computer*, **35**, 66.

Fortuin, C. M. and Kasteleyn, P. W. (1972). On the random-cluster model: I. Introduction and relation to other models. *Physica*, **57**, 536.

Fortunato, S. and Barthélemy, M. (2007). Resolution limit in community detection. *PNAS*, **104**, 36.

Fortunato, S., Boguñá, M., Flammini, A., and Menczer, F. (2006). Approximating PageRank from in-degree. In *International Workshop on Algorithms and Models for the Web-Graph*, pp. 59–71. Springer, Berlin.

Friedrich, T., Sauerwald, T., and Stauffer, A. (2013). Diameter and broadcast time of random geometric graphs in arbitrary dimensions. *Algorithmica*, **67**, 65.

Fronczak, A., Fronczak, P., and Hołyst, J. A. (2003). Mean-field theory for clustering coefficients in Barabási-Albert networks. *Phys. Rev. E*, **68**, 046126.

Fronczak, A. and Fronczak, P. (2009). Biased random walks in complex networks: The role of local navigation rules. *Phys. Rev. E*, **80**, 016107.

Gani, J. (2000). The Maki–Thompson rumour model: A detailed analysis. *Environ. Model. Softw.*, **15**, 721.

Gao, J., Buldyrev, S. V., Havlin, S., and Stanley, H. E. (2011). Robustness of a network of networks. *Phys. Rev. Lett.*, **107**, 195701.

Gao, J., Buldyrev, S. V., Havlin, S., and Stanley, H. E. (2012*a*). Robustness of a network formed by n interdependent networks with a one-to-one correspondence of dependent nodes. *Phys. Rev. E*, **85**, 066134.

Gao, J., Buldyrev, S. V., Stanley, H. E., and Havlin, S. (2012*b*). Networks formed from interdependent networks. *Nature Phys.*, **8**, 40.

Gao, J., Buldyrev, S. V., Stanley, H. E., Xu, X., and Havlin, S. (2013). Percolation of a general network of networks. *Phys. Rev. E*, **88**, 062816.

Garlaschelli, D., Battiston, S., Castri, M., Servedio, V. D. P., and Caldarelli, G. (2005). The scale-free topology of market investments. *Phys. A: Stat. Mech. Appl.*, **350**, 491.

REFERENCES

Garlaschelli, D. and Loffredo, M. I. (2009). Generalized Bose-Fermi statistics and structural correlations in weighted networks. *Phys. Rev. Lett.*, **102**, 038701.

Gauvin, L., Génois, M., Karsai, M., Kivelä, M., Takaguchi, T., Valdano, E., and Vestergaard, C. L. (2018). Randomized reference models for temporal networks. *arXiv:1806.04032*.

Gilbert, E. N. (1961). Random plane networks. *SIAM Rev. Soc. Ind. Appl. Math.*, **9**, 533.

Girvan, M. and Newman, M. E. J. (2002). Community structure in social and biological networks. *PNAS*, **99**, 7821.

Gleeson, J. P. (2013). Binary-state dynamics on complex networks: Pair approximation and beyond. *Phys. Rev. X*, **3**, 021004.

Gleeson, J. P. and Cahalane, D. J. (2007). Seed size strongly affects cascades on random networks. *Phys. Rev. E*, **75**, 056103.

Gleeson, J. P., Cellai, D., Onnela, J.-P., Porter, M. A., and Reed-Tsochas, F. (2014). A simple generative model of collective online behavior. *PNAS*, **111**, 10411.

Gleeson, J. P., Melnik, S., and Hackett, A. (2010). How clustering affects the bond percolation threshold in complex networks. *Phys. Rev. E*, **81**, 066114.

Gleeson, J. P., O'Sullivan, K. P., Baños, R. A., and Moreno, Y. (2016). Effects of network structure, competition and memory time on social spreading phenomena. *Phys. Rev. X*, **6**, 021019.

Gleiss, P. M., Stadler, P. F., Wagner, A., and Fell, D. A. (2001). Relevant cycles in chemical reaction networks. *Adv. Complex Syst.*, **4**, 207.

Goh, K.-I., Kahng, B., and Kim, D. (2001). Universal behavior of load distribution in scale-free networks. *Phys. Rev. Lett.*, **87**, 278701.

Goltsev, A. V., Dorogovtsev, S. N., and Mendes, J. F. F. (2003). Critical phenomena in networks. *Phys. Rev. E*, **67**, 026123.

Goltsev, A. V., Dorogovtsev, S. N., and Mendes, J. F. F. (2006). k-core (bootstrap) percolation on complex networks: Critical phenomena and nonlocal effects. *Phys. R. E*, **73**, 056101.

Goltsev, A. V., Dorogovtsev, S. N., and Mendes, J. F. F. (2008). Percolation on correlated networks. *Phys. Rev. E*, **78**, 051105.

Goltsev, A. V., Dorogovtsev, S. N., Oliveira, J. G., and Mendes, J. F. F. (2012). Localization and spreading of diseases in complex networks. *Phys. Rev. Lett.*, **109**, 128702.

Goltsev, A. V., Timár, G., and Mendes, J. F. F. (2017). Sensitivity of directed networks to the addition and pruning of edges and vertices. *Phys. Rev. E*, **96**, 022317.

Gómez, S., Arenas, A., Borge-Holthoefer, J., Meloni, S., and Moreno, Y. (2010). Discrete-time Markov chain approach to contact-based disease spreading in complex networks. *EPL*, **89**, 38009.

Gomez, S., Diaz-Guilera, A., Gómez-Gardeñes, J., Perez-Vicente, C. J., Moreno, Y., and Arenas, A. (2013). Diffusion dynamics on multiplex networks. *Phys. Rev. Lett.*, **110**, 028701.

Gómez-Gardenes, J., Gómez, S., Arenas, A., and Moreno, Y. (2011). Explosive synchronization transitions in scale-free networks. *Phys. Rev. Lett.*, **106**, 128701.

Granell, C., Gomez, S., and Arenas, A. (2012). Hierarchical multiresolution method to overcome the resolution limit in complex networks. *Int. J. Bifurc. Chaos Appl. Sci. Eng.*, **22**, 1250171.

Granovetter, M. S. (1973). The strength of weak ties. *Amer. J. Sociology*, **78**, 1360.

Gribov, V. N. (1968). A reggeon diagram technique. *Sov. Phys. JETP*, **26**, 27.

Guimerà, R. and Sales-Pardo, M. (2009). Missing and spurious interactions and the reconstruction of complex networks. *PNAS*, **106**, 22073.

Gumbel, E. J. (2004). *Statistics of Extremes*. Courier Corporation, Chelmsford, MA.

Hackett, A., Cellai, D., Gómez, S., Arenas, A., and Gleeson, J. P. (2016). Bond percolation on multiplex networks. *Phys. Rev. X*, **6**, 021002.

Hakimi, S. L. (1962). On realizability of a set of integers as degrees of the vertices of a linear graph. I. *J. SIAM*, **10**, 496.

Hardy, G. H. and Wright, E. M. (1979). *An Introduction to the Theory of Numbers*. Oxford University Press, Oxford.

Harris, J. K. (2013). *An Introduction to Exponential Random Graph Modeling*. Sage Publications, Los Angeles, CA.

Harris, T. E. (1974). Contact interactions on a lattice. *Ann. Probab.*, 969.

Hartmann, A. K. (2011). Large-deviation properties of largest component for random graphs. *Eur. Phys. J. B*, **84**, 627.

Hartmann, A. K. (2017). Large-deviation properties of the largest 2-core component for random graphs. *Eur. Phys. J. Spec. Top.*, **226**, 567.

Hartmann, A. K. and Weigt, M. (2006). *Phase Transitions in Combinatorial Optimization Problems: Basics, Algorithms and Statistical Mechanics*. John Wiley & Sons, Weinheim.

Hashimoto, K. (1989). Zeta functions of finite graphs and representations of p-adic groups. In *Automorphic Forms and Geometry of Arithmetic Varieties* (ed. K. Hashimoto and Y. Namikawa), pp. 211–280. Elsevier, Amsterdam.

Hauert, C. (2001). Fundamental clusters in spatial 2×2 games. *Proc. Royal Soc. London. Ser. B: Biol. Sci.*, **268**, 761.

Havel, V. (1955). A remark on the existence of finite graphs (czech). *Časopis Pest. Mat.*, **80**, 477.

Henkel, M., Hinrichsen, H., and Lübeck, S. (2008). *Non-equilibrium Phase Transitions: Vol. 1. Absorbing Phase Transitions*. Springer, Dordrecht.

Hethcote, H. W. and Yorke, J. A. (1984). *Gonorrhea Transmission Dynamics and Control*. Springer, Berlin.

Hoffmann, T., Porter, M. A. and Lambiotte, R. (2013). Random walks on stochastic temporal networks. In *Temporal Networks* (ed. P. Holme and J. Saramäki), pp. 295–313. Springer, Berlin.

Holland, P. W., Laskey, K. B., and Leinhardt, S. (1983). Stochastic blockmodels: First steps. *Soc. Netw.*, **5**, 109.

Holme, P. (2003). Congestion and centrality in traffic flow on complex networks. *Adv. Complex Syst.*, **6**, 163.

Holme, P. (2015). Modern temporal network theory: A colloquium. *Eur. Phys. J. B*, **88**, 1.

Holme, P. and Kim, B. J. (2002). Growing scale-free networks with tunable clustering. *Phys. Rev. E*, **65**, 026107.

Holme, P. and Litvak, N. (2017). Cost-efficient vaccination protocols for network epidemiology. *PLoS Comput. Biol.*, **13**, e1005696.

Holme, P. and Saramäki, J. (2012). Temporal networks. *Phys. Rep.*, **519**, 97.

Hołyst, J. A., Sienkiewicz, J., Fronczak, A., Fronczak, P., and Suchecki, K. (2005). Universal scaling of distances in complex networks. *Phys. Rev. E*, **72**, 026108.

Hong, H., Ha, M., and Park, H. (2007). Finite-size scaling in complex networks. *Phys. Rev. Lett.*, **98**, 258701.

Hu, Y., Ksherim, B., Cohen, R., and Havlin, S. (2011). Percolation in interdependent and interconnected networks: Abrupt change from second-to first-order transitions. *Phys. Rev. E*, **84**, 066116.

Hu, Y., Zhou, D., Zhang, R., Han, Z., Rozenblat, C., and Havlin, S. (2013). Percolation of interdependent networks with intersimilarity. *Phys. Rev. E*, **88**, 052805.

Huang, X., Gao, J., Buldyrev, S. V., Havlin, S., and Stanley, H. E. (2011). Robustness of interdependent networks under targeted attack. *Phys. Rev. E*, **83**, 065101.

Hwang, S., Choi, S., Lee, D., and Kahng, B. (2015). Efficient algorithm to compute mutually connected components in interdependent networks. *Phys. Rev. E*, **91**, 022814.

Hwang, S., Lee, D.-S., and Kahng, B. (2012*a*). Effective trapping of random walkers in complex networks. *Phys. Rev. E*, **85**, 046110.

Hwang, S., Lee, D.-S., and Kahng, B. (2012*b*). First passage time for random walks in heterogeneous networks. *Phys. Rev. Lett.*, **109**, 088701.

Hwang, S., Lee, D. S., and Kahng, B. (2014). Effective spectral dimension in scale-free networks. In *First-passage Phenomena and Their Applications* (ed. R. Metzler, G. Oshanin, and S. Redner), pp. 122–144. World Scientific Publishing, London.

Hwang, S., Yun, C.-K., Lee, D.-S., Kahng, B., and Kim, D. (2010). Spectral dimensions of hierarchical scale-free networks with weighted shortcuts. *Phys. Rev. E*, **82**, 056110.

Itzkovitz, S., Milo, R., Kashtan, N., Ziv, G., and Alon, U. (2003). Subgraphs in random networks. *Phys. Rev. E*, **68**, 026127.

Iwata, M. and Sasa, S.-I. (2009). Dynamics of k-core percolation in a random graph. *J. Phys. A: Math. Theor.*, **42**, 075005.

Janson, S. (2008). The largest component in a subcritical random graph with a power law degree distribution. *Ann. Appl. Probab.*, **18**, 1651.

Janssen, A. J. E. M., van Leeuwaarden, J. S. H., and Shneer, S. (2019). Counting cliques and cycles in scale-free inhomogeneous random graphs. *J. Stat. Phys.*, **175**, 161.

Janssen, H.-K., Müller, M., and Stenull, O. (2004). Generalized epidemic process and tricritical dynamic percolation. *Phys. Rev. E*, **70**, 026114.

Jonasson, J. (1998). On the cover time for random walks on random graphs. *Combin. Probab. Comput.*, **7**, 265.

Kac, M. (1947). Random walk and the theory of Brownian motion. *Am. Math. Mon.*, **54**(7P1), 369.

Kalisky, T. and Cohen, R. (2006). Width of percolation transition in complex networks. *Phys. Rev. E*, **73**, 035101.

Karp, B. and Kung, H.-T. (2000). GPSR: Greedy perimeter stateless routing for wireless networks. In *Proceedings of the 6th Annual International Conference on Mobile Computing and Networking*, pp. 243–254.

Karp, R. M. and Sipser, M. (1981). Maximum matching in sparse random graphs. In *22nd Annual Symposium on Foundations of Computer Science (SFCS 1981)*, pp. 364–375. IEEE, New York.

Karrer, B., Newman, M. E. J., and Zdeborová, L. (2014). Percolation on sparse networks. *Phys. Rev. Lett.*, **113**, 208702.

Karrer, B. and Newman, M. E. J. (2010a). Message passing approach for general epidemic models. *Phys. Rev. E*, **82**, 016101.

Karrer, B. and Newman, M. E. J. (2010b). Random graphs containing arbitrary distributions of subgraphs. *Phys. Rev. E*, **82**, 066118.

Karrer, B. and Newman, M. E. J. (2011). Stochastic blockmodels and community structure in networks. *Phys. Rev. E*, **83**, 016107.

Kasteleyn, P. W. and Fortuin, C. M. (1969). Phase transitions in lattice systems with random local properties. *J. Phys. Soc. Japan*, **26** (Suppl.), 11.

Kawamoto, T. (2016). Localized eigenvectors of the non-backtracking matrix. *J. Stat. Mech.: Theor. Exp.*, **2016**, 023404.

Kenyon, R., Radin, C., Ren, K., and Sadun, L. (2016). Bipodal structure in oversaturated random graphs. *Int. Math. Res. Not.*, **2018**, 1009.

Kenyon, R., Radin, C., Ren, K., and Sadun, L. (2017a). Multipodal structure and phase transitions in large constrained graphs. *J. Stat. Phys.*, **168**, 233.

Kenyon, R., Radin, C., Ren, K., and Sadun, L. (2017b). The phases of large networks with edge and triangle constraints. *J. Phys. A: Math. Theor.*, **50**, 435001.

Khajeh, E., Dorogovtsev, S. N., and Mendes, J. F. F. (2007). Berezinskii–Kosterlitz–Thouless-like transition in the Potts model on an inhomogeneous annealed network. *Phys. Rev. E*, **75**, 041112.

Kharel, S. R., Mezei, T. R., Chung, S., Erdős, P. L., and Toroczkai, Z. (2022). Degree-preserving network growth. *Nature Phys.*, **18**, 100.

Kim, D.-H. and Motter, A. E. (2007). Ensemble averageability in network spectra. *Phys. Rev. Lett.*, **98**, 248701.

Kitsak, M., Gallos, L. K., Havlin, S., Liljeros, F., Muchnik, L., Stanley, H. E., and Makse, H. A. (2010). Identification of influential spreaders in complex networks. *Nature Phys.*, **6**, 888.

Kleinberg, J. (2000a). The small-world phenomenon: An algorithmic perspective. In *Proceedings of the Thirty-Second Annual ACM Symposium on Theory of Computing*, pp. 163–170.

Kleinberg, J. (2006). Complex networks and decentralized search algorithms. In

Proceedings of the International Congress of Mathematicians (ICM), Volume 3, pp. 1019–1044.

Kleinberg, J. M. (2000*b*). Navigation in a small world. *Nature*, **406**, 845.

Kleinberg, R. (2007). Geographic routing using hyperbolic space. In *IEEE INFO-COM 2007-26th IEEE International Conference on Computer Communications*, pp. 1902–1909. IEEE, New York.

Kosterlitz, J. M. and Thouless, D. J. (1973). Ordering, metastability and phase transitions in two-dimensional systems. *J. Phys. C*, **6**, 1181.

Krapivsky, P. L. (2021). Infection process near criticality: Influence of the initial condition. *J. Stat. Mech.: Theor. Exp.*, **2021**, 013501.

Krapivsky, P. L., Redner, S., and Ben-Naim, E. (2010). *A Kinetic View of Statistical Physics*. Cambridge University Press, Cambridge.

Krapivsky, P. L., Redner, S., and Leyvraz, F. (2000). Connectivity of growing random networks. *Phys. Rev. Lett.*, **85**, 4629.

Krapivsky, P. L. and Redner, S. (2001). Organization of growing random networks. *Phys. Rev. E*, **63**, 066123.

Krapivsky, P. L. and Redner, S. (2002). Finiteness and fluctuations in growing networks. *J. Phys. A: Math. Gen.*, **35**, 9517.

Krasnytska, M., Berche, B., Holovatch, Y., and Kenna, R. (2016). Partition function zeros for the Ising model on complete graphs and on annealed scale-free networks. *J. Phys. A: Math. Theor.*, **49**, 135001.

Krings, G., Karsai, M., Bernhardsson, S., Blondel, V. D., and Saramäki, J. (2012). Effects of time window size and placement on the structure of an aggregated communication network. *EPJ Data Sci.*, **1**, 1.

Krioukov, D., Papadopoulos, F., Kitsak, M., Vahdat, A., and Boguñá, M. (2010). Hyperbolic geometry of complex networks. *Phys. Rev. E*, **82**, 036106.

Krivelevich, M. and Sudakov, B. (2003). The largest eigenvalue of sparse random graphs. *Combin. Probab. Comput.*, **12**, 61.

Krot, A. and Prokhorenkova, L. O. (2015). Local clustering coefficient in generalized preferential attachment models. In *International Workshop on Algorithms and Models for the Web-Graph*, pp. 15–28. Springer, Berlin.

Krzakala, F., Moore, C., Mossel, E., Neeman, J., Sly, A., Zdeborová, L., and Zhang, P. (2013). Spectral redemption in clustering sparse networks. *PNAS*, **110**, 20935.

Kumar, R., Raghavan, P., Rajagopalan, S., Sivakumar, D., Tomkins, A., and Upfal, E. (2000). Stochastic models for the web graph. In *Proceedings 41st Annual Symposium on Foundations of Computer Science*, pp. 57–65. IEEE, New York.

Kuramoto, Y. (1975). Self-entrainment of a population of coupled non-linear oscillators. In *International Symposium on Mathematical Problems in Theoretical Physics*, pp. 420–422. Springer, New York.

Kuramoto, Y. (1984). *Chemical Oscillations, Waves, and Turbulence*. Springer, Berlin.

Lambiotte, R., Krapivsky, P. L., Bhat, U., and Redner, S. (2016). Structural transitions in densifying networks. *Phys. Rev. Lett.*, **117**, 218301.

Lancichinetti, A., Fortunato, S., and Kertész, J. (2009). Detecting the overlapping and hierarchical community structure in complex networks. *New J. Phys.*, **11**, 033015.

Lancichinetti, A. and Fortunato, S. (2011). Limits of modularity maximization in community detection. *Phys. Rev. E*, **84**, 066122.

Lancichinetti, A., Radicchi, F., Ramasco, J. J., and Fortunato, S. (2011). Finding statistically significant communities in networks. *PloS One*, **6**, e18961.

Landau, L. D. and Lifshitz, E. M. (2013). *Statistical Physics*. Elsevier, Amsterdam.

Lau, H. W. and Szeto, K. Y. (2010). Asymptotic analysis of first passage time in complex networks. *EPL*, **90**, 40005.

Lauritsen, K. B., Zapperi, S., and Stanley, H. E. (1996). Self-organized branching processes: Avalanche models with dissipation. *Phys. Rev. E*, **54**, 2483.

Lee, C. and Wilkinson, D. J. (2019). A review of stochastic block models and extensions for graph clustering. *Appl. Netw. Sci.*, **4**, 1.

Lee, D., Choi, S., Stippinger, M., Kertész, J., and Kahng, B. (2016a). Hybrid phase transition into an absorbing state: Percolation and avalanches. *Phys. Rev. E*, **93**, 042109.

Lee, D., Jo, M., and Kahng, B. (2016b). Critical behavior of k-core percolation: Numerical studies. *Phys. Rev. E*, **94**, 062307.

Lee, D., Kahng, B., Cho, Y. S., Goh, K.-I., and Lee, D.-S. (2018). Recent advances of percolation theory in complex networks. *J. Korean Phys. Soc.*, **73**, 152.

Leicht, E. A. and D'Souza, Raissa M. (2009). Percolation on interacting networks. *arXiv:0907.0894*.

Leone, M., Vázquez, A., Vespignani, A., and Zecchina, R. (2002). Ferromagnetic ordering in graphs with arbitrary degree distribution. *Eur. Phys. J. B*, **28**, 191.

Leskovec, J., Adamic, L. A., and Huberman, B. A. (2007a). The dynamics of viral marketing. *ACM Trans. Web*, **1**, 5–es.

Leskovec, J., Kleinberg, J., and Faloutsos, C. (2007b). Graph evolution: Densification and shrinking diameters. *ACM Trans. Knowl. Discov. Data*, **1**, 2–es.

Levin, D. A. and Peres, Y. (2017). *Markov Chains and Mixing Times*. Volume 107. American Mathematical Society, Providence, RI.

Leyva, I., Sevilla-Escoboza, R., Buldú, J. M., Sendina-Nadal, I., Gómez-Gardeñes, J., Arenas, A., Moreno, Y., Gómez, S., Jaimes-Reátegui, R., and Boccaletti, S. (2012). Explosive first-order transition to synchrony in networked chaotic oscillators. *Phys. Rev. Lett.*, **108**, 168702.

Leyvraz, F. (2003). Scaling theory and exactly solved models in the kinetics of irreversible aggregation. *Phys. Rep.*, **383**, 95.

Li, A., Cornelius, S. P., Liu, Y.-Y., Wang, L., and Barabási, A.-L. (2017). The fundamental advantages of temporal networks. *Science*, **358**, 1042.

Li, M., Deng, Y., and Wang, B.-H. (2015). Clique percolation in random graphs. *Phys. Rev. E*, **92**, 042116.

Li, Z., Zhang, S., Wang, R.-S., Zhang, X.-S., and Chen, L. (2008). Quantitative function for community detection. *Phys. Rev. E*, **77**, 036109.

Liben-Nowell, D. and Kleinberg, J. (2007). The link-prediction problem for social networks. *J. Assoc. Inf. Sci. Technol.*, **58**, 1019.

Liu, S., Perra, N., Karsai, M., and Vespignani, A. (2014). Controlling contagion processes in activity driven networks. *Phys. Rev. Lett.*, **112**, 118702.

Liu, X., Stanley, H. E., and Gao, J. (2016). Breakdown of interdependent directed networks. *PNAS*, **113**, 1138.

Liu, Y.-Y., Csóka, E., Zhou, H., and Pósfai, M. (2012). Core percolation on complex networks. *Phys. Rev. Lett.*, **109**, 205703.

Liu, Y.-Y., Slotine, J.-J., and Barabási, A.-L. (2011). Controllability of complex networks. *Nature*, **473**, 167.

Liu, Z., Lai, Y.-C., and Ye, N. (2003). Propagation and immunization of infection on general networks with both homogeneous and heterogeneous components. *Phys. Rev. E*, **67**, 031911.

Lokhov, A. Y., Mézard, M., Ohta, H., and Zdeborová, L. (2014). Inferring the origin of an epidemic with a dynamic message-passing algorithm. *Phys. Rev. E*, **90**, 012801.

Lovász, L. (2012). *Large Networks and Graph Limits*. American Mathematical Society, Providence, RI.

Lü, L., Medo, M., Yeung, C. H., Zhang, Y.-C., Zhang, Z.-K., and Zhou, T. (2012). Recommender systems. *Phys. Rep.*, **519**, 1.

Łuczak, T. (1991). Size and connectivity of the k-core of a random graph. *Discrete Math.*, **91**, 61.

Ludwig, D. (1975). Final size distribution for epidemics. *Math. Biosci.*, **23**, 33.

Lyons, R. (1989). The Ising model and percolation on trees and tree-like graphs. *Commun. Math. Phys.*, **125**, 337.

Magner, A., Grama, A., Sreedharan, J., and Szpankowski, W. (2017). Recovery of vertex orderings in dynamic graphs. In *2017 IEEE International Symposium on Information Theory (ISIT)*, pp. 1563–1567. IEEE, New York.

Maier, B. F. and Brockmann, D. (2017). Cover time for random walks on arbitrary complex networks. *Phys. Rev. E*, **96**, 042307.

Maki, D. P and Thompson, M. (1973). *Mathematical Models and Applications: With Emphasis on the Social, Life, and Management Sciences*. Prentice-Hall Englewood Cliffs.

Marro, J. and Dickman, R. (2005). *Nonequilibrium Phase Transitions in Lattice Models*. Cambridge University Press, Cambridge.

Martens, E. A., Barreto, E., Strogatz, S. H., Ott, E., So, P., and Antonsen, T. M. (2009). Exact results for the Kuramoto model with a bimodal frequency distribution. *Phys. Rev. E*, **79**, 026204.

Martin, T., Zhang, X., and Newman, M. E. J. (2014). Localization and centrality in networks. *Phys. Rev. E*, **90**, 052808.

Maslov, S., Sneppen, K., and Alon, U. (2003). Correlation profiles and motifs in complex networks. In *Handbook of Graphs and Networks: From the Genome to the Internet* (ed. S. Bornholdt and H. G. Schuster), pp. 168–198. Wiley-VCH Germany, Weinheim.

Maslov, S. and Sneppen, K. (2002). Specificity and stability in topology of protein networks. *Science*, **296**, 910.

Masuda, N. and Hiraoka, T. (2020). Waiting-time paradox in 1922. *NEJCS*, **2**, 1.

Masuda, N., Klemm, K., and Eguíluz, V. M. (2013). Temporal networks: Slowing down diffusion by long lasting interactions. *Phys. Rev. Lett.*, **111**, 188701.

Masuda, N. and Lambiotte, R. (2016). *A Guide to Temporal Networks*. World Scientific, Singapore.

Mata, A. S. and Ferreira, S. C. (2013). Pair quenched mean-field theory for the susceptible-infected-susceptible model on complex networks. *EPL*, **103**, 48003.

Matousek, J. (2013). *Lectures on Discrete Geometry*. Springer, Berlin.

Matula, D. W. (1970). On the complete subgraphs of a random graph. In *Proceedings of the Second Chapel Hill Conference on Combinatorial Mathematics and its Applications*, pp. 356–369. University of North Carolina Press, Chapel Hill, NC.

McKay, B. D. (1981). The expected eigenvalue distribution of a large regular graph. *Linear Algebra Appl.*, **40**, 203.

Meester, R. and Roy, R. (1996). *Continuum Percolation*. Cambridge University Press, Cambridge.

Mehta, M. L. (2004). *Random Matrices*. Elsevier, Amsterdam.

Mezard, M. and Montanari, A. (2009). *Information, Physics, and Computation*. Oxford University Press, Oxford.

Milgram, S. (1967). The small world problem. *Psychol. Today*, **2**, 60.

Miller, J. C. (2016). Equivalence of several generalized percolation models on networks. *Phys. Rev. E*, **94**, 032313.

Milo, R., Shen-Orr, S., Itzkovitz, S., Kashtan, N., Chklovskii, D., and Alon, U. (2002). Network motifs: Simple building blocks of complex networks. *Science*, **298**, 824.

Min, B. (2018). Identifying an influential spreader from a single seed in complex networks via a message-passing approach. *Eur. Phys. J. B*, **91**, 1.

Min, B. and Castellano, C. (2020). Message-passing theory for cooperative epidemics. *Chaos*, **30**, 023131.

Min, B., Lee, S., Lee, K.-M., and Goh, K.-I. (2015). Link overlap, viability, and mutual percolation in multiplex networks. *Chaos Solit. Fract.*, **72**, 49.

Mirchev, M. J. (2017). On the spectra of scale-free and small-world networks. In *Proc. of XXV Conference 'Telecom 2017' 26–27 October, NSTC, Sofia, Bulgaria*, pp. 5–12.

Mizutaka, S. and Hasegawa, T. (2018). Disassortativity of percolating clusters in random networks. *Phys. Rev. E*, **98**, 062314.

Mohar, B. (1991). The Laplacian spectrum of graphs. In *Graph Theory, Combinatorics, and Applications* (ed. Y. Alavi, G. Chartrand, O. Ollermann, and A. Schwenk), pp. 871–898. Wiley, New York.

Molloy, M. and Reed, B. (1995). A critical point for random graphs with a given degree sequence. *Random Struct. Algor.*, **6**, 161.

Molloy, M. and Reed, B. (1998). The size of the giant component of a random graph with a given degree sequence. *Combin. Probab. Comput.*, **7**, 295.

Montbrió, E., Kurths, J., and Blasius, B. (2004). Synchronization of two inter-acting populations of oscillators. *Phys. Rev. E*, **70**, 056125.

Moon, J. W. and Moser, L. (1965). On cliques in graphs. *Isr. J. Math.*, **3**, 23.

Moore, C. (2017). The computer science and physics of community detection: Landscapes, phase transitions, and hardness. *arXiv:1702.00467*.

Moore, C., Ghoshal, G., and Newman, M. E. J. (2006). Exact solutions for models of evolving networks with addition and deletion of nodes. *Phys. Rev. E*, **74**, 036121.

Moore, S. and Rogers, T. (2020). Predicting the speed of epidemics spreading in networks. *Phys. Rev. Lett.*, **124**, 068301.

Moreno, Y., Nekovee, M., and Pacheco, A. F. (2004a). Dynamics of rumor spread-ing in complex networks. *Phys. Rev. E*, **69**, 066130.

Moreno, Y., Nekovee, M., and Vespignani, A. (2004b). Efficiency and reliability of epidemic data dissemination in complex networks. *Phys. Rev. E*, **69**, 055101.

Moreno, Y., Pastor-Satorras, R., and Vespignani, A. (2002). Epidemic outbreaks in complex heterogeneous networks. *Eur. Phys. J. B*, **26**, 521.

Morone, F., Del Ferraro, G., and Makse, H. A. (2019). The *k*-core as a predictor of structural collapse in mutualistic ecosystems. *Nature Phys.*, **15**, 95.

Motter, A. E., Zhou, C., and Kurths, J. (2005). Network synchronization, diffu-sion, and the paradox of heterogeneity. *Phys. Rev. E*, **71**, 016116.

Murray, J. D. (2007). *Mathematical Biology: I. An Introduction*. Springer, Berlin.

Nadakuditi, R. R. and Newman, M. E. J. (2012). Graph spectra and the de-tectability of community structure in networks. *Phys. Rev. Lett.*, **108**, 188701.

Nadakuditi, R. R. and Newman, M. E. J. (2013). Spectra of random graphs with arbitrary expected degrees. *Phys. Rev. E*, **87**, 012803.

Navlakha, S. and Kingsford, C. (2011). Network archaeology: Uncovering ancient networks from present-day interactions. *PLoS Comput. Biol.*, **7**, e1001119.

Nekovee, M., Moreno, Y., Bianconi, G., and Marsili, M. (2007). Theory of rumour spreading in complex social networks. *Phys. A: Stat. Mech. Appl.*, **374**, 457.

Newman, M. E. J. (2000). Models of the small world. *J. Stat. Phys.*, **101**, 819.

Newman, M. E. J. (2001a). Clustering and preferential attachment in growing networks. *Phys. Rev. E*, **64**, 025102.

Newman, M. E. J. (2001b). Scientific collaboration networks. II. Shortest paths, weighted networks, and centrality. *Phys. Rev. E*, **64**, 016132.

Newman, M. E. J. (2002a). Assortative mixing in networks. *Phys. Rev. Lett.*, **89**, 208701.

Newman, M. E. J. (2002b). Spread of epidemic disease on networks. *Phys. Rev. E*, **66**, 016128.

Newman, M. E. J. (2003). Mixing patterns in networks. *Phys. Rev. E*, **67**, 026126.

Newman, M. E. J. (2004a). Analysis of weighted networks. *Phys. Rev. E*, **70**, 056131.

Newman, M. E. J. (2004b). Fast algorithm for detecting community structure in networks. *Phys. Rev. E*, **69**, 066133.

Newman, M. E. J. (2006). Finding community structure in networks using the eigenvectors of matrices. *Phys. Rev. E*, **74**, 036104.

Newman, M. E. J. (2007). Component sizes in networks with arbitrary degree distributions. *Phys. Rev. E*, **76**, 045101.

Newman, M. E. J. (2009). Random graphs with clustering. *Phys. Rev. Lett.*, **103**, 058701.

Newman, M. E. J. (2010). *Networks: An Introduction*. Oxford University Press, Oxford.

Newman, M. E. J. (2011). Complex systems: A survey. *arXiv:1112.1440*.

Newman, M. E. J. (2018a). Estimating network structure from unreliable measurements. *Phys. Rev. E*, **98**, 062321.

Newman, M. E. J. (2018b). Network structure from rich but noisy data. *Nature Phys.*, **14**, 542.

Newman, M. E. J. (2019). Spectra of networks containing short loops. *Phys. Rev. E*, **100**, 012314.

Newman, M. E. J. and Clauset, A. (2016). Structure and inference in annotated networks. *Nature Commun.*, **7**, 1.

Newman, M. E. J. and Ghoshal, G. (2008). Bicomponents and the robustness of networks to failure. *Phys. Rev. Lett.*, **100**, 138701.

Newman, M. E. J. and Girvan, M. (2004). Finding and evaluating community structure in networks. *Phys. Rev. E*, **69**, 026113.

Newman, M. E. J., Strogatz, S. H., and Watts, D. J. (2001). Random graphs with arbitrary degree distributions and their applications. *Phys. Rev. E*, **64**, 026118.

Newman, M. E. J., Zhang, X., and Nadakuditi, R. R. (2019). Spectra of random networks with arbitrary degrees. *Phys. Rev. E*, **99**, 042309.

Nicosia, V., Tang, J., Mascolo, C., Musolesi, M., Russo, G., and Latora, V. (2013). Graph metrics for temporal networks. In *Temporal Networks* (ed. P. Holme and J. Saramäki), pp. 15–40. Springer, Berlin.

Nicosia, V., Tang, J., Musolesi, M., Russo, G., Mascolo, C., and Latora, V. (2012). Components in time-varying graphs. *Chaos*, **22**, 023101.

Noh, J. D. and Rieger, H. (2004). Random walks on complex networks. *Phys. Rev. Lett.*, **92**, 118701.

Nowak, M. A. (2006). *Evolutionary Dynamics: Exploring the Equations of Life*. Harvard University Press, Cambridge, MA.

Nowak, M. A. and May, R. M. (1992). Evolutionary games and spatial chaos. *Nature*, **359**, 826.

O'Connell, N. (1997). From laws of large numbers to large deviation principles. Technical Report HPL-BRIMS-97-17, Hewlett-Packard Laboratories.

O'Connell, N. (1998). Some large deviation results for sparse random graphs. *Probab. Theory Relat. Fields*, **110**, 277.

Oh, E., Lee, D.-S., Kahng, B., and Kim, D. (2007). Synchronization transition of heterogeneously coupled oscillators on scale-free networks. *Phys. Rev. E*, **75**, 011104.

Ohno, S. (2013). *Evolution by Gene Duplication*. Springer, Berlin.

Olson, W. H. and Uppuluri, V. R. R. (1972). Asymptotic distribution of eigenvalues of random matrices. In *Proceedings of the Sixth Berkeley Symposium on Mathematical Statistics and Probability, Vol. III*, pp. 615–644. University of

California Press, Berkeley, CA.

Onnela, J.-P., Saramäki, J., Hyvönen, J., Szabó, G., De Menezes, M. A., Kaski, K., Barabási, A.-L., and Kertész, J. (2007*a*). Analysis of a large-scale weighted network of one-to-one human communication. *New J. Phys.*, **9**, 179.

Onnela, J.-P., Saramäki, J., Hyvönen, J., Szabó, G., Lazer, D., Kaski, K., Kertész, J., and Barabási, A.-L. (2007*b*). Structure and tie strengths in mobile communication networks. *PNAS*, **104**, 7332.

Opsahl, T. and Panzarasa, P. (2009). Clustering in weighted networks. *Soc. Netw.*, **31**, 155.

Ott, E. and Antonsen, T. M. (2008). Low dimensional behavior of large systems of globally coupled oscillators. *Chaos*, **18**, 037113.

Otter, R. (1948). The number of trees. *Ann. Math*, **49**, 583.

Ouvrard, X., Goff, J.-M. L., and Marchand-Maillet, S. (2017). Adjacency and tensor representation in general hypergraphs. Part 1: *e*-adjacency tensor uniformisation using homogeneous polynomials. *arXiv:1712.08189*.

Pacheco, J. M., Traulsen, A., and Nowak, M. A. (2006). Coevolution of strategy and structure in complex networks with dynamical linking. *Phys. Rev. Lett.*, **97**, 258103.

Pachner, U. (1991). PL homeomorphic manifolds are equivalent by elementary shellings. *Eur. J. Comb.*, **12**, 129.

Pade, J. P. and Pereira, T. (2015). Improving network structure can lead to functional failures. *Sci. Rep.*, **5**, 1.

Page, L., Brin, S., Motwani, R., and Winograd, T. (1999). The PageRank citation ranking: Bringing order to the web. Technical report, Stanford InfoLab.

Palla, G., Derényi, I., and Vicsek, T. (2007). The critical point of *k*-clique percolation in the Erdős–Rényi graph. *J. Stat. Phys.*, **128**, 219.

Palla, G., Derényi, I., Farkas, I., and Vicsek, T. (2005). Uncovering the overlapping community structure of complex networks in nature and society. *Nature*, **435**, 814.

Panagiotou, K., Spöhel, R., Steger, A., and Thomas, H. (2011). Explosive percolation in Erdős-Rényi-like random graph processes. *Electron. Notes Discrete Math.*, **38**, 699.

Papadopoulos, F., Kitsak, M., Serrano, M. Á., Boguñá, M., and Krioukov, D. (2012). Popularity versus similarity in growing networks. *Nature*, **489**, 537.

Papadopoulos, F., Krioukov, D., Boguñá, M., and Vahdat, A. (2010). Greedy forwarding in dynamic scale-free networks embedded in hyperbolic metric spaces. In *2010 Proceedings IEEE INFOCOM*, pp. 1–9. IEEE, New York.

Parisi, G. (2002). Complex systems: A physicist's viewpoint. *arXiv:cond-mat/0205297*.

Park, J. and Newman, M. E. J. (2004). Statistical mechanics of networks. *Phys. Rev. E*, **70**, 066117.

Park, J., Yi, S., Choi, K., Lee, D., and Kahng, B. (2019). Interevent time distribution, burst, and hybrid percolation transition. *Chaos*, **29**, 091102.

Parshani, R., Dickison, M., Cohen, R., Stanley, H. E., and Havlin, S. (2010). Dynamic networks and directed percolation. *EPL*, **90**, 38004.

Pastor-Satorras, R. and Castellano, C. (2016). Distinct types of eigenvector localization in networks. *Sci. Rep.*, **6**, 18847.

Pastor-Satorras, R. and Castellano, C. (2018). Eigenvector localization in real networks and its implications for epidemic spreading. *J. Stat. Phys.*, **173**, 1110.

Pastor-Satorras, R. and Castellano, C. (2020). The localization of non-backtracking centrality in networks and its physical consequences. *Sci. Rep.*, **10**, 1.

Pastor-Satorras, R., Castellano, C., Van Mieghem, P., and Vespignani, A. (2015). Epidemic processes in complex networks. *Rev. Mod. Phys.*, **87**, 925.

Pastor-Satorras, R., Smith, E., and Solé, R. V. (2003). Evolving protein interaction networks through gene duplication. *J. Theor. Biol.*, **222**, 199.

Pastor-Satorras, R., Vázquez, A., and Vespignani, A. (2001). Dynamical and correlation properties of the internet. *Phys. Rev. Lett.*, **87**, 258701.

Pastor-Satorras, R. and Vespignani, A. (2001a). Epidemic dynamics and endemic states in complex networks. *Phys. Rev. E*, **63**, 066117.

Pastor-Satorras, R. and Vespignani, A. (2001b). Epidemic spreading in scale-free networks. *Phys. Rev. Lett.*, **86**, 3200.

Pastor-Satorras, R. and Vespignani, A. (2002). Immunization of complex networks. *Phys. Rev. E*, **65**, 036104.

Pastor-Satorras, R. and Vespignani, A. (2007). *Evolution and Structure of the Internet: A Statistical Physics Approach.* Cambridge University Press, Cambridge.

Pazó, D. (2005). Thermodynamic limit of the first-order phase transition in the Kuramoto model. *Phys. Rev. E*, **72**, 046211.

Pearson, K. J. and Zhang, T. (2015). The Laplacian tensor of a multi-hypergraph. *Discrete Math.*, **338**, 972.

Pecora, L. M. and Carroll, T. L. (1998). Master stability functions for synchronized coupled systems. *Phys. Rev. Lett.*, **80**, 2109.

Peixoto, T. P. (2013a). Eigenvalue spectra of modular networks. *Phys. Rev. Lett.*, **111**, 098701.

Peixoto, T. P. (2013b). Parsimonious module inference in large networks. *Phys. Rev. Lett.*, **110**, 148701.

Peixoto, T. P. (2018). Reconstructing networks with unknown and heterogeneous errors. *Phys. Rev. X*, **8**, 041011.

Penrose, M. (2003). *Random Geometric Graphs.* Oxford University Press, Oxford.

Perra, N., Gonçalves, B., Pastor-Satorras, R., and Vespignani, A. (2012). Activity driven modeling of time varying networks. *Sci. Rep.*, **2**, 1.

Pittel, B. (1990). On tree census and the giant component in sparse random graphs. *Random Struct. Algor.*, **1**, 311.

Pittel, B. (1994). Note on the heights of random recursive trees and random m-ary search trees. *Random Struct. Algor.*, **5**, 337.

Pittel, B., Spencer, J., and Wormald, N. (1996). Sudden emergence of a giant k-core in a random graph. *J. Comb. Theory Ser. B*, **67**, 111.

Plotkin, J. M. and Rosenthal, J. W. (1994). How to obtain an asymptotic expansion of a sequence from an analytic identity satisfied by its generating function. *J. Austral. Math. Soc. Ser. A*, **56**, 131.

REFERENCES

Pósfai, M., Liu, Y.-Y., Slotine, J.-J., and Barabási, A.-L. (2013). Effect of correlations on network controllability. *Sci. Rep.*, **3**, 1067.

Privman, V. (1990). *Finite Size Scaling and Numerical Simulation of Statistical Systems*. World Scientific, Singapore.

Pržulj, N., Corneil, D. G., and Jurisica, I. (2004). Modeling interactome: Scale-free or geometric? *Bioinformatics*, **20**, 3508.

Quintanilla, J., Torquato, S., and Ziff, R. M. (2000). Efficient measurement of the percolation threshold for fully penetrable discs. *J. Phys. A: Math. Gen.*, **33**, L399.

Radicchi, F. (2015). Predicting percolation thresholds in networks. *Phys. Rev. E*, **91**, 010801.

Radicchi, F. and Bianconi, G. (2017). Redundant interdependencies boost the robustness of multiplex networks. *Phys. Rev. X*, **7**, 011013.

Radicchi, F. and Bianconi, G. (2020). Epidemic plateau in critical susceptible-infected-removed dynamics with nontrivial initial conditions. *Phys. Rev. E*, **102**, 052309.

Radicchi, F., Castellano, C., Cecconi, F., Loreto, V., and Parisi, D. (2004). Defining and identifying communities in networks. *PNAS*, **101**, 2658.

Radin, C., Ren, K., and Sadun, L. (2014). The asymptotics of large constrained graphs. *J. Phys. A: Math. Theor.*, **47**, 175001.

Ravasz, E. and Barabási, A.-L. (2003). Hierarchical organization in complex networks. *Phys. Rev. E*, **67**, 026112.

Ravasz, E., Somera, A. L., Mongru, D. A., Oltvai, Z. N., and Barabási, A.-L. (2002). Hierarchical organization of modularity in metabolic networks. *Science*, **297**, 1551.

Razborov, A. A. (2008). On the minimal density of triangles in graphs. *Combin. Probab. Comput.*, **17**, 603.

Redner, S. (2001). *A Guide to First-passage Processes*. Cambridge University Press, Cambridge.

Reichardt, J. and Bornholdt, S. (2004). Detecting fuzzy community structures in complex networks with a Potts model. *Phys. Rev. Lett.*, **93**, 218701.

Reichardt, J. and Bornholdt, S. (2006). Statistical mechanics of community detection. *Phys. Rev. E*, **74**, 016110.

Rintoul, M. D. and Torquato, S. (1997). Precise determination of the critical threshold and exponents in a three-dimensional continuum percolation model. *J. Phys. A: Math. Gen.*, **30**, L585.

Riordan, O. and Warnke, L. (2011). Explosive percolation is continuous. *Science*, **333**, 322.

Robson, J. M. (2001). Finding a maximum independent set in time $o(2n/4)$. Technical report, LaBRI, Université Bordeaux I.

Rodgers, G. J., Austin, K., Kahng, B., and Kim, D. (2005). Eigenvalue spectra of complex networks. *J. Physics A: Math. Gen.*, **38**, 9431.

Rodgers, G. J. and Bray, A. J. (1988). Density of states of a sparse random matrix. *Phys. Rev. B*, **37**, 3557.

Rodrigues, F. A., Peron, T. K. DM., Ji, P., and Kurths, J. (2016). The Kuramoto model in complex networks. *Phys. Rep.*, **610**, 1.

Rosvall, M. and Bergstrom, C. T. (2008). Maps of random walks on complex networks reveal community structure. *PNAS*, **105**, 1118.

Saade, A., Krzakala, F., and Zdeborová, L. (2014). Spectral clustering of graphs with the Bethe Hessian. In *Advances in Neural Information Processing Systems 27 (NIPS 2014)*, pp. 406–414. Curran Associates Inc., Red Hook, NY.

Samukhin, A. N., Dorogovtsev, S. N., and Mendes, J. F. F. (2008). Laplacian spectra of, and random walks on, complex networks: Are scale-free architectures really important? *Phys. Rev. E*, **77**, 036115.

Santos, F. C. and Pacheco, J. M. (2005). Scale-free networks provide a unifying framework for the emergence of cooperation. *Phys. Rev. Lett.*, **95**, 098104.

Schawe, H. and Hartmann, A. K. (2019). Large-deviation properties of the largest biconnected component for random graphs. *Eur. Phys. J. B*, **92**, 73.

Schenk, K., Drossel, B., Clar, S., and Schwabl, F. (2000). Finite-size effects in the self-organized critical forest-fire model. *Eur. Phys. J. B*, **15**, 177.

Schwab, K. and Malleret, T. (2020). *Covid-19: The Great Reset*. World Economic Forum, Geneva.

Schwarz, J. M., Liu, A. J., and Chayes, L. Q. (2006). The onset of jamming as the sudden emergence of an infinite k-core cluster. *EPL*, **73**, 560.

Seidman, S. B. (1983). Network structure and minimum degree. *Soc. Netw.*, **5**, 269.

Semerjian, G. and Cugliandolo, L. F. (2002). Sparse random matrices: The eigenvalue spectrum revisited. *J. Phys. A: Math. Gen.*, **35**, 4837.

Serrano, M. Á., Boguñá, M., and Díaz-Guilera, A. (2005). Competition and adaptation in an internet evolution model. *Phys. Rev. Lett.*, **94**, 038701.

Serrano, M. Á., Boguñá, M., and Pastor-Satorras, R. (2006). Correlations in weighted networks. *Phys. Rev. E*, **74**, 055101.

Serrano, M. Á. and Boguñá, M. (2006*a*). Clustering in complex networks. I. General formalism. *Phys. Rev. E*, **74**, 056114.

Serrano, M. Á. and Boguñá, M. (2006*b*). Clustering in complex networks. II. Percolation properties. *Phys. Rev. E*, **74**, 056115.

Serrano, M. Á., Krioukov, D., and Boguñá, M. (2008). Self-similarity of complex networks and hidden metric spaces. *Phys. Rev. Lett.*, **100**, 078701.

Shah, D. and Zaman, T. (2011). Rumors in a network: Who's the culprit? *IEEE Trans. Inf. Theory*, **57**, 5163.

Shah, D. and Zaman, T. (2016). Finding rumor sources on random trees. *Oper. Res.*, **64**, 736.

Shao, J., Buldyrev, S. V., Havlin, S., and Stanley, H. E. (2011). Cascade of failures in coupled network systems with multiple support-dependence relations. *Phys. Rev. E*, **83**, 036116.

Shinomoto, S. and Kuramoto, Y. (1986). Phase transitions in active rotator systems. *Progr. Theor. Phys.*, **75**, 1105.

Shklovskii, B. I. and Efros, A. L. (1984). *Electronic Properties of Doped Semiconductors*. Springer, Berlin.

Shutters, S. T. (2013). Towards a rigorous framework for studying 2-player continuous games. *J. Theor. Biol.*, **321**, 40.

Small, M. and Tse, C. K. (2005). Small world and scale free model of transmission of sars. *Int. J. Bifurc. Chaos Appl. Sci. Eng.*, **15**, 1745.

Söderberg, B. (2002). General formalism for inhomogeneous random graphs. *Phys. Rev. E*, **66**, 066121.

Solé-Ribalta, A., De Domenico, M., Kouvaris, N. E., Díaz-Guilera, A., Gómez, S., and Arenas, A. (2013). Spectral properties of the Laplacian of multiplex networks. *Phys. Rev. E*, **88**, 032807.

Solomonoff, R. (1952). An exact method for the computation of the connectivity of random nets. *Bull. Math. Biophys.*, **14**, 153.

Solomonoff, R. and Rapoport, A. (1951). Connectivity of random nets. *Bull. Math. Biophys.*, **13**, 107.

Son, S.-W., Bizhani, G., Christensen, C., Grassberger, P., and Paczuski, M. (2012). Percolation theory on interdependent networks based on epidemic spreading. *EPL*, **97**, 16006.

Song, J. U., Um, J., Park, J., and Kahng, B. (2020). Effective-potential approach to hybrid synchronization transitions. *Phys. Rev. E*, **101**, 052313.

Sood, V., Antal, T., and Redner, S. (2008). Voter models on heterogeneous networks. *Phys. Rev. E*, **77**, 041121.

Sood, V. and Grassberger, P. (2007). Localization transition of biased random walks on random networks. *Phys. Rev. Lett.*, **99**, 098701.

Sood, V. and Redner, S. (2005). Voter model on heterogeneous graphs. *Phys. Rev. Lett.*, **94**, 178701.

Speidel, L., Klemm, K., Eguíluz, V. M., and Masuda, N. (2016). Temporal interactions facilitate endemicity in the susceptible-infected-susceptible epidemic model. *New J. Phys.*, **18**, 073013.

Speidel, L., Klemm, K., Eguíluz, V. M., and Masuda, N. (2017). Epidemic threshold in temporally-switching networks. In *Temporal Network Epidemiology* (ed. N. Masuda and P. Holme), pp. 161–177. Springer, Berlin.

Speidel, L., Lambiotte, R., Aihara, K., and Masuda, N. (2015). Steady state and mean recurrence time for random walks on stochastic temporal networks. *Phys. Rev. E*, **91**, 012806.

Squartini, T., de Mol, J., den Hollander, F., and Garlaschelli, D. (2015). Breaking of ensemble equivalence in networks. *Phys. Rev. Lett.*, **115**, 268701.

Starnini, M., Baronchelli, A., Barrat, A., and Pastor-Satorras, R. (2012). Random walks on temporal networks. *Phys. Rev. E*, **85**, 056115.

Stauffer, D. and Aharony, A. (1991). *Introduction to Percolation Theory*. Taylor and Francis, London.

Strauss, D. (1986). On a general class of models for interaction. *SIAM Review*, **28**, 513.

Strauss, D. J. (1975). A model for clustering. *Biometrika*, **62**, 467.

Strogatz, S. H. (2000). From Kuramoto to Crawford: Exploring the onset of synchronization in populations of coupled oscillators. *Physica D*, **143**, 1.

Strogatz, S. H., Mirollo, R. E., and Matthews, P. C. (1992). Coupled nonlinear oscillators below the synchronization threshold: Relaxation by generalized Landau damping. *Phys. Rev. Lett.*, **68**, 2730.

Strogatz, S. H. and Mirollo, R. E. (1991). Stability of incoherence in a population of coupled oscillators. *J. Stat. Phys.*, **63**, 613.

Suchecki, K., Eguíluz, V. M., and San Miguel, M. (2004). Conservation laws for the voter model in complex networks. *EPL*, **69**, 228.

Suchecki, K., Eguíluz, V. M., and San Miguel, M. (2005). Voter model dynamics in complex networks: Role of dimensionality, disorder, and degree distribution. *Phys. Rev. E*, **72**, 036132.

Sudbury, A. (1985). The proportion of the population never hearing a rumour. *J. Appl. Probab.*, 443.

Susca, V. A. R., Vivo, P., and Kühn, R. (2021). Cavity and replica methods for the spectral density of sparse symmetric random matrices. *SciPost Phys. Lecture Notes*, **33**, 1.

Szabó, G. and Fath, G. (2007). Evolutionary games on graphs. *Phys. Rep.*, **446**, 97.

Szymański, J. (1987). On a nonuniform random recursive tree. *Ann. Discrete Math.*, **33**, 297.

Tadić, B., Thurner, S., and Rodgers, G. J. (2004). Traffic on complex networks: Towards understanding global statistical properties from microscopic density fluctuations. *Phys. Rev. E*, **69**, 036102.

Tao, T. (2012). *Topics in Random Matrix Theory*. American Mathematical Society, Providence, RI.

Timár, G., da Costa, R. A., Dorogovtsev, S. N., and Mendes, J. F. F. (2017a). Nonbacktracking expansion of finite graphs. *Phys. Rev. E*, **95**, 042322.

Timár, G., da Costa, R. A., Dorogovtsev, S. N., and Mendes, J. F. F. (2020). Choosing among alternative histories of a tree. *Phys. Rev. E*, **102**, 032304.

Timár, G., da Costa, R. A., Dorogovtsev, S. N., and Mendes, J. F. F. (2021). Approximating nonbacktracking centrality and localization phenomena in large networks. *arXiv:2103.04655*.

Timár, G., Dorogovtsev, S. N., and Mendes, J. F. F. (2016). Scale-free networks with exponent one. *Phys. Rev. E*, **94**, 022302.

Timár, G., Goltsev, A. V., Dorogovtsev, S. N., and Mendes, J. F. F. (2017b). Mapping the structure of directed networks: Beyond the bow-tie diagram. *Phys. Rev. Lett.*, **118**, 078301.

Tishby, I., Biham, O., and Katzav, E. (2021). Analytical results for the distribution of first return times of random walks on random regular graphs. *J. Phys. A: Math. Theor.*, **54**, 325001.

Touchette, H. (2009). The large deviation approach to statistical mechanics. *Phys. Rep.*, **478**, 1.

Travers, J. and Milgram, S. (1969). An experimental study of the small world problem. *Sociometry*, **32**, 425.

Valdano, E., Ferreri, L., Poletto, C., and Colizza, V. (2015). Analytical computation of the epidemic threshold on temporal networks. *Phys. Rev. X*, **5**, 021005.

Van den Driessche, P. (2017). Reproduction numbers of infectious disease models.

Infect. Dis. Model., **2**, 288.

Van Mieghem, P., Omic, J., and Kooij, R. (2008). Virus spread in networks. *IEEE/ACM Trans. Netw.*, **17**, 1.

Vasiliev, A. N. (2019). *Functional Methods in Quantum Field Theory and Statistical Physics*. Routledge, Abingdon.

Vazquez, A., Dobrin, R., Sergi, D., Eckmann, J.-P., Oltvai, Z. N., and Barabási, A.-L. (2004). The topological relationship between the large-scale attributes and local interaction patterns of complex networks. *PNAS*, **101**, 17940.

Vázquez, A. and Moreno, Y. (2003). Resilience to damage of graphs with degree correlations. *Phys. Rev. E*, **67**, 015101.

Vitali, S., Glattfelder, J. B., and Battiston, S. (2011). The network of global corporate control. *PloS One*, **6**, e25995.

Von Luxburg, U. (2007). A tutorial on spectral clustering. *Stat. Comput.*, **17**, 395.

Vukadinović, D., Huang, P., and Erlebach, T. (2002). On the spectrum and structure of Internet topology graphs. In *International Workshop on Innovative Internet Community Systems, Lecture Notes in Computer Science*, Volume 2346, pp. 83–95. Springer, Berlin.

Wang, Y., Chakrabarti, D., Wang, C., and Faloutsos, C. (2003). Epidemic spreading in real networks: An eigenvalue viewpoint. In *22nd International Symposium on Reliable Distributed Systems, 2003*, pp. 25–34. IEEE, New York.

Watts, D. J. (2002). A simple model of global cascades on random networks. *PNAS*, **99**, 5766.

Watts, D. J., Dodds, P. S., and Newman, M. E. J. (2002). Identity and search in social networks. *Science*, **296**, 1302.

Watts, D. J. and Strogatz, S. H. (1998). Collective dynamics of 'small-world' networks. *Nature*, **393**, 440.

Waxman, B. M. (1988). Routing of multipoint connections. *IEEE Journal on Selected Areas in Communications*, **6**, 1617.

Weigt, M. and Hartmann, A. K. (2000). Number of guards needed by a museum: A phase transition in vertex covering of random graphs. *Phys. Rev. Lett.*, **84**, 6118.

Weiss, Y. and Freeman, W. T. (2001). On the optimality of solutions of the max-product belief-propagation algorithm in arbitrary graphs. *IEEE Trans. Inf. Theory*, **47**, 736.

Wigner, E. P. (1955). Characteristic vectors of bordered matrices with infinite dimensions. I. *Ann. Math.*, **62**, 548.

Wigner, E. P. (1957*a*). Characteristic vectors of bordered matrices with infinite dimensions. II. *Ann. Math.*, **65**, 203.

Wigner, E. P. (1957*b*). *Statistical Properties of Real Symmetric Matrices with Many Dimensions*. Princeton University Press, Princeton, NJ.

Wilf, H. S. (2013). *Generatingfunctionology*. Elsevier, Amsterdam.

Wilkinson, R. R., Ball, F. G., and Sharkey, K. J. (2017). The relationships between message passing, pairwise, Kermack–McKendrick and stochastic SIR epidemic models. *J. Math. Biol.*, **75**, 1563.

Wu, Z., Menichetti, G., Rahmede, C., and Bianconi, G. (2015). Emergent complex network geometry. *Sci. Rep.*, **5**, 1.

Yook, S.-H., Jeong, H., Barabási, A.-L., and Tu, Y. (2001). Weighted evolving networks. *Phys. Rev. Lett.*, **86**, 5835.

Yoon, S., Goltsev, A. V., and Mendes, J. F. F. (2018). Structural stability of interaction networks against negative external fields. *Phys. Rev. E*, **97**, 042311.

Yoon, S., Goltsev, A. V., Dorogovtsev, S. N., and Mendes, J. F. F. (2011). Belief-propagation algorithm and the ising model on networks with arbitrary distributions of motifs. *Phys. Rev. E*, **84**, 041144.

Yoon, S., Sindaci, M. S., Goltsev, A. V., and Mendes, J. F. F. (2015). Critical behavior of the relaxation rate, the susceptibility, and a pair correlation function in the Kuramoto model on scale-free networks. *Phys. Rev. E*, **91**, 032814.

Young, J.-G., St-Onge, G., Laurence, E., Murphy, C., Hébert-Dufresne, L., and Desrosiers, P. (2019). Phase transition in the recoverability of network history. *Phys. Rev. X*, **9**, 041056.

Zachary, W. W. (1977). An information flow model for conflict and fission in small groups. *J. Anthropol. Res.*, **33**, 452.

Zdeborová, L. and Krzakala, F. (2016). Statistical physics of inference: Thresholds and algorithms. *Adv. Phys.*, **65**, 453.

Zdeborová, L. and Mézard, M. (2006). The number of matchings in random graphs. *J. Stat. Mech.: Theor. Exp.*, **2006**, P05003.

Zhang, X., Hu, X., Kurths, J., and Liu, Z. (2013). Explosive synchronization in a general complex network. *Phys. Rev. E*, **88**, 010802.

Zhang, Z.-K., Liu, C., Zhan, X.-X., Lu, X., Zhang, C.-X., and Zhang, Y.-C. (2016). Dynamics of information diffusion and its applications on complex networks. *Phys. Rep.*, **651**, 1.

Zhao, J.-H., Zhou, H.-J., and Liu, Y.-Y. (2013). Inducing effect on the percolation transition in complex networks. *Nature Commun.*, **4**, 1.

Zhao, J.-H. and Zhou, H.-J. (2019). Controllability and maximum matchings of complex networks. *Phys. Rev. E*, **99**, 012317.

Zhou, D., Bashan, A., Cohen, R., Berezin, Y., Shnerb, N., and Havlin, S. (2014). Simultaneous first- and second-order percolation transitions in interdependent networks. *Phys. Rev. E*, **90**, 012803.

Zhou, H. and Ou-Yang, Z.-c. (2003). Maximum matching on random graphs. *arXiv:cond-mat/0309348*.

Zhou, S. and Mondragón, R. J. (2004). The rich-club phenomenon in the Internet topology. *IEEE Commun. Lett.*, **8**, 180.

Zhou, T., Lü, L., and Zhang, Y.-C. (2009). Predicting missing links via local information. *Eur. Phys. J. B*, **71**, 623.

Zlatić, V., Bianconi, G., Díaz-Guilera, A., Garlaschelli, D., Rao, F., and Caldarelli, G. (2009). On the rich-club effect in dense and weighted networks. *Eur. Phys. J. B*, **67**, 271.

Index